Locating Right to the City in the Global South

Despite the fact that virtually all urban growth is occurring, and will continue to occur, in the cities of the Global South, the conceptual tools used to study cities are distilled disproportionately from research on the highly developed cities of the Global North. With urban inequality widely recognized as central to many of the most pressing challenges facing the world, there is a need for a deeper understanding of cities of the South on their own terms.

Locating Right to the City in the Global South marks an innovative and far-reaching effort to document and make sense of urban transformations across a range of cities, as well as the conflicts and struggles for social justice these are generating. The volume contains empirically rich, theoretically informed case studies focused on the social, spatial and political dimensions of urban inequality in the Global South. Drawing from scholars with extensive fieldwork experience, this volume covers sixteen cities in fourteen countries across a belt stretching from Latin America, to Africa and the Middle East, and into Asia. Central to what binds these cities are deeply rooted, complex and dynamic processes of social and spatial division that are being actively reproduced. These cities are not so much fracturing as they are being divided by governance practices informed by local histories and political contestation, and refracted through or infused by market-based approaches to urban development. Through a close examination of these practices and resistance to them, this volume provides perspectives on neoliberalism and right to the city that advance our understanding of urbanism in the Global South.

In mapping the relationships between space, politics and populations, the volume draws attention to variations shaped by local circumstances, while simultaneously elaborating a distinctive transnational Southern urbanism. It provides indepth research on a range of practical and policy-oriented issues, from housing and slum redevelopment to building democratic cities that include participation by lower-income and other marginal groups. It will be of interest to students and practitioners alike studying Urban Studies, Globalization and Development.

Tony Roshan Samara is Associate Professor, Department of Sociology and Anthropology, George Mason University, USA.

Shenjing He is Professor and Assistant Dean at the School of Geography and Planning, interdisciplinary Urban Research Center, Sun Yat-Sen University, China.

Guo Chen is Assistant Professor of Geography and Global Urban Studies at Michigan State University, USA.

Routledge studies in human geography

This series provides a forum for innovative, vibrant and critical debate within human geography. Titles will reflect the wealth of research that is taking place in this diverse and ever-expanding field.

Contributions will be drawn from the main sub-disciplines and from innovative areas of work that have no particular sub-disciplinary allegiances.

Published:

1 **A Geography of Islands**
Small island insularity
Stephen A. Royle

2 **Citizenships, Contingency and the Countryside**
Rights, culture, land and the environment
Gavin Parker

3 **The Differentiated Countryside**
Jonathan Murdoch, Philip Lowe, Neil Ward and Terry Marsden

4 **The Human Geography of East Central Europe**
David Turnock

5 **Imagined Regional Communities**
integration and sovereignty in the global south
James D Sidaway

6 **Mapping Modernities**
Geographies of Central and Eastern Europe 1920–2000
Alan Dingsdale

7 **Rural Poverty**
Marginalisation and exclusion in Britain and the United States
Paul Milbourne

8 **Poverty and the Third Way**
Colin C. Williams and Jan Windebank

9 **Ageing and Place**
Edited by Gavin J. Andrews and David R. Phillips

10 **Geographies of Commodity Chains**
Edited by Alex Hughes and Suzanne Reimer

11 **Queering Tourism**
Paradoxical performances at Gay Pride parades
Lynda T. Johnston

12 **Cross-Continental Food Chains**
Edited by Niels Fold and Bill Pritchard

13 **Private Cities**
Edited by Georg Glasze, Chris Webster and Klaus Frantz

14 **Global Geographies of Post Socialist Transition**
Tassilo Herrschel

15 **Urban Development in Post-Reform China**
Fulong Wu, Jiang Xu and Anthony Gar-On Yeh

16 **Rural Governance**
international perspectives
Edited by Lynda Cheshire, Vaughan Higgins and Geoffrey Lawrence

17 **Global Perspectives on Rural Childhood and Youth**
Young rural lives
Edited by Ruth Panelli, Samantha Punch, and Elsbeth Robson

18 **World City Syndrome**
Neoliberalism and inequality in Cape Town
David A. McDonald

19 **Exploring Post Development**
Aram Ziai

20 **Family Farms**
Harold Brookfield and Helen Parsons

21 **China on the Move**
Migration, the state, and the household
C. Cindy Fan

22 **Participatory Action Research Approaches and Methods**
Connecting people, participation and place
Edited by Sara Kindon, Rachel Pain and Mike Kesby

23 **Time-Space Compression**
Historical geographies
Barney Warf

24 **Sensing Cities**
Monica Degen

25 **international Migration and Knowledge**
Allan Williams and Vladimir Baláž

26 **The Spatial Turn**
interdisciplinary perspectives
Edited by Barney Warf and Santa Arias

27 **Whose Urban Renaissance?**
An international comparison of urban regeneration policies
Edited by Libby Porter and Katie Shaw

28 **Rethinking Maps**
Edited by Martin Dodge, Rob Kitchin and Chris Perkins

29 **Rural–Urban Dynamics**
Livelihoods, mobility and markets in African and Asian frontiers
Edited by Jytte Agergaard, Niels Fold and Katherine V. Gough

30 **Spaces of Vernacular Creativity**
Rethinking the cultural economy
Edited by Tim Edensor, Deborah Leslie, Steve Millington and Norma Rantisi

31 **Critical Reflections on Regional Competitiveness**
Gillian Bristow

32 **Governance and Planning of Mega-City Regions**
An international comparative perspective
Edited by Jiang Xu and Anthony G.O. Yeh

33 **Design Economies and the Changing World Economy**
Innovation, production and competitiveness
John Bryson and Grete Rustin

34 **Globalization of Advertising**
Agencies, cities and spaces of creativity
James Faulconbridge, Peter J. Taylor, J.V. Beaverstock and C. Nativel

35 **Cities and Low Carbon Transitions**
Edited by Harriet Bulkeley, Vanesa Castán Broto, Mike Hodson and Simon Marvin

36 **Globalization, Modernity and The City**
John Rennie Short

37 **Climate Change and the Crisis of Capitalism**
A chance to reclaim, self, society and nature
Edited by Mark Pelling, David Manual Navarette and Michael Redclift

38 **New Economic Spaces in Asian Cities**
From industrial restructuring to the cultural turn
Edited by Peter W. Daniels, Kong Chong Ho and Thomas A. Hutton

39 **Cities, Regions and Flows**
Edited by Peter V. Hall and Markus Hesse

40 **The Politics of Urban Cultural Policy**
Global perspectives
Edited by Carl Grodach and Daniel Silver

41 **Ecologies and Politics of Health**
Edited by Brian King and Kelley Crews

42 **Producer Services in China**
Economic and urban development
Edited by Anthony G.O. Yeh and Fiona F. Yang

43 **Locating Right to the City in the Global South**
Tony Roshan Samara, Shenjing He and Guo Chen

Forthcoming:

44 **Fieldwork in the Global South**
Ethical challenges and dilemmas
Edited by Jenny Lunn

Locating Right to the City in the Global South

Edited by Tony Roshan Samara, Shenjing He and Guo Chen

Routledge
Taylor & Francis Group
LONDON AND NEW YORK

First published 2013
by Routledge
2 Park Square, Milton Park, Abingdon, Oxon OX14 4RN

Simultaneously published in the USA and Canada
by Routledge
711 Third Avenue, New York, NY 10017

Routledge is an imprint of the Taylor & Francis Group, an informa business

British Library Cataloguing in Publication Data
A catalogue record for this book is available from the British Library

Library of Congress Cataloging in Publication Data
Locating right to the city in the Global South / edited by Tony Roshan Samara, Shenjing He and Guo Chen.
p. cm.
Includes bibliographical references and index.
1. Urbanization–Southern Hemisphere. 2. Urbanization–Developing countries. 3. Urban sociology–Southern Hemisphere. 4. Urban sociology–Developing countries. 5. Urban policy–Southern Hemisphere. 6. Urban policy–Developing countries. I. Samara, Tony Roshan. II. He, Shenjing. III. Chen, Guo.
HT384.D44L63 2013
307.76–dc23 2012025386

ISBN: 978-0-415-63564-6 (hbk)
ISBN: 978-0-203-09110-4 (ebk)

Typeset in Times New Roman
by Wearset Ltd, Boldon, Tyne and Wear

Printed and bound in the United States of America by Publishers Graphics, LLC on sustainably sourced paper.

Contents

Illustrations

Figures

Tables

Boxes

Contributors

Rachel Berney is an Assistant Professor at the School of Architecture, University of Southern California.

Wouter Bervoets is an architect and urban planner, currently enrolled as a PhD student in Architecture at the Department of Architecture, Urbanism and Planning of the University of Leuven, Belgium.

Koenraad Bogaert is a member of the Middle East and North Africa Research Group (MENARG) at Ghent University.

Pablo S. Bose is an Assistant Professor at the Department of Geography, University of Vermont.

Ryan Centner is an Assistant Professor at the Department of Sociology, Tufts University.

Guo Chen is an Assistant Professor of Geography and Global Urban Studies at Michigan State University.

Jia-Ching Chen is a PhD candidate researching City and Regional Planning at the department of Global Metropolitan Studies, University of California, Berkeley.

Wael Salah Fahmi is an Associate Professor of Urbanism at Helwan University, Cairo.

Mona Fawaz is an Associate Professor for the graduate program in Urban Planning, Policy and Design at the American University of Beirut.

Shenjing He is an Associate Professor at the School of Geography and Planning, Sun Yat-Sen University.

Christine Hentschel is a Postdoctoral Fellow at the Department of Urban and Regional Sociology, Humboldt University, Berlin.

Maarten Loopmans is an Assistant Professor in Urban and Social Geography at the Department for Earth and Environmental Sciences, University of Leuven, Belgium.

Ernesto López-Morales is an Assistant Professor at the Department of Urban Planning, University of Chile.

Martin J. Murray is Professor of Urban Planning and Professor of Sociology at the Taubman College of Architecture and Urban Planning, University of Michigan.

Xuefei Ren is an Assistant Professor at the Department of Sociology and Global Urban Studies, Michigan State University.

Tony Roshan Samara is an Associate Professor at the Department of Sociology and Anthropology, George Mason University.

David M. Walker is an Assistant Professor at the Department of Geology and Geography, Ohio Wesleyan University.

Liza Weinstein is an Assistant Professor at the Department of Sociology and Anthropology, Northeastern University.

Introduction

Locating Right to the City in the Global South

Tony Roshan Samara, Shenjing He and Guo Chen

> You don't see yourself as part of the city – there are no places that you relate to, that you love to go. No corner, no area touched by a certain kind of light. You have no memory of any material, texture, shape. Everything is constantly changing, according to somebody else's will, somebody else's power.
>
> Al Weiwei 2011

Cities in the Global South have been moving steadily from the margins to the center of the global community of urban scholars. For far too long cities in the North played an outsized role in thinking about cities on a global scale, contributing to a structural neglect of research on "other" cities. While interest in these cities as objects of study has certainly grown, so too has an understanding that a wide range of challenges that are global in scope can only be properly understood if viewed through an urban lens. At the same time, cities themselves are easier to decipher if we understand them as global spaces. As this awareness increases, it is difficult to hold on to many explicit and implicit assumptions about how we should think about cities, and which cities we should be thinking about. Specifically, global demographics, world events and what appears to be a shifting geopolitical terrain demand that more attention be paid to those cities and metropolitan regions across the globe where the majority of the urban population is located.

Recent empirical and theoretical work is quickly taking us past the era when cities in the South were either ignored or interpreted through analytical lenses produced by research on very different cities, upending much of the conventional thinking about cities and global urbanism in the process (Murray, Chapter 12 of this volume; Roy and Ong 2011; Roy 2009; Mayaram 2008; Huyssen 2008; Amen *et al.* 2006; Simone and Abouhani 2005; Robinson 2005, 2002). Roughly sketched, the recent interest in Southern urbanism, as distinct from urbanism "in general", began with challenges to research, policy and discourse that attempted to understand or make comprehensible cities in the South based on the experiences of and research on cities of the North. This was followed by efforts to challenge empirically and theoretically the knowledge produced about cities of the South, and to generate new approaches derived from studies that engaged these cities directly and, as much as possible, on their own terms. Work on the South has now progressed to the point where there is a distinct body of scholarship, with its own questions, approaches and contributions to the urban studies literature. This scholarship will have relevance for the North as well, as the two poles of the urban world increasingly bend towards each other in some important ways (Hirt 2012; Smith 2002). It is perhaps premature to say so, but we can see in these recent shifts the emergence of a truly global urban studies, one that is firmly rooted in a changing understanding of what constitutes urban centrality.

The present volume contributes to these ongoing efforts through empirically rich, theoretically informed case studies focused on the social, spatial and political dimensions of urban inequality. Its goal is not to draw from these an impregnable theory or singular perspective for studying, or even defining, cities of the South: instead, the research published here represents an effort to document and make sense of urban transformations in the South, as well as the conflicts and struggles for social justice these are generating. Drawing from scholars with extensive fieldwork experience, the volume covers sixteen cities in fourteen countries across a belt stretching from Latin America to Africa and the Middle East and into Asia. Central to what binds these cities are deeply rooted, complex and dynamic processes of social and spatial division that are actively reproduced. These cities are not so much fracturing as they are being strategically divided by governance practices informed by local histories and political contestation, and refracted through or infused by market-based approaches to urban development.

This volume advances our understanding of these processes by drawing attention to three defining aspects of the city, according to which the book is organized: first, the increasing social polarization and spatial division of the city, and the local expressions of transnational governance driving these developments; second, the refashioning of certain city quarters of the divided city into cosmopolitan landscapes and the integration of unevenly developed spaces into an unstable and conflict-prone "whole"; finally, many of the chapters gathered here reveal the complicated politics arising from and feeding into the changes that cities are experiencing. Divided cities are sites of competing claims and oppositional forms of governance. These give rise to distinctive political struggles seeking to exert influence across city spaces, and to remake, defend or control those spaces. Taken together these contributions constitute a powerful argument that there is indeed an identifiable, if evolving and contested, transnational urbanism distinctive to the Global South (Smith 2002).

The remainder of this introduction elaborates these themes and concludes with a discussion of the two interrelated concepts whose meaning they illuminate, and which occupy a central position in the production of contemporary urban spaces in the Global South: neoliberalism and right to the city. If these have distinctive Southern expressions – and we believe they do – then it is important that we try to build them up from a foundation of fieldwork based in the South. The themes are meant to capture important constitutive elements of the concepts as they are emerging on the ground in the South, and to inform further theorization. We offer the collected chapters here as a contribution to that effort, rather than as a definitive statement of their meaning outside the Global North, where most of the field and theoretical work informing them to date has been produced. The end goal, of course, is to develop further a critical urban theory (Brenner 2009) that allows us to make some headway in addressing the admittedly overwhelming challenges posed by inequality across the vast urban regions of the South. In the conclusion to this volume, Martin Murray addresses this challenge, situating the questions and contributions offered here in the wider global urban studies literature. This, we feel, is an important and useful way to end, as he highlights key issues and questions for future research that complement and complicate the discussion offered here.

A city divided against itself

Under present urban governance regimes, cities of the South are experiencing pressure towards greater and relatively entrenched socio-spatial distance between groups of residents who become increasingly foreign to each other and to other places in the city. The specific outcomes of this pressure may vary greatly, but the differences can be thought of in terms of degrees of apartness. The divided city remains, as it has been for ages, a preeminent urban form, segregation a relative constant rather than an exception in the social and spatial life of cities (Garrido 2012; Nightingale 2012; Elate 2004; Marcuse 2003). Yet, while socio-spatial segregation may be a defining feature of cities, the forms it has assumed, and will assume, are far from uniform across time and space. What we believe to be constant here is the relational aspect; inequality is fundamental to all cities and this inequality has both social and spatial components. The oft-evoked concepts of core and periphery, in this sense, refer to a relationship through which to grasp the urban at its root.

At the same time, the principle of integration and the even less tangible vision of an "inclusive city" retain a certain popularity among many policymakers, planners, officials and scholars, while actual policy and practice often produce (or reproduce) quite the opposite effect. Indeed, the more critical literature notes that the neoliberal or market-driven development approaches that have gained prominence in the past two decades are directly linked to the making or deepening of social and spatial divisions, particularly in the context of the "world city" aspirations which hold a significant proportion of urban elites across the Global South in their grip (Samara 2011; Batra 2008; McDonald 2008; Broudehoux 2007).

The evocation of inclusion and integration in this context of divisions with widening and hardening tendencies is at best naïve and at worst intentionally misleading. The question is not really one of integration or exclusion, at least not without some substantial qualification. The urban poor have been and will continue to be integrated into the city, just not as equal participants politically or as equal beneficiaries of city resources and government policy. Their centrality to the social reproduction of the city as a whole, and of the lifestyles of more affluent classes, is a direct function of their deep integration; but it is an integration that assumes inequality rather than one which seeks to overcome it, that bespeaks of affluent residents resigned to being dependent on the labor of the poor, yet determined to share the city with them as residents as little as possible. Our understanding of the shape and substance of today's divided city of the South hinges to a great extent on this distinction, and is therefore an essential lens to employ in interrogating urban inequality.

None of this should be mistaken for an argument that urban divisions look the same across vast differences of geography, history and culture, or across the range of integration with the global, or that they are ever absolute and impermeable. While we hold that the principle of divided cities as outlined above is an important one for making sense of inequality and urban politics in the present, there is great variety in how these divisions appear in cities around the world (Abaza 2011; Labbé and Bourdreau 2011; Samara 2011; Kneebone and Garr 2010; Murray 2010, 2004; Roberts and Wilson 2009; Rodgers 2004; Connell 1999).

The making of divided cities is accomplished by a variety of means, but a number of contributions to this volume confirm findings elsewhere that displacement, demolition and other forms of coercion and violence quite often play a central role, and can often be tied directly or indirectly to the introduction or intensification of commodified and financialized land and housing markets (Goldman 2011; UN-HABITAT 2011; Searle 2010; Birkinshaw and Harris 2009; COHRE 2006). There is nothing "natural" about the divided city, nor is it simply an aggregation of individual choices, whether mediated by markets or other institutional forms. Rather, agents of the state, of private security or more shadowy actors play a key role here, often operating as proxies for particular coalitions of interests linked to accumulation and/or class. Great force is often involved in the (re)making of the divided city, and a great deal of force is required for its governance once it has been (re)made (Samara 2011; Amar 2010; Wacquant 2008).

Taking the divided city as a starting point, the chapters in Part I of the book make a number of contributions that fill out these cursory observations, and hopefully provoke useful debate. Mona Fawaz, in her chapter on Beirut, draws our attention to some key overarching themes: the adoption of market-based approaches to "integrate" informal neighborhoods, the uses of violence and the centrality of local conditions in shaping governance practice. Importantly, her research shows that efforts at a particular kind of integration relied on extensive securitization and produced a new kind of marginalization, one that undermines the citymaking power residents had accumulated and created outside of the formal regulatory gaze of state and market. She closes her piece with the argument that a meaningful right to the city must be disentangled from state and market and embedded in the political struggle by residents for secure land and housing.

In the second chapter of Part I, Koenraad Bogaert draws our attention to a similar set of issues in Morocco. The role of slum eradication measures as a form of remarginalization is detailed through a discussion of how access to informal areas by the market is facilitated by the state. Here again, state and market circumscribe the political space of urban residents through efforts at formalizing housing and civic participation in cities across the country. The resulting governance regime does not simply reflect an imposed neoliberal logic; rather, it represents a complex interplay between local and extra-local forces that in turn constitute the substance of neoliberal globalization. As represented here, this includes as primary features the "rediscovery" of neglected urban hinterlands by the state and market – including "new" populations to regulate, an intensification of land politics and struggles and a reconfiguration of preexisting socio-spatial divisions. Bogaert concludes by evoking Lefebvre's critique of the formal empowerment of marginal urban residents, noting that social and political tensions generated by inequality (along with potential disorder) have not been defused through the slum eradication program, but have simply been displaced.

The next contribution to the volume, by Wouter Bervoets and Maarten Loopmans, is an insightful case study of Ouagadougou, Burkina Faso. While also focused on the recent wave of slum redevelopment schemes, Bervoets and Loopmans detail the differences between current market-driven versions and the socialist programs of the postcolonial period. They situate redevelopment in Ouagadougou within a multiscalar context, from the broader "reforms" at the national level mandated by the IMF and World Bank in the 1980s, to the desire of local elites to turn the capital city into a cultural and political center of West Africa, and the subsequent emphasis on downtown and slum redevelopment. Their research demonstrates how local dynamics and global forces combine to form pernicious governance regimes that fail to develop marginal areas, reinforce and expand existing divisions and, importantly, undermine mobilization among marginalized residents. Their insights into the absence of a sustained and popular movement against unpopular reforms are valuable for understanding how divided cities are reproduced and why care must be taken in ascribing inherent radical potential to objectively oppressed communities.

In the final chapter of Part I, Jia-Ching Chen examines a master-planned ecocity in Yixing, China. Chen's analysis builds upon the previous chapters, particularly in illuminating the state-driven process of, quite literally, producing urban space. His study of green development driven by the solar photovoltaic industry in Jiangsu Province cuts across a number of central concerns, situating regional development within both national and global contexts. Chen shows that the development of Yixing into a green economy is accompanied by large-scale land enclosure and displacement of agricultural villages, as newly minted urban areas are brought under direct state administration. As Chen shows, the combination of state control, market-based redevelopment and a discourse of clean energy and rural modernization integrates Yixing into a transnational green economy, while simultaneously extending and entrenching social and spatial inequality across this newly expanded urban region.

Taken together, the chapters in Part I make a strong argument that neoliberalism remains an important analytical lens for understanding urban change in the

South, even as they reveal how complex the interactions are between markets, market ideology, the state and other elements that comprise urban governance regimes. They also provide compelling evidence that policies and practices nominally intended to promote integration through redevelopment of informal and rural areas fail to overcome existing social and spatial distance. Instead market-based approaches carried out by networks of public and private actors are re-inscribing division across unevenly developed urban landscapes, but through perhaps more palatable discourses linked to markets, growth, ownership, participation and clean energy. Within this context conflicts are emerging between increasingly differentiated populations throughout the city, around the use of and access to space, belonging and citizenship and efforts to remake the image of city. It is to this set of issues that we turn in Part II.

Governance and cosmopolitanism: escaping the South

We have seen in recent years growing attention to those cities of the South that have managed to scale the global urban hierarchy based on the histories and current forms of cities in the North. Scholarly research and popular media coverage of Beijing, Shanghai, Mumbai, Rio de Janeiro, Dubai and a handful of other cities underscore the extent to which large sections of these cities have undergone massive transformations, such that their new façades would be familiar to most visitors from the North. However, while the vast majority of cities in the South will never be prominent world cities, recognized as global brands or specialty destinations, this does not mean the idiom of the world city is without significant influence in the reshaping of their urban spaces. Whatever the degree of success a particular city achieves with these aspirations, we argue that the project itself can engender, reanimate and reframe conflict over visions for and of the city.

Perhaps the most striking and immediate manifestation of this phenomenon is the change in built environment. Locally based growth coalitions seek both to emulate and distinguish themselves from prominent Northern cityscapes, and in doing so make their own contribution to the shaping of a hegemonic transnational urban aesthetic (Ren 2011). Even in the grandest new urban centers, however, the transformation is always partial; this partiality is itself a reflection of the very divisions that shape transformation. The renovated built environment serves to highlight the nature of these divisions under market-driven redevelopment, and the widening socio-spatial inequalities that underpin them. It also, by virtue of its limited reach, draws attention to its own status as exception within the local context (Ong 2006). Far from being seamlessly integrated into the broader urban terrain, this cosmopolitan center is notable for how foreign and even hostile a space it can be for most local residents, even as it provides comfort and leisure to moneyed international visitors (Binnie *et al.* 2006).

The partiality of transformation is not unique to the Global South. Wealthy cities in the wealthiest nations have not succeeded in erasing their own "off the map" neighborhoods, even, in the case of the United States, after three decades of gentrification. What is different is the size of the marginalized multitudes and the extent of their deprivation, neither of which can ever be entirely escaped no matter how many layers of security and distance more affluent groups put between themselves and everyone else. From formal policy and governance to the rhythms and disjuncture of daily life, the reality of the divided city infuses local politics and culture. The ways in which it does so, and the consequences, will vary greatly between cities undergoing massive, planned redevelopment, such as Shanghai at one end of the scale, to those like Luanda and Managua on the other, where the local elite are too small, and the scope of transformation too narrow to sustain more than connected enclaves in what is otherwise (perceived to be) a hostile terrain (Chen 2009; Rodrigues 2009; Rodgers 2004). What binds them despite their differences are the disjointed, oppositional spaces that result and a politics of social distancing, both deeply shaped by the constraints of being a city of the South.

The creation of cosmopolitan, transnational spaces of affluence and exclusivity is closely linked to emergent identities and notions of belonging (Samara 2010; Young *et al.* 2006). There is more going on here than an attempt to create spectacular and attractive places to live, shop and recreate. Newly built or refurbished urban centers are also markers of territory and points of reference in what are rapidly changing and expanding environments. More affluent interest groups seek to create and cordon off their cosmopolitan enclaves as much as possible from the disorder around them. Through these efforts, the intimacy often shared between the urban poor and middle class becomes a site of intensified conflict. The new urban geography in many cities has given rise to an anti-poor politics often spearheaded by new and more established middle-class groups of residents (Baviskar and Ray 2011; Davis 2010; Swanson 2007; Ferndandes and Heller 2006; Mawdsley 2004).

The chapters in Part II advance our understanding of these dynamics considerably, and pose important questions for future research. In the first chapter, Xuefei Ren and Liza Weinstein approach the question of transformation and governance from the perspective of mega-project development. The authors seek to explain the differences in planning and implementation between two iconic cities in the South: Shanghai and Mumbai. In explaining how different landscapes emerge, their contribution reveals not only that the state remains central to the implementation of market-based urban transformations, but also how variations in landscape transformation between divided cities are substantially shaped by different inter-scalar state articulations, which are themselves embedded in highly specific local histories. Their chapter draws our attention to the broader networks – of which the state forms one central node, through which these projects are initiated, facilitated, obstructed and unevenly implemented – and how these operate effectively to exclude or reduce participation by local residents in the process.

In the second chapter, Pablo Bose presents an indepth study of cosmopolitan transformation in Kolkata by engaging three seeming contradictions: the rise of neoliberal urbanism in a center of Marxist power; the reemergence of transnational forms in a city that many believed had been left behind by globalization;

and the invocation by more affluent groups of an environmental aesthetic at odds with the social and ecological justice agendas of the urban poor. Bose provides a careful and detailed account of how city spaces are shaped by tensions between what a city can and should be in the eyes of more affluent groups and what the city is. Importantly, he reveals the uses of violence in this process, as an accusation hurled against the urban poor when they infringe upon certain spaces, as in the case of hawkers, and as a method of eviction and removal against these very same populations in the name of beautification and modernization. This view, of the street-level reality of remaking city spaces and into the political coalitions behind them, builds on the work of Ren and Weinstein to complicate our understanding of how city divisions are made and remade. Further, it also emphasizes the distance that can separate reigning governance regimes and networks, regardless of expressed ideology, from the politics of the marginalized.

Rachel Berney offers an insightful analysis of how public spaces become sites for the production of citizenship and the regulation of behavior. Drawing from her extensive fieldwork in Bogotá, she examines efforts by city leaders to use public spaces to integrate residents who are both socially and spatially divided. Adding to the analysis provided by Bose, she shows that from the start the process was shaped by two related constraints: the perceived need for the city to be an attractive place for investment, and the desire of middle- and upper-income residents and officials for an "orderly" city. At the outset, then, the meaning of public space was caught between two competing visions, as a space for interaction between citizens of equal status, or as a space of discipline and differentiation. Public space and citizenship, rather than representing institutions of integration, much less political liberty, become intertwined regulatory mechanisms deployed by exclusionary networks of urban actors. A policy oriented around integration through the use of public space to facilitate the creation of more representative publics had become instead a mechanism of governance linked to existing divisions.

David Walker confronts a similar set of issues in his work on Mexico City and the efforts to redevelop the city's historic center. Walker analyses the connections between the network of actors involved in the redevelopment, the outcomes for the area's physical spaces and the people who live and work there, and the different forms of resistance it provokes. He shows that here too perceptions by more affluent residents and local officials of appropriate uses of public and quasi-public spaces, in combination with specific market imperatives, act to constrain the lives of marginalized workers, in this case street vendors. Walker also draws our attention to the less than straightforward relationship between the middle class and neoliberalism, showing that more affluent residents are quite vocal and organized in rejecting redevelopment when it is seen to impinge on their quality of life, yet are actively involved in supporting policies that increase insecurity for more precarious workers.

In the final chapter of Part II Christine Hentschel compares struggles against segregation in Durban with those against gentrification in Berlin. She finds that in the former, historically rooted concern over overt and repressive forms of segregation often miss how in a post-apartheid context micro-scaled "soft" forms of segregation linked to aesthetic and lifestyle markers can function to reproduce exclusivity and apartness. In contrast, Hentschel finds in the gentrifying Berlin neighborhood of Neukölln that the narrow focus by white activists on conventional forms of gentrification fails to appreciate the broader context of racial and ethnic segregation and undermines alliances with working-class immigrant communities. In each case, possibilities for transformative struggles over urban space could be expanded through a more critical view of the processes through which divisions are being reproduced.

The chapters in Part II detail how a particular vision of cosmopolitan urbanism, driven by networks of affluent actors and expressed through redeveloped built environments, figures prominently in creating the texture of urban spaces and divisions. They show how different articulations of this vision appear, but also support the argument that there is a transnational dimension to the environments themselves, the actors involved and the consequences for urban marginality and the lives of less powerful and affluent residents. In doing so they contribute to ongoing discussions over the respective roles of and relationships between state, market and civil society in urban governance, while simultaneously drawing attention to spaces outside these institutions, and forces which cut across and through them. Of particular note is the prominent role of middle-income groups, and their posture towards the urban poor with whom they often, and it seems unwillingly, share urban space. In this relationship we can see a city pulled between two trajectories that are not only separate but mutually exclusive, while remaining intimately and materially bound to each other. These opposing forces will shape contests over space, resources and power in the South for some time to come, and in the final section of the book we turn to the politics of the divided city.

Governance and counter-governance – the shape of urban conflict and the urban future

The politics arising from divided cities in the South are located squarely within a wider global democracy deficit and the concomitant spread of governance regimes in which market-based economies are tightly bound to a variety of formal political systems. The phenomenon is global in scope and transnational in nature, but may be concentrated and at its most severe in the world's urban centers. While up to this point we have been primarily concerned with the city's socio-spatial production, the issue of political territory has been at or near the surface the entire time. The aims of distinct actors that form a part of local regimes, coalitions or networks may be driven variously by economic incentives, aesthetic and moral concerns or desire for particular kinds of (secure) lifestyles. Regardless of which of these is implicated the resulting policies and practices are at root expressions of particular systems of governance that function to create and control territories. Further, a defining feature of these systems in relation to territory, and in this discussion to political territory, is a movement towards enclosure.

The enclosure of political space under market-driven forms of urban governance is a rational development given the general drift of the control over a range of resources from the public domain to private and public-private networks; it is indeed a movement that has been central in reinvigorating the interest in right to the city and urban democracy more generally (Purcell 2002). In the Global South there appears to be a rough match between the division of city spaces and the political disenfranchisement of the majority of urban residents, as local, national and transnational actors combine to form highly effective governance networks that even the most organized local opposition will struggle to locate and access, much less impact. Yet, local struggles to defend, retake and remake urban spaces abound, and in many cities the so-called peripheries are sites of social, cultural, economic, political and technical innovations that somehow and at least for some time evade the reach of governance regimes (Roy 2010; Simone 2010; Holston 2009; Lazar 2008; Neuwirth 2004). Enclosure may be a powerful repressive tendency, but it is not one that can ever be fully realized in practice.

Urban politics arise from the struggles to survive, live and thrive within the constraints of the divisions the city imposes on its inhabitants. In most if not all cities of the South these constraints include dependence on the labor of the poor by more affluent groups who do not necessarily want to live side by side with them. We can refer to the politics arising from this as expressions of governance and counter-governance. If the former can be characterized as attempts to exert decisive control over territories in pursuit of specific agendas, then the latter refers to efforts to oppose this control and, beyond that, to assert alternative control over the same territory. These do not encapsulate the entirety of politics in the city, either of affluent or marginal residents, but they do speak to a central political dynamic. The politics of counter-governance play out across and outside formal institutions and processes (as do those of governance), they can engage with, subvert or ignore (at least to some extent) existing governance structures, as well as produce their own. What binds them is that they in some way represent responses to political exclusion, whether through a lack of legal protections/rights, inadequate enforcement of existing rights or the imposition of disempowering "development" schemes (Huchzermeyer 2011; Brown *et al.* 2010; Kranthi and Rao 2010; Chatterjee 2004).

Taking up these issues directly in the first chapter of Part III, Ernesto López-Morales uses the real estate-led expansion of middle- and upper-income housing in Santiago to explore resistance by working-class residents to gentrification and displacement. Although the immediate target of the activists was the redrafting of the master plan for the Pedro Aguirre Cerda district, López-Morales shows that this opposition was at the same time a contestation of power along class lines, in this case over the remaking of a working-class community, and of the relatively unaccountable and unreachable institutions of governance that characterize the Chilean planning system. In response, an increasingly organized and active social movement has emerged to project its own power in the contestation of urban space and maneuver against top-down planning processes. Mobilizing and disseminating technical knowledge at the grassroots level, the movement was able to counter plans generated by powerful local actors through innovative engagement with and shaping of the political process.

Returning to the role of the middle class in the city, Ryan Centner examines a similar set of issues from quite a different perspective: how the middle classes in Buenos Aires, Istanbul, Rio de Janeiro and São Paulo negotiate and construct identity while producing and mobilizing around particular notions of the right kind of city. He draws attention to the relationship between urban governance, the shifting, malleable meanings of "middleclassness" and the complex politics of distinction that result, with great consequences for city spaces. Centner locates these phenomena in the growth of the middle class in many countries of the South and the intensification of the politics of urban space that has accompanied it. He finds that although class remains central to urban struggles across the cases, class itself is less a fixed category than a complex and contested one linked to other dimensions of identity and status that draw legitimacy from a range of sources, not all rooted in the local. What emerges in all cases, though, are exclusionary notions of citizenship and rights that complicate attempts to link rights-based movements and discourses facilely to the production of more just cities for all urban residents.

Finally, Wael Fahmi presents research on the innovative interpenetrating, hybrid spaces being created by bloggers in Cairo. Constrained by government repression and surveillance in the city's public squares, bloggers and "hacktivists" are creating new spaces within which to claim and exercise rights, moving between real and virtual worlds. In doing so, protestors have been able to re-empower themselves in the city, and remake the spaces from which they had either been excluded or within which their activities were being constrained. Fahmi's piece provides crucial insights into how spaces of freedom are constructed and continuously evolving in response to hegemonic, and in this case authoritarian, governance structures. Further, the activists in Cairo show that even in the most repressive conditions possibilities exist for producing, expanding and inhabiting political spaces of resistance.

These chapters touch on a number of key concerns that are useful in orienting our thinking on these issues. In Santiago and Cairo, mobilized residents intrude into existing spaces and create new ones within which to enact improvisational forms of urban citizenship that overcome deficiencies or obstacles in existing citizenship practices, and themselves have the potential to become institutionalized. Cairo bloggers in particular demonstrate the possibilities of autoconstruction in the context of political spaces (Holston 2009). Centner, however, provides an important reminder that both the definition and outcomes of rights-based struggles are indeterminate, linked to complex local dynamics and shifting terrains that can produce competing and contradictory rights claims. In the final section of this introduction we draw from the volume's contributions and offer some thoughts on urban political struggles and, more specifically, right to the city across the Global South, as well as the hostile terrain that generates them.

Neoliberalism versus right to the city

To close, we offer some thoughts on how the work in this volume contributes to building an understanding of neoliberalism and right to the city. It might be useful to begin by positing that each of these concepts capture and isolate opposing time- and place-specific tendencies that push cities towards two fundamentally different urban futures: one characterized by various forms of enclosure; the other by various forms of justice. We must stress that these are tendencies, and that they are subject to alterations and mutations that cannot be anticipated and may render the terms themselves anachronistic. In that sense, we see this discussion as one of many that can take place, and has been taking place, about the making of cities of the South, while we hold that the research presented here offers compelling evidence that at present the concepts remain important for thinking about these cities from a critical social justice perspective. Developing a distinct and independent Southern urbanism is still in the early, basic research phase; consequently we offer early reflections on these concepts that will hopefully continue to undergo critique and refinement. To begin, we suggest that neoliberalism functions as a mode of economic inequality and marginalization that articulates with three other interlocking modes of exclusion: physical, social and political.

The exclusive city

While the socio-spatial division of urban space in not new, the chapters here suggest that neoliberalism is linked to the production of particular kinds of cores and peripheries through the commodification of land and housing markets, playing a central, historically grounded role in the deepening and hardening of divided cityscapes. These processes produce important differences shaped by local context – as in those between Beirut and Ouagadougou, for example – but important aspects that bind them include an articulation or rearticulation of land value through the real estate market that is directly linked to enclosure, displacement, eviction, commodification, redevelopment, piecemeal upgrading, containment or neglect, depending on assigned values. These in turn shape socio-spatial marginalization and help to define core and peripheral spaces, whether through increased distance from economic, political and cultural "centers" or various forms of in situ regulation when physical distancing would be an onerous undertaking. The primary constraints on how far market-led redevelopment can penetrate into and monetize "disorderly" spaces in the South are the power and organization of state actors and the networks in which they participate for these purposes, the size and durability of "troublesome" neighborhoods and populations, the extent of urban poverty and the degree of inequality.

The rise and expansion of affluent classes in the South has been attributed to the adoption of market reforms by many nations. The disproportionate concentration of these populations in cities has, of course, had profound implications for urban space and politics. Kolkata, Durban and Mexico City certainly differ in many ways, but there is a clear convergence in the carving out of certain spaces from their hostile and disorderly surroundings, and not only redeveloping these sequestered spaces but also infusing them with a certain status and class-based morality qua civic virtue. Belonging is coded here in terms that betray a differentiated citizenship, linking social status to territory. From here an entire politics of place becomes possible, which functions across numerous cities as a mechanism of exclusion from or regulation in public and quasi-public spaces. Again, the constraints imposed by cities in the South are important, particularly the dependence of affluent groups on the labor of the poor and the proximity of unequal groups to each other in many places. Affluent and poor are more tightly bound together because of the central role of cheap labor in underwriting affluence, and more distant because of the often glaring contrasts between living space, lifestyles and life trajectories. It makes sense that a particularly intense spatial politics of belonging emerges with clear markers given this cheek-by-jowl existence, whether these result in fortification, repression or simply dispersion as affluent groups move away.

Neoliberalism in the South is implicated not just in physical and social exclusion, but also in the enclosure of political space. The splitting of social space and subsequent distancing of the urban poor, as well as elite control over nominally democratic institutions and processes, contribute to the unequal distribution of political power and access. The rise and expansion of governance networks to which most urban residents have limited access, comprised of affluent citizens, local officials and representatives of the private sector seeking to extend control over territories, can limit and even effectively disenfranchise large numbers of residents. From efforts in Casablanca and Ouagadougou to eliminate slums, to mobilizations by middle-class residents in Rio, Buenos Aires or Istanbul to assert their own vision of the right kind of city, we see policies and practices that are often about the urban poor, but rarely from them. As Ren and Weinstein suggest, while variations in local and national political context can play an important role in determining pace and scale of transformations, spatial and political marginalization mark mega-project development in both Shanghai and Mumbai, despite the very different political systems in each case.

Neoliberalism in the South is associated with the (re)production of urban divisions and the emergence of redeveloped cosmopolitan spaces of affluence that increasingly are in tension with, even as they are bound to, the peripheral spaces of the urban majority. From this emerges a politics of physical, social and political distancing, as affluent citizens seek to remove themselves from the poverty around them, and the populations upon whose labor their affluence largely depends. Central to this politics are efforts to enclose and commandeer political spaces themselves. While these dynamics are not absent in the North, and may in fact be increasing in importance, they operate at a scale and intensity in the urban South that produces a distinct category of urbanism and, as we turn to next, a distinct set of political questions and politics.

Right to the city as counter-governance

For now, right to the city cannot be understood separately from neoliberalism, it must be located within the context of and as a response to the substantial rescaling of inequality, crisis and, to a great extent, global governance (Harvey 2012; Samara 2012; O'Connor 2008; Purcell 2002). If neoliberalism is viewed as an innovation in the reproduction and recreation of socio-spatial inequality, then right to the city is a political response and challenge to this development. If neoliberalism is a central force driving the enclosure of political spaces, then right to the city, to start, can be understood as both resistance and an attempt to create or open existing political spaces. Taken in aggregate, the daily pushes against various sub-populations in cities across the South, just as cities have reemerged as spaces of citizenship, amount to a collective effort to deny urban majorities the *right* to be *of* the city, or at least those parts of the divided city in which certain forms of power (linked to governance) and types of resources (linked to affluence) are concentrated (Isin 2002; see also Sassen 2004).

Writings on right to the city relevant to the South cover a relatively broad spectrum of thought, underscoring the challenges of conceptual and political specification. From these a number of key issues and debates emerge. In a useful distinction, Peter Marcuse separates the demand for rights *in* cities (cities as sites of politics) from a right *to* the city (the city as the object of politics), the latter a potentially more transformative approach which can include a constructive power (of space, for example) that goes beyond mere participation in existing structures and processes (Marcuse 2010; see also Harvey 2008; Isin 2002). Developing this line of thought in his work on water provision in South Africa, Patrick Bond distinguishes between what he calls the limited scope of consumption rights and a more expansive and radical right to the commons (Bond 2010). Other writers have taken up the critique of right-to-the-city approaches that center state, law, civil society and other institutional sites of reform, and marginalize more radical and politicized iterations (Kipfer *et al.* 2012; Samara 2012; Mayer 2009; Fernandez 2007). This focus is matched by attention to the meaning of right to the city for traditionally excluded social groups, in particular women, racial and ethnic minorities and undocumented migrants (Taylor 2011; Gilbert and Dikeç 2008; Varsanyi 2008; Fenster 2005; McEwan 2005).

Looming behind many of these arguments and discussions is the meta-issue of socio-spatial segregation, which as we have argued here is a defining feature of cities and, from the point of view of inequality and social justice, an especially salient one. Calls for urban policy and practice that integrate divided spaces and populations, or for building inclusive cities, remain popular among many non-governmental organizations, scholars and community organizations (Zerah *et al.* 2011; Parnell and Pieterse 2010; UN-HABITAT 2010; Brown and Kristiansen 2009). A common thread uniting many of these calls is a belief that integration is possible and necessary for equitable urban development, and desirable as a matter of principle. The massive challenges facing cities of the South, emerging from such scale and depth of inequality, should caution us against too readily accepting any of these beliefs as essential features of what right to the city can mean in the South. Indeed, we should consider whether right to the city may require new conceptions of centrality and more radical forms of politicization of urban conflict than the integrationist approaches offer (Kipfer *et al.* 2012; Turkmen 2011). At the very least, the reality on the ground suggests that we reconsider the parameters of possibility. What does the divided city tell us about right to the city as a particular intervention into urban politics, and, then, what can right to the city tell us about the possible futures of divided cities?

There is much to suggest that divided cities will be the predominant urban form for some time to come; David Harvey (2003), Bervoets and Loopmans (Chapter 3, this volume), Chen (Chapter 4, this volume) and Fawaz (Chapter 1, this volume) all show urbanization across the South operating as a mechanism of enclosure and separation. Before positing that right to the city is dependent of some form of integration, we must ask whether the divided city is an obstacle to be overcome or a relatively fixed urban configuration from which new politics (and polities) will emerge and new, unanticipated urban forms produced? Is the goal an integrated city or is the integrated city a proxy concept for other goals whose achievement is not dependent on (a highly unlikely) integration? Inherent in these questions are two assertions: that, whatever the answers to them may be, right to the city is fundamentally linked to place; and that the city alone is too broad a spatial lens and can make it difficult to see – and center – the important fault lines upon which it is constructed.

An additional phenomenon that work in this volume also documents is that affluent groups, networks and coalitions seem to pursue, often quite vigorously, exclusive cities, and it seems a conception of right to the city would need to contend directly with this political reality. A conception which is in any way reliant upon petitioning (state, the "public") to "let the poor in" or building partnerships with these groups for various development schemes seems to misrecognize the problem as one of historical inertia, resource scarcity, bureaucratic inefficiency or some other explanation that discounts the centrality of self-aware agents who are very much certain of the kind of city they want. The thinness of such an approach becomes clear in light of the contributions here by Bose (Chapter 6), Walker (Chapter 8), Hentschel (Chapter 9), Centner (Chapter 11) and Berney (Chapter 7), each in its own way detailing the power of cosmopolitan urbanism in remaking places but also politics, as new and established affluent groups confront the perceived disorder that often surrounds them and, for some, the fragility of their own positions. The point is not that such approaches are doomed to failure or should never be pursued, but that analysis of their potential must take the political limitations they often entail more seriously. At present, right to the city in the South has to be built upon the antinomic political foundation that divided cities produce and are produced by.

Divided cities in which more powerful groups collaborate in enclosing political spaces and excluding urban majorities raise the question of where a politics of the excluded is located, if not in the formal institutions and processes that, at least in liberal democracies, are meant to afford all legal residents with access to political power. Partha Chatterjee offers a useful orientation to this question through his discussion of popular politics. Although not writing in the idioms of urbanism or geography, the civil society that appears through his critique resonates with both: "Civil society ... will appear as the closed association of modern elite groups, sequestered from the wider popular life of the communities, walled up within enclaves of civic freedom and rational law" (Chatterjee 2004: 4). Political exclusion, to the extent we see this as an effective exclusion from civil society and state, then raises two related questions: where is the actual or potential political power of the excluded centered, and how it is expressed?

The loose correlation between social, spatial and political exclusion requires closer examination in future work. To the extent that marginalized populations construct homes, communities and economies in peripheral spaces, further exploration of how political power and right to the city are produced from this position is needed, from exploring how working-class communities in Santiago mobilize in innovative ways to enter and control institutions and processes from which they have been excluded, to Fahmi's study of how activists in Cairo construct new – in this case virtual – political spaces from which they orchestrate efforts to claim downtown public spaces for political protest. What binds these efforts, those that are successful as well as those that are not, is the labor of building political centers outside of enclosed spaces, from which a range of options present themselves. While many of these are certainly driven, at least in part, by a desire for participation, they also open up the possibility of a politics beyond integration and inclusion.

Conclusion

The aim of this introduction has not been to provide readymade definitions of neoliberalism and right to the city in a Global South context, but to explore the terrain from which conceptions of each emerge. This is what is meant by locating the concepts: situating them in specific spaces and as aspects of distinctive processes and dynamics. If we privilege them here it is only because they appear to have a value in capturing certain important and common aspects of urban life in the South related to the global challenges of inequality and deprivation. We add our voices to those who have warned against claiming exaggerated, and unsubstantiated, explanatory power for either concept, and liberatory power for right to the city. At the same time, we hold to the continued significance of and potential in each concept, and in their interrelationship. More broadly, in our focus on the urban South, we hope this volume marks a contribution to efforts at situating what have for too long been considered marginal and peripheral cities and spaces as centers, both longstanding and emerging, while also avoiding the pitfall of attributing to these inherent or essential social, spatial or political qualities. We do believe, however, that a fundamental shift is occurring and that it will be increasingly untenable to refer to as marginal or peripheral those areas of the world where most people live, where most economic activity takes place, where multiple forms of global urban culture are forged and from which the strongest political storms of the future may well emerge.

References

Abaza, M. (2011) "Cairo's Downtown Imagined: Dubaisation or Nostalgia?", *Urban Studies*, 48 (6): 1075–87.

Amar, P. (2010) *New Racial Missions of Policing: International Perspectives on Evolving Law-Enforcement Politics*, New York: Routledge.

Amen, M. M., Archer, K. and Bosman, M. M. (eds) (2006) *Relocating Global Cities: From the Center to the Margins*, New York: Rowman & Littlefield Publishers.

Batra, L. (2008) "Deconstructing the 'World Class' City", *Seminar*, 582. Online: www.india-seminar.com/semframe.html (accessed 24 March 2012).

Baviskar, A. and Ray, R. (eds) (2011) *Elite and Everyman: The Cultural Politics of the Indian Middle Classes*, New Delhi: Routledge India.

Binnie, J., Holloway, J., Millington, S. and Young, C. (eds) (2006) *Cosmopolitan Urbanism*, New York: Routledge.

Birkinshaw, M. and Harris, V. (2009) "The Right to the 'World Class City'? City Visions and Evictions in Mumbai", *The Urban Reinventors*, special issue 3: "The Right to the City: the Entitled and the Excluded".

Bond, P. (2010) "Fighting for the Right to the City: Discursive and Political Lessons from the Right to Water", paper presented at the Right to Water Conference, Syracuse, NY, March.

Brenner, N. (2009) "What is Critical Urban Theory?", *City*, 13 (2–3): 198–207.

Broudehoux, A. (2007) "Spectacular Beijing: The Conspicuous Construction of an Olympic Metropolis", *Journal of Urban Affairs*, 29 (4): 383–99.

Brown, A. and Kristiansen, A. (2009) *Urban Policies and the Right to the City: Rights, Responsibilities and Citizenship*, Nairobi: UN Habitat.

Brown, A., Lyons, M. and Dankoco, I. (2010) "Street Traders and the Emerging Spaces for Urban Voice and Citizenship in African Cities", *Urban Studies*, 47 (3): 666–83.

Chatterjee, P. (2004) *The Politics of the Governed: Reflections on Popular Politics in Most of the World*, New York: Columbia University Press.

Chen, X. (ed.) (2009) *Shanghai Rising: State Power and Local Transformations in a Global Megacity*, Minneapolis, MN: University of Minnesota Press.

COHRE Global Forced Evictions Programme (2006) *Forced Evictions: Violations of Human Rights 2003–2006*, Geneva: Centre on Housing Rights and Evictions. Online: www.cohre.org/globalsurvey (accessed 14 March 2010).

Connell, J. (1999) "Beyond Manila: Walls, Malls, and Private Spaces", *Environment and Planning*, A 31 (3): 417–39.

Davis, D. (2010) "The Sociospatial Reconfiguration of Middle Classes and their Impact on Politics and Development in the Global South: Preliminary Ideas for Future Research", *Political Power & Social Theory*, 21: 241–67.

Elate, S. (2004) "African Urban History in the Future", in T. Falola and S. J. Salm (eds) *Globalization and Urbanization in Africa*, Asmara, Eritrea: Africa World Press.

Fenster, T. (2005) "The Right to the Gendered City: Different Formations of Belonging in Everyday Life", *Journal of Gender Studies*, 14 (3): 217–31.

Fernandes, L. and Heller, P. (2006) "Hegemonic Aspirations: New Middle Class Politics and India's Democracy in Comparative Perspective", *Critical Asian Studies*, 38 (4): 495–522.

Fernandez, E. (2007) "Constructing the 'Right to the City' in Brazil", *Social & Legal Studies*, 16 (2): 201–19.

Garrido, M. (2012) "The Ideology of the Dual City: The Modernist Ethic in the Corporate Development of Makati City, Metro Manila", *International Journal of Urban and Regional Research*, Online (restricted access; accessed 7 March 2012).

Gilbert, L. and Dikeç, M. (2008) "Right to the City: Politics of Citizenship", in Goodewardena, K., Kipfer, S., Milgrom, R. and Schmid, C. (eds) *Space, Difference, Everyday Life: Reading Henri Lefebvere*, New York: Routledge.

Goldman, M. (2011) "Speculative Urbanism and the Making of the Next World City", *International Journal of Urban and Regional Research*, 35 (3): 555–81.

Harvey, D. (2012) *Rebel City: From Right to the City to the Urban Revolution*, London: Verso.

Harvey, D. (2008) "The Right to the City", *New Left Review*, 53 (September–October). Online: http://newleftreview.org/II/53/david-harvey-the-right-to-the-city (accessed 1 November 2010).

Harvey, D. (2003) *The New Imperialism*, Oxford: Oxford University Press.

Hirt, S. A. (2012) *Iron Curtains: Gates, Suburbs and Privatization of Space in the Post-Socialist City*, Malden, MA: John Wiley & Sons, Ltd.

Holston, J. (2009) *Insurgent Citizenship: Disjunctions of Democracy and Modernity in Brazil*, Princeton, NJ: Princeton University Press.

Huchzermeyer, M. (2011) *Cities with "Slums": from Informal Settlement Eradication to a Right to the City in Africa*, Cape Town: University of Cape Town Press.

Huyssen, A. (ed.) (2008) *Other Cities, Other Worlds: Urban Imaginaries in a Globalizing Age*, Durham, NC: Duke University Press.

Isin, E. (2002) *Being Political: Genealogies of Citizenship*, Minneapolis, MN: University of Minnesota Press.

Kipfer, S., Saberi, P. and Wieditz, T. (2012) "Henri Lefebvre: Debates and controversies", *Progress in Human Geography*, 29 May. Online: http://phg.sagepub.com/content/early/2012/05/29/0309132512446718 (accessed 1 June 2012).

Kneebone, E. and Garr, E. (2010) *The Suburbanization of Poverty: Trends in Metropolitan America, 2000–2008*, Washington, DC: Brookings Institute. Online: www.brookings.edu/~/media/Files/rc/papers/2010/0120_poverty_kneebone/0120_poverty_paper.pdf (accessed 15 January 2012).

Kranthi, N. and Rao, K. D. (2010) "Security of Tenure and Protection Against Evictions of Slum Dwellers: a Case of Hyderabad", Institute of Town Planners, *India Journal*, 7 (2): 41–9.

Labbé, D. and Bourdreau, J. (2011) "Understanding the Causes of Urban Fragmentation in Hanoi: the Case of New Urban Areas", *International Development Planning Review*, 33 (3): 273–91.

Lazar, S. (2008) *El Alto, Rebel City: Self and Citizenship in Andean Bolivia*, Durham, NC: Duke University Press.

Marcuse, P. (2010) "Rights in Cities and Right to the City", in A. Sugranyes and C. Mathivet (eds) *Cities for All: Proposals and Experiences towards the Right to the City*, Santiago: Habitat International Coalition.

Marcuse, P. (2003) "What's So New About Divided Cities?", *International Journal of Urban and Regional Research*, 17 (3): 355–65.

Mawdsley, E. (2004) "India's Middle Classes and the Environment", *Development and Change*, 35 (1): 79–103.

Mayaram, S. (ed.) (2008) *The Other Global City*, New York: Routledge.

Mayer, M. (2009) "The 'Right to the City' in the Context of Shifting Mottos of Urban Social Movements", *City*, 13 (2–3): 362–74.

McDonald, D. A. (2008) *World City Syndrome: Neoliberalism and Inequality in Cape Town*, New York: Routledge.

McEwan, C. (2005) "New Spaces of Citizenship? Rethinking Gendered Participation and Empowerment in South Africa", *Political Geography*, 24 (8): 969–91.

Murray, M. J. (2010) *City of Extremes: The Spatial Politics of Johannesburg*, Durham, NC: Duke University Press.

Murray, M. J. (2004) "The Evolving Spatial Form of Cities in a Globalising World Economy: Johannesburg and São Paulo", Democracy and Governance Research Programme, Occasional Paper 5, Cape Town: HSRC Publishers.

Neuwirth, R. (2004) *Shadow Cities: A Billion Squatters, A New Urban World*, New York: Routledge.

Nightingale, C. H. (2012) *Segregation: a Global History of Divided Cities*, Chicago, IL: University of Chicago Press.

O'Conner, A. (2008) "The Privatized City: The Manhattan Institute, the Urban Crisis, and the Conservative Counterrevolution in New York", *Journal of Urban History*, 34 (2): 333–53.

Ong, A. (2006) *Neoliberalism as Exception: Mutations in Citizenship and Sovereignty*, Durham, NC: Duke University Press.

Parnell, S. and Pieterse, E. (2010) "The 'Right to the City': Institutional Imperatives of a Developmental State", *International Journal of Urban and Regional Research*, 34 (1): 146–62.

Purcell, M. (2002) "Excavating Lefebvre: the Right to the City and its Urban Politics of the Inhabitant", *Geojournal*, 58 (2–3): 99–108.

Ren, X. (2011) *Building Globalization: Transnational Architecture Production in Urban China*, Chicago, IL: University of Chicago Press.

Roberts, B. R. and Wilson, R. W. (eds) (2009) *Urban Segregation and Governance in the Americas*, New York: Palgrave Macmillan.

Robinson, J. (2005) "Urban Geography: World Cities, or a World of Cities", *Progress in Human Geography*, 29 (6): 757–65.

Robinson, J. (2002) "Global and World Cities: a View from Off the Map", *International Journal of Urban and Regional Research*, 26 (3): 53–4.

Rodgers, D. (2004) "'Disembedding' the City: Crime, Insecurity and Spatial Organization in Managua, Nicaragua", *Environment & Urbanization*, 16 (2):113–23.

Rodrigues, C. U. (2009) "Angolan Cities: Urban (Re)Segregation?", in F. Locatelli and P. Nugent (eds) *African Cities: Competing Claims on Urban Spaces*, Leiden: Brill NV.

Roy, A. (2010) *Poverty Capital: Microfinance and the Making of Development*, New York: Routledge.

Roy, A. (2009) "Why India Cannot Plan its Cities: Informality, Insurgence and the Idiom of Urbanization", *Planning Theory*, 8 (1): 76–87.

Roy, A. and Ong, A. (eds) (2011) *Worlding Cities: Asian Experiments and the Art of Being Global*, Malden, MA: Blackwell.

Samara, T. R. (2012) "Citizens in Search of a City: Towards a New Infrastructure of Political Belonging", in M. Smith and M. McQuarrie (eds) *Remaking Urban Citizenship: Organizations, Institutions and Right to the City*, Piscataway, NJ: Transaction Publishers.

Samara, T. R. (2011) *Cape Town After Apartheid: Crime and Governance in the Divided City*, Minneapolis, MN: University of Minnesota Press.

Samara, T. R. (2010) "Order and Security in the City", *Ethnic and Racial Studies*, 33 (4): 637–55.

Sassen, S. (2004) "The Repositioning of Citizenship", in A. Brysk and G. Shafir (eds) *People Out of Place: Globalization, Human Rights and the Citizenship Gap*, New York: Routledge.

Searle, L. (2010) "Making Space for Capital: The Production of Global Landscapes in Contemporary India", Publicly accessible Penn dissertations, Paper 160. Online: http://repository.upenn.edu/edissertations/160 (accessed 24 January 2012).

Simone, A. M. (2010) *City Life from Jakarta to Dakar: Movements at the Crossroads*, New York: Routledge.

Simone, A. and Abouhani, A. (eds) (2005) *Urban Africa: Changing Contours of Survival in the City*, London: Zed Books.

Smith, M. P. (2000) *Transnational Urbanism: Locating Globalization*, Malden, MA: Wiley-Blackwell.

Smith, N. (2002) "New Globalism, New Urbanism: Gentrification as Global Urban Strategy", *Antipode*, 34 (3): 427–50.

Swanson, K. (2007) "Revanchist Urbanism Heads South: the Regulation of Indigenous Beggars and Street Vendors in Ecuador", *Antipode*, 39 (4): 708–28.

Taylor, A. (2011) *Women and the City: Examining the Gender Impact of Violence and Urbanization*, Johannesburg: Action Aid.

Turkmen, H. (2011) "Debates on Right to the City in Istanbul", paper presented at the International RC21 Conference, Amsterdam, August.

Varsanyi, M. (2008) "Immigration Policing Through the Backdoro: City Ordinances, the 'Right to the City,' and the Exclusion of Undocumented Day Laborers", *Urban Geography*, 29 (1): 29–52.

United Nations-Habitat (2011) "Forced Evictions: Global Crisis, Global Solutions", Nairobi, Kenya. Online: www.unhabitat.org/pmss/listItemDetails.aspx?publicationID=3187 (accessed 12 May 2012).

United Nations-Habitat (2010) *State of the World's Cities 2010/2011: Bridging the Urban Divide*, London: Earthscan.

United Nations-Habitat (2008) "Secure Land Rights for All", Nairobi, Kenya. Online: www.responsibleagroinvestment.org/rai/sites/responsibleagroinvestment.org/files/Secure%20land%20rights%20for%20all-UN%20HABITAT.pdf (accessed 30 April 2012).

Wacquant, L. (2008) "The Militarization of Urban Marginality: Lessons from the Brazilian Metropolis", *International Political Sociology*, 2 (1): 56–74.

Weiwei, A. (2011) "The City: Beijing", *Newsweek*, August 28. Online: www.thedailybeast.com/newsweek/2011/08/28/ai-weiwei-on-beijing-s-nightmare-city.html (accessed 20 March 2012).

Young, C., Diep, M. and Drabble, S. (2006) "Living with Difference?: the 'Cosmopolitan City' and the Urban Reimagining in Manchester, UK", *Urban Studies*, 43 (10): 1687–714.

Zerah, M., Dupont, V. and Lama-Rewal, S. (eds) (2011) *Urban Policies and the Right to the City in India: Rights, Responsibilities and Citizenship*, New Delhi: UNESCO. Online: http://unesdoc.unesco.org/images/0021/002146/214602e.pdf (accessed 10 April 2012).

Part I

A city divided against itself

1 Towards the right to the city in informal settlements

Mona Fawaz

Throughout the months of April and May 2011, a surge of illegal constructions occupied the limelight of Lebanese politics, and news coverage reported daily confrontations between police forces, sometimes backed by the national army, and self-help builders. In multiple instances, violence was uncontrolled on both ends: police cars were torched and several civilians were severely wounded, others killed. A common thread in the public responses of policymakers and the news coverage was the criminalization of those who had built in violation of property, building and/or urban regulations. Most official responses stemmed from the Ministry of Interior and adopted the security discourse of law enforcement and curtailing illegality. These materialized in actual military strategies, such as the deployment of army tanks to bulldoze recently built structures.

Neither public officials nor the media drew connections between this short-lived building boom and national housing and land policies. One is hard pressed, however, to ignore the housing dimension in this illegal construction saga. It is first evidenced in the narratives of those who have engaged in processes of illegal construction: families sold jewelry, cars and other personal valuables when the opportunity to add a floor or even a room to their existing shelter was provided. The unmet housing demand is also poignantly revealed by skyrocketing property values, triggered over the past two decades by development policies that conceptualize of real estate investments as a main source of national economic growth. Since the 1990s, incentives have been deployed to encourage national and foreign capital into this the real estate sector[1] and property values have risen severalfold, rendering the smallest apartment in Beirut unaffordable to the vast majority of the country's population.

One can read in this short narrative the same old song of neoliberal urbanization as it has materialized in many contexts: an ever more exclusive city where the efforts of low-income dwellers to circumvent difficulties are often criminalized and met with militarized reactions (Wacquant 2008; Samara 2010). Zooming in on details, one can also identify many of the ingredients associated with this form of urbanization: property values pegged to global (regional and diaspora) interests and demands (Harvey 2007; Sassen 2000; Smith 2002; Shatkin 2004), urban regulations and taxations organized to facilitate foreign investments and high-end commercial and residential development and, conversely, dwindling tenure security for low-income dwellers whose presence in the city becomes even more superfluous as their once needed labor is replaced by cheaper goods and/or labor imported from nearby and faraway locations, and numerous population displacements caused by highway and other so-called development projects (Deboulet and Fawaz 2011).

But what about the informal settlements of these cities, the areas that traditionally housed the working classes of most, if not all, Third World cities? While much of the literature listed above concurs on an assessment of "forgotten" and "neglected" places (Davis 2007; Shatkin 2004), I argue that the contemporary moment betrays a neoliberal policy vis-à-vis informal settlements, one that goes beyond the traditional laissez faire and the notorious reluctance of Lebanese policymakers to initiate regularization policies (Fawaz and Peillen 2002) in order to work actively towards the erasure of these neighborhoods in what I will show is a dual strategy of criminalization and de facto integration in the citywide housing market. It is, in fact, in this convergence of direct and indirect strategies to reinforce market-type relations in these neighborhoods that one sees best the effects of neoliberalism, defined throughout this chapter as an active set of interventions aimed at delegating to market institutions areas of social and political life that are regulated by other institutions (Dikeç 2006; Brenner and Theodore 2002). Neoliberalism, hence, is not mere laissez faire. It requires policymakers to work actively towards facilitating the expansion of the market. This chapter argues that such an active neoliberal policy is at work in Beirut. It materializes, on the one hand, with the marketization of the process of housing and service provision in informal settlements, and on the other hand by deploying the necessary policing to protect market operations.

To what extent is the Lebanese case reflective of other contexts and what, then, should be advocated for those who still find in informal settlements a reasonable form or shelter provision in increasingly exclusive cities? Based on a comparative analysis of the Lebanese context and other cases documented in the literature, I argue that it is possible to outline common patterns across geographic contexts that converge towards the delegation of processes of housing provision and exchange to market forces. In order to respond to current conditions, I further argue that it will be important to reframe the current context of public policymaking vis-à-vis informal settlements in political rather than market terms. I offer the concept of the "right to the city" as a powerful approach in which reframing can be articulated. Recognizing, however, that neoliberalism has materialized differently across regional and national contexts (Robinson 2002), I zoom in on the Lebanese case to outline its specificity and reflect on the formulation of the "right to the city" that should be articulated for this context.

This chapter is based on over a decade of engagement in informal settlements in Lebanon. I began looking at the historical production of informal settlements in Lebanon back in 1999, when I started my dissertation work (Fawaz 2004). Since then, I have conducted numerous projects that have sought, through a variety of qualitative research approaches, to reconstruct the history of spatial production in Beirut's informal settlements and understand how the position of these neighborhoods has changed over time vis-à-vis the urbanization of the city. Although I didn't conduct any fieldwork specifically for this chapter, I was able to access interviews conducted with dwellers in Tyre (South Lebanon) through a joint initiative by the American University of Beirut's Center for Civil Engagement and UN-Habitat in June 2011. I also build on an extensive and long-term engagement and draw on several earlier researches (Fawaz and Peillen 2002; Fawaz 2009) as well as detailed literature and newspaper reviews reporting on recent changes in Lebanon and the region.

In the next section, I outline the materialization of neoliberal policies on informal settlements, beginning by documenting the Lebanese case and moving to a more general review of the literature. In the second part of the chapter, I move to a discussion of the right to the city and argue for the necessity to activate concepts such as this in order to repoliticize the discourse on housing policy.

Informal settlements: then and now, here and there

Beirut's informal settlements and the advent of neoliberalism

A small city of some 180,000 inhabitants under the French Mandate (1919–42), Beirut rapidly grew after Lebanon gained its independence in 1942, establishing itself as a main regional tourist and commercial center. Between 1950 and 1975, the capital city of the newly established State of Lebanon witnessed a real population explosion, going from 300,000 inhabitants to some 1.2 million inhabitants. By 1975 it housed about half of Lebanon's population (Kassir 2003). Most images that circulate of this period flaunt the beautiful sea-front resorts, illustrating well the widely celebrated liberal lifestyle, as well as experimentations in architecture that placed Beirut at the forefront of regional explorations of modernity. Efforts at nation-building and the elaboration of a social and economic infrastructure that would support the development of the young nation state lagged behind, however, creating eventually a huge gap between, the wealthy neighborhoods of the city on the one hand, and impoverished rural areas on the other (Traboulsi 2007). It is therefore not surprising that much of the population growth described above was fueled by rural-to-urban migration, rural areas suffering from the notorious disinterest of national planners. It was also fueled by an influx of foreign workers (mostly Syrians and Egyptians) as well as refugees coming from nearby countries (particularly Palestinians in 1948).

It was during this period, particularly in the mid-1960s, that despite a laissez faire attitude and staunch aversion to any form of development constraints, urban and building regulations were adopted, imposing restrictions on building practices, especially in Beirut and its suburbs. The natural result of these regulations is that while some neighborhoods indeed benefited from improvements in their physical infrastructure, a sizable portion of the urbanization occurred informally, that is in violation of property, urban and/or building regulations. This was how low-income dwellers could carve spaces for themselves in the form of pockets in the city and mostly in its fringes, accessing land outside prime areas, at times purchased in installments and at other times squatted, and building and servicing homes incrementally. Since most of their practices did not abide by public regulations, they were rapidly tagged as "illegal" and judged "undesirable". On the eve of the Lebanese civil war (1975), scholars echoed the dismay of the local press by decrying the "misery cordon" on the peripheries of the city (Bourgey and Phares 1972). It was also possible, however, to see in many of these settlements neighborhoods in the making, areas which – following the widely repeated dictum at the time – reflected dwellers' capacities to take charge of their own access to shelter and incrementally improve their living conditions as resources became available to them (Turner 1972). Indeed, in many of these neighborhoods, dwellers had consolidated their housing with permanent materials, installed self-help sewer systems, and obtained, through political mobilization and/or clientelism, connection to water and electricity networks.

The process of illegal land occupation and building was further facilitated by the outbreak of the civil war in 1975. Throughout the years of civil war (1975–90), a large section of the city's urbanization occurred illegally, within and outside informal settlements, as squatting and violations of various regulations became largely the norm (Fawaz and Peillen 2002). By 1990, informal settlements had gained considerable importance in the city, particularly in its southern suburbs where vast tracts of land were occupied and buildings exceeded severalfold the allowable exploitation ratios. Their dwellers were organized in neighborhood committees where processes of building and exchanging housing relied on social and political forms of organizations and were regulated by a set of informal rules that often did not abide by state laws (Fawaz 2008; Clerc-Huybrechts 2008).

Since 1990, and despite various moments of heightened violence, Lebanon has entered a so-called period of post-war reconstruction. During this phase, the country has renewed its commitment to liberal policymaking and has strived to reaffirm the position of its capital city as the service and tourist hub of the Middle East. Over the past two decades, policies were adopted in order to attract capitals and investors from the region and beyond, creating a framework of facilities and incentives akin to the neoliberal reforms witnessed in the region and elsewhere (Makdisi 1996). Sizable public investments were also placed in large-scale infrastructure projects, particularly the development of a new international airport with a capacity of six million visitors per year, the renovation and enlargement of the city's harbor and the implementation of a network of highways superposed on the fabric of the historic city to facilitate circulation, particularly across strategic points such as the redeveloped historic core and the international airport. To date, these policies have not materialized in the projected economic growth but they have caused widening social inequalities and translated spatially into pockets of exclusive consumption and luxurious residence amidst an impoverished city.

Throughout the modern and contemporary histories, planning agencies failed to recognize informal housing as a form of housing acquisition. Not only have there been no regularization policies of the type observed in many other contexts (Fernandes and Varley 1998), but most policymakers continue to describe neighborhoods that existed decades prior to the civil war as "war-generated, temporary settlements" that should be dismantled once war-displaced populations "return" to their areas of origin. In some instances, services were extended to these areas – mostly as a pragmatic decision from the water and electricity agencies, which had to insure the financial sustainability of their operations (Fawaz 2004) – but these have never been accompanied with an acknowledgment of this form of housing as a legitimate or at least permanent form of city-making. Housing agencies have also played a very limited role, mostly confined to the provision of housing loans, but they have consistently failed to recognize informal housing as part of their jurisdictions.

It is in this backdrop that the current policies of marketization and securitization should be read. Earlier investigations of their implications in the informal settlements of Greater Beirut indicated that these neighborhoods had witnessed a rapid deterioration of living conditions, with increasing density, poor environmental conditions and deficient services (Fawaz 2009; Deboulet and Fawaz 2011). One of the most important changes in informal settlements, however, is the fact that land in these neighborhoods has become attractive to developers working in the formal sector because of skyrocketing land prices everywhere in the city. Through the operations of these developers, processes of housing production and exchange in informal settlements have been further integrated in the citywide housing market, rendering housing production in informal settlements closer in form and process to other segments of the market. This role is facilitated by an array of public interventions that have facilitated the provision of lump-sum capital to be used as a down payment for acquiring housing, often disbursed as indemnities for those displaced by the development projects of the post-war reconstruction (Bou Akar 2005). Thus, what is accessed or exchanged in today's informal settlements of Beirut is mostly a finished apartment within a multistory apartment building, much like the rest of the city market, and not a plot of land on which one can incrementally develop one's housing as the family's needs grow and its financial capacities improve. The channels of housing production have also been considerably modified, as developers who operate at the scale of a particular suburb or region now intervene within informal settlements, building housing complexes with personal savings and bank loans, and relying on design firms and their contracting companies. These developers advertise their projects in venues that are likely to attract reliable clients, the most desired being the impoverished members of the city's middle class who can no longer afford to live in the city's traditional formal quarters.[2] The outcomes of these changes have been twofold. They have first lead to direct population displacement, which is organized through market channels, in order to leave space for new developments and clients. They have also lead to the degradation of the old structures as the pressure to increase available space – notably for rent – has lead to additional ad hoc floors and the re-subdivision of existing housing units. Elsewhere, in the old squatter settlements of the city, in a combination of private developments and public projects, blocks are being dissected and reduced in size and capacity, forcing out dwellers who had occupied them for decades. In short, we observe empirically that the housing and land markets of informal settlements in the city are increasingly integrated into the rest of the citywide housing market, with the marked consequences of degrading living and environmental conditions and reducing the chances of dwellers from these neighborhoods to participate in the production of the city, either as builders, developers or homeowners.

It is in this integration of informal settlements with housing markets in the rest of the city that I believe one can best read the marks of neoliberalism, conducted with a blend of the traditional Lebanese laissez faire. As described, above, the higher integration of informal settlements in the wider housing market of the city has ushered a more powerful role for market institutions such as bank loans, advertisement channels, and formal contracts in the making and exchange of housing in these neighborhoods. Such channels are more restrictive than those of informal housing production, however, and outside the reach of a large section of low-income city dwellers, for whom informal channels had traditionally secured shelter. The new apartments developed in informal settlements often target members of the impoverished middle classes, who are pushed away from the rapidly gentrifying middle-income neighborhoods of the city. Not surprisingly, higher reliance on market channels has also necessitated the increased deployment of the police forces to control and monitor the production of space and its organization within market channels. In a perverse set of circumstances in which members of the police force can also be bribed, and hence policing can be reduced to financial exchanges, every aspect of the making of space, from access to property to legal protection, has become monetized. In the process, the possibilities of accessing housing outside market channels have been severely reduced, if not completely erased.

Informal settlements in the neoliberal era: an overview of other contexts

A reading of the literature exploring changes in informal settlements in several national contexts during the past decade confirms these trends in other cities, such as Delhi, Cairo and Casablanca to name a few (Benjamin and Raman 2011; Bogaert 2011; Denis 2011; Michel *et al.* 2011; Durant-Lasserve 2006). However, what is achieved in Lebanon indirectly, as the de facto outcome of a policy framework that facilitates speculative land strategies, has materialized elsewhere as the outcome of public policy directly facilitating the marketization of properties in informal settlements. This is indeed, at least in large part, the outcome of decades of so-called regularization policies that have strived to "formalize the informal" and integrate "illegal neighborhoods" in wider processes of housing production. However, as housing production was delegated almost entirely to the market in most national contexts, formalization and integration have increasingly materialized in policies that facilitate the penetration of market forces in informal settlements. Chief among these policies are titling projects, interventions that seek to clarify property rights in informal settlements through the disbursement of individual property titles to home dwellers. These interventions were granted wide visibility in 2000, notably with the work of the Peruvian economist Hernando de Soto, who argued that the regularization of property rights not only improves tenure security but also places an "asset" in the hands of low-income dwellers and increases their participation in the "economy", by giving them the opportunity either to exchange their property or to use their property as collateral to access loans (de Soto 2000). Whatever the profit that is realistically achieved from the legal recognition of property, the main lesson that was retained from the work of de Soto and others is the emphasis within public interventions in informal settlements on the disbursement of property titles. This recommendation has consistently dominated the academic literature and policy prescriptions forwarded by international organizations such as the World Bank and various United Nations agencies for over a decade. It has materialized with an array of titling policies that swept aside other approaches to regularization and informal settlement regularization throughout the Third World (Payne *et al.* 2009; Denis 2011; Huchzermeyer 2008). These are complemented with other interventions, which also look to facilitate the integration of the poor in larger housing markets and their access to title – most notably an array of financial instruments such as microcredit loans that are deployed in order to re-involve the poor in the city "on market terms" (Bogaert 2011; Berner and Phillips 2003).

One should also mention here the proliferation of so-called security policies that are seen as necessary for the protection of the "market". Thus, claims for social redistribution are met with disciplinary responses that criminalize poverty, of the type described in the introduction of this chapter. These responses feed on the alarmist tone in which many scholars have addressed degrading living conditions around the world – epitomized by the reintroduction of the term "slum" to refer to these neighborhoods (Davis 2007; UN-Habitat 2003). These responses blur the boundaries between "fearing the poor" and "fearing for the poor", fueling alarmist discourses that describe the dwellers of informal settlements as "virtual enemies of the nation" or, more generally, criminalizing poverty (Wacquant 2008). Their outcome is higher investment in security mechanisms and their deployment vis-à-vis these low-income urban neighborhoods and their dwellers whose activities must be monitored and controlled (Dikeç 2007; Wacquant 2008), while a blind eye is turned to growing corporate crime and corruption. It ushers in two distinct, yet interrelated forms of security interventions in informal settlements. In their better version, these representations justify interventions on informal settlements – such as titling or servicing – as a strategy to curtail violence in other sections of the city (Bogaert 2011). Their more common version, however, justifies the violent repression of informal housing practices in the forms we witnessed in Beirut last spring, for example, where the army is deployed with its full military apparatus to prevent the production of housing outside market channels.

It is important to note at this stage that if neoliberalism ushers in higher levels of integration for informal settlements within the market mechanisms of the city, it does not mean that these neighborhoods ever operated entirely "outside" state and/or market institutions. To the contrary, ample evidence has emerged over the past decades to indicate that the production of informal housing happens in direct relation to the frameworks of the capitalist land market and state-sanctioned regulations (Benton 1994; Fawaz 2008; Razzaz 1998; Varley 2002). On the one hand, processes of land acquisition do not happen "freely", meaning that dwellers always pay some cost for squatting or accessing land, although the monetized value remains well below that ascribed by the capitalist land market, hence tilting the balance between the relative influence of use and exchange values. Furthermore, the development of these neighborhoods generally contributes to reducing property values in their vicinities, challenging capitalist interests in real estate markets. On the other hand, these neighborhoods are produced according to regulations and principles that deviate from and bend state-sanctioned urban regulations, without ever completely stepping outside them (Azuela de la Cueva 1987; Fawaz 2009).

Until recently, nonetheless, informal settlements still presented an alternative model of spatial organization in which, for example, dwellers relied on social networks and other informal institutions in order to build their spaces and organize their services (Turner and Fichter 1972). This was manifested in the sequences in which their homes were built and consolidated, which rarely followed those dictated by state regulations (such as land purchase, permitting, servicing, building, occupation) and abided instead by the logic of household needs and the pattern of negotiation with public authorities (Razzaz 1994; Ward and Macoloo 1992). The physical form and organization of these neighborhoods respected social principles such as local notions of privacy or safety, rather than building codes sanctioned by state agencies. The channels of financing depended on informal credit organizations (Pamuk 2000), family relations and informal exchanges (Varley 2002) rather than formal lending agencies. Most important, the management of these neighborhoods relied on local neighborhood committees and social forms of organization and, ultimately, their production presents a model of local control over the production of space, without ever being definitive or completely independent of state or market authorities (Berry-Chikhanoui and Deboulet 2000; Fernandes and Varley 1998; Ward 1982).

The ongoing integration of informal settlements in the city is therefore in some sense an intensification of already existing commercial relations that worked towards the production of these neighborhoods. But it is also, in other ways, a reversal of these trends – most notably in terms of the loss of the margin of maneuver that allowed for these neighborhoods to act as viable channels of housing acquisition. This integration should not be confused with further social integration across formal and informal neighborhoods, since market integration does not necessarily foster social bridges across sections of the city. Finally, this integration within market channels, or rather the entrenchment of the market in the making of these neighborhoods, can also be observed in other aspects of the daily life of informal settlements – most notably in terms of the urban services that have increasingly depended on commercial forms of provision, which compensate for weakening public agencies and poor forms of social organization. This is clearly visible in the informal settlements of Beirut and beyond, where the documented effects on the poor have been quite alarming (Huzchemeyer 2008).

A policy of erasure: neglect, displacement, and criminalization

What does this review of the literature and the cases taken from the Lebanese context tell us? Not surprisingly, the past decades have witnessed a severe degradation of living conditions in the informal settlements of Beirut and other cities (Davis 2007). Despite the resonance of images of dilapidated neighborhoods and difficult living conditions with representations of informal settlements as "neglected" or "forgotten spaces" (Shatkin 2004), I argue that contemporary policymaking towards informal settlements amounts to more than sheer neglect, and that it is possible to speak of actual erasure operating through the entrenchment of market-type relations and the security that is deployed to defend and protect the functioning of these markets. This erasure happens through the displacement of large population groups that is triggered by the rising land prices that can result from titling projects (Durand-Lasserve 2006; Denis 2011), or, more generally, from a strictly market-based conceptualization of land as real estate property. Erasure also happens more significantly through the transformation of the place and significance of informal settlements in the city. By displacing the opportunities for small-scale developers to take part in the production of housing and by reducing the possibilities and forms in which the acquisition of housing occurs, what is erased are the very possibilities or potentials of the informal, or, in other words, the significance of the informal settlement as a place of dissent. Reduced to its value as a real estate asset, land is stripped of its social significance. More generally, informal settlements are stripped from their political significance as a form of dissent (Leontidou 2010) and as the embodiment of alternatives to the market. They are thus reduced in many contexts to "slums", places where low-income (and increasingly middle-income) city dwellers can be stored. Once again, it is not that low-income dwellers' shelters are neglected – they always have been, to a large extent. The fact, however, that it has become necessary to inscribe the language of public policymaking in a market-based set of options that appear as the *sine qua non* of the ability of the poor to acquire shelter is symptomatic of the way neoliberalism works: the penetration and entrenchment of market relations in sectors of life until then delegated to other institutions – notably the public and the social (Dikeç 2006; Harvey 2007). It is also the takeover, by the forces of the market, of those areas of the city that had managed to maintain some margin of maneuver outside market and state institutions, without ever being separate from either sphere. It is here that the importance of the "right to the city" as a political project that reclaims the political nature of "space" or the "urban" becomes clearest. This is what I will turn to in the next section.

Informal settlements and the right to the city

The right to the city

The concept of the "right to the city" was put forth by Henri Lefebvre in the 1960s as a political program that aimed to strengthen the ability of "city-zens" to take charge of processes of spatial production (Lefebvre 1968, 1974; see also Dikeç 2001; Purcell 2002, 2003; Shields 1998). It entailed two intertwined rights: (1) the right to participate in the conception, design and implementation of the production of urban spaces, shifting decisions about the production of these spaces away from the state, towards urban residents; and (2) the right to appropriate – through access, occupation and use – urban spaces, and produce them in ways that would meet the needs of urban inhabitants, hence shifting control away from capital that conceptualizes space as a commodity that can be exchanged in the form of property (Lefebvre 1974; Logan and Molotch 1987). Central to these two claims is the recognition of the political potential of "space", both as the embodiment of previous struggles and as the enabler for the possibilities of future collective and individual social action (Lefebvre 1974).

Many scholars have considered "informal settlements" to be the embodiment of the "right to the city", neighborhoods where dwellers are producing their living quarters in ways that challenge the dominant (state-sanctioned and market-dictated) modes of spatial production. Through processes of land acquisition, building construction, servicing and/or exchange, they challenge and reshape the existing norms in which space is produced elsewhere in the same cities within the capitalist order (Bey 1991; Shields 1998; see also Dikeç 2001; Harvey 2008; Lefebvre 1968). For example, by occupying land without obtaining property titles, dwellers are forwarding a conceptualization of land as shelter that challenges the state-sanctioned conception of land as real estate property. Through negotiations and mobilization, they succeed in securing a relatively safe tenure and building livelihoods in the city without going through the channels of property markets. Similarly, low-income dwellers in informal settlements often rely on their own labor, on shared networks of social relations, and on other non-market-based forms of homemaking that challenge the production of home as market.

It is worth noting that Lefebvre's formulation of the "right to the city" differs considerably from the way this concept has been used by theorists and policy-makers since the 1970s within a normative framework of redistributive justice. In the latter formulation, "the right to the city" is evoked to support the entitlement of informal settlement dwellers to access public services (for example, water and electricity), regardless of violations of property rights or urban rules and building codes and/or to "regularize" their settlements and integrate them in wider urban markets (Miraftab and Wills 2005). It is also the latter formulation of the "right to the city" which dominates the contemporary urban literature that has decried the growing disenfranchisement of urban inhabitants in the context of contemporary neoliberal policymaking (Friedman 1988; Harvey 2008; Isin

2000; McCann 2002, 2003; Mitchell 2003; Purcell 2002, 2003; Simone 2005). In most cases, authors have identified the potentials of Lefebvre's formulation of the concept to reform traditional models of public participation by, for example, replacing national citizenship as a prerequisite for political participation by other forms of entitlement, such as inhabitance (Martin *et al.* 2003). Most of these investigations are, however, directed towards reforming existing state models of governance and thinking of strategies and legal frameworks that can render them more inclusive, while Lefebvre's right to the city is not derived from the state per se, but rather from challenging its authority. Applied in the context of Western democracies, researchers are using the right to the city in order to react, for example, against the growing powers of multinational corporations and to argue for the need for forums where inhabitants are able to influence the decisions that big capital takes when it shapes their spaces (Martin *et al.* 2003; Purcell 2002). Similar arguments have been deployed in reference to urban planning reforms in lower-income countries, especially in Colombia and Brazil where more inclusive and participatory processes of planning are invoking the "right to the city" in their formulation (Lopez de Souza 2001). The United Nations has also put forth a proposal for a "World Charter of the Right to the City" that relies on the legal system in order to secure inclusiveness in urban planning practices (Fernandes 2007) or, more generally, to promote a "rights-based" approach to the challenges of urbanization (Faetani 2011). Although these formulations somewhat limit the "right to the city" to distributive justice and conceptualization of participation only in state-sanctioned public forums, they also mark an essential contribution to thinking about informal settlements in relation to the "right to the city" and neoliberal policymaking: the search for inclusive frameworks in which decisions about the production of space can be taken, and the recognition of the social/use value of property that needs to be strengthened vis-à-vis the dominance of exchange value (Dikeç 2001; Purcell 2003).

Given the current pressures to strip space down to its (abstracted) property value and to reduce processes of spatial acquisition and negotiation to sanctioned forms of market exchange, the "right to the city" speaks to the "imperative" of the political, the necessity to re-inscribe the quest for a livable city within larger questions of social justice that account for the political nature of space. By infusing the notion of "rights" within the discourse on housing and/or space, this formulation activates a discourse of entitlement where housing can and should be conceptualized beyond questions of "affordability" and access can occur through other venues than market and/or state channels. Moreover, through Lefebvre's wider body of work, "space" itself cannot be reduced to an abstract exchangeable unit because it is defined as the product of a particular historic and geographic moment, as the embodiment of particular modes of production and reproduction, and as the determinant of the social and political possibilities. As a result, key questions about shelter in relation to informal settlements can neither be posed nor addressed solely in market terms, and the discussion of "shelter policy" and public policy vis-à-vis informal settlements is reframed outside of the plethora of market-based tools in which international organizations nowadays specialize. More importantly, there is a need to acknowledge that the market is not a fair arbitrator of spatial allocation, as has been widely proved by experiences around the world. Instead, dwellers can resort to a multitude of political venues in order to express dissent. An acknowledgment to the right to the city therefore amounts to recognizing the importance of political organization, including conflictive forms such as protests, sit-ins, temporary and permanent occupations of public and private space and others as part of the strategies and approaches to shape the city (Routledge 2010).

Tenure security and housing policy

In order to illustrate how a shift from a market-based analysis to a right-to-the city framework can change our approach to informal settlement, let us take "tenure security", which I described above as the main area of concern for international organizations for at least a decade. Within the framework of the market, tenure security is widely equated with titling programs where property rights are clarified and claims over particular space consolidated through the disbursement of (mostly freehold) property titles. Evidence has nonetheless emerged from around the world to signal that property titles are neither a necessary nor a sufficient source of tenure security. Already in the early 1990s, Razzaz (1994) showed that dwellers in informal settlements in Amman (Jordan) asserted tenure security through the consolidation of houses where they felt that a well-finished house was the best protection from bulldozing. This evidence is now widely endorsed, as reported in the review paper for the World Bank that looked at such policies in thirty-five countries (Payne *et al.* 2009). Research also shows that tenure security is not necessarily secured by titles. To begin with, titling is known to increase land prices, forcing tenants who constitute the most vulnerable social groups in informal settlements to leave the premises for more affordable areas (Denis 2011; Payne *et al.* 2009). Increased land prices also often act as a strong incentive for low-income urban dwellers to sell their property when they need money, causing them to be in a more vulnerable position vis-à-vis economic downturns and homelessness. Finally, titles do not necessarily prevent displacement: in numerous instances, social and political backing determines whether dwellers are evicted and the amount of compensation they will receive. Thus, numerous city dwellers holding property titles have been forced out of their residences when development needs clashed with their interests, while dwellers without legal property claims, having nonetheless secured sufficient political support, have resisted displacements and/or secured appropriate financial compensations (Deboulet and Fawaz 2011). In a restrictive, market-based conceptualization of access to housing, however, titling is the only possible form of accessing secure housing because titling limits the types of claims that can be laid on space to an economic exchange which is concluded with the transfer of a formal property title sanctioned by national land authorities – as happens in the formal market. If, however, a framework of right to the city is adopted, it could recognize the multiplicity of arrangements through which people access property and articulate larger policies that can indeed provide security of tenure, such as limiting displacements triggered by large-scale planning projects or controlling land prices to reduce the incentives for real estate speculations. It also acknowledges the agency that low-income dwellers have in occupying spaces, reclaiming areas in the city as theirs, violating urban and building regulations, and hence imagining alternative forms in which the city can be made (Leontidou 2010). More generally, at this critical juncture, the right to the city can be used as a framework to repoliticize the discourse of housing in ways that decry the marketization of land and reaffirm its social value (Blomley 2004), that recognize the multiplicity of meanings a home can take and the necessity for more flexible and expandable frameworks to organize and allocate homemaking than the market.

Localizing the right to the city

Claims for re-inscribing the political, however, cannot go without accounting for the actually existing structures of organization in the geographic and historical context in which the claim is made. In most countries of the Arab region, political unrest and social movements have since January 2011 challenged dictatorships and ushered in new eras of political activism. Embedded in those movements is the resistance to neoliberal trends, which – most notably in Egypt (Denis 2011; Mitchell 2007), Morocco (Berry-Chikaoui and Deboulet 2000; Bogaert 2011), Jordan (Parker 2009) and Lebanon (Makdisi 1996; Fawaz 2009) to name only a few cases in the region – have radically transformed traditional systems of social welfare and reshaped cities in line with the demands of capital. Within this same region, however, the configurations of political systems differ considerably from one national context to another and, as a result, the forms that neoliberalism and its oppositions take also differ.

Let me more specifically talk of the materialization of neoliberalism – or "actually existing neoliberalism" in the Lebanese context – and reflect on the possibilities to activate the "right to the city" in this specific context (Brenner and Theodore 2002). I have outlined above how neoliberalism was indeed endorsed across the spectrum by Lebanese policymakers through a combination of privatization, delegation of planning to the private, and other measures. I have also described the deployment of security mechanisms that are consistently used to "secure the market" and provide it with the type of safety that it needs to maintain its operations. Lest these descriptions recall, however, the usual set-ups observed in other contexts where centralized state authorities and all-too-powerful public bodies coherently work to restrain the poor through the rise of a "penal state" (Wacquant 2008; Dikeç 2007), it should be noted that Lebanon displays a different understanding of the "public".

In Lebanon, the state is a fragmented and weak entity, where one arm fights another, where what is public and what is not are continuously redefined and rarely clearly delineated, where citizens often prioritize social and politico-religious forms of self-identification over an abstract national belonging. Things are further complicated as political parties, self-appointed representatives of these low-income, vulnerable groups, have themselves been in office for at least a decade and work from their capacity as public officials to organize social groups outside the state, often in antagonism to communal notions of the public (Harb 2011; Bou Akar 2010). As a result, the "state" or rather a particular public agency or institution (e.g., ministry, municipality) can be weak or strong, depending on whether the person heading the agency is "well backed" by a political party or not. Even the criminalization of the poor to which I pointed earlier in the chapter emerges as a strategy between contending political groups rather than a coherent public response against the poor.

This fragmentation forecloses the development of a rights-based discourse in line with what we have seen elsewhere (e.g., Leontidou 2010), because the latter would require a coherent public interlocutor. It confines dwellers further to the protection of political parties, sometimes militias and sometimes militias disguised as public representatives. It leads to a scenario where informal settlement dwellers have ceased to look for the recognition they once sought from public agencies. Thus, local neighborhood commissions which once struggled for regularizing interventions, or at least for the recognition of public agencies, have been largely inactive for decades. Interviewed dwellers expressed skepticism vis-à-vis the necessity to regularize or the actual protection it provides, often pointing to political parties as their protectors (Fawaz 2009). Their decision to disinvest from the authority of the public is quite logical in post-war Lebanon. Suffice it to say that those who look for formal titles and regular status to secure their tenure are quickly disillusioned: a large number of low-income dwellers with property titles were displaced when the demands for infrastructure provided by capital necessitated this displacement while others occupying land without titles maintained their prime locations in the city, when they were sufficiently 'protected' by any of the local forces (Deboulet and Fawaz 2010). As a result, in today's Beirut, informal settlement dwellers do not struggle for the regularization of their status, as they had done during the 1960s when most of their social movements sought recognition from public authorities. Neither is the right to the city, in terms of a potential entitlement to imagine and/or take part in the making of the city, in the framework of their claims. They instead strive for the protection of one arm of the "public" sector against another, finding in these contestations over sovereignty the space to negotiate some kind of entitlement to services and shelter.

The fragmentation of the "public" in Lebanon and its deeply divisive and divided nature suggests the necessity to work with an understanding of the right to the city that goes back to the origins of the concept, as I outlined it above. Rather than trying to reform existing channels of political participation and working through housing policies that limit the ceiling to their activities to the market, as much of the cotemporary discourses of the right to the city had operated, it would be important to articulate visions and imaginaries that re-infuse political intentionality in acts of squatting. Thus, the significance of contemporary processes of illegal construction should go beyond the dire response for a few inches of land where one can access shelter, to reclaim instead one's right to be part of conceptualizing and building the city, of accessing and shaping its public spaces – rejecting hence the imposition of the market as a fair arbitrator of spatial allocation. Such a process seems quite far-fetched in today's Beirut, where the entire landscape of the country is conceptualized in dollarized property values. It also fails to resonate with any of the contemporary discourses forwarded by the contending political parties. It nonetheless responds to a different imaginary of what constitutes the urban – one that is worth striving for.

Conclusions

I have argued in this chapter for the need to infuse in the reading of contemporary public policy vis-à-vis informal settlements a sense of intentionality that counters common descriptions of "laissez faire" or "neglect", to recognize deliberate attempts to erase these neighborhoods. I have further argued that these attempts are conducted, on the one hand, by seeking to integrate these neighborhoods in the rest of the city's housing market and, on the other, by protecting their realm through intensive security measures that prevent the development of housing outside market channels. I have also shown that these policies have materialized in Beirut with negative repercussions for the informal settlements of the city, whose position in the general process of urbanization has moved from possible sites of hope or "neighborhoods in the making", to paraphrase the classical writings on informal settlements, to mere "slums" or high-density areas with degrading living conditions. More generally, I have shown that, despite variations, the general outcomes of a heightened role for market institutions in low-income housing markets have led to further impoverishment of low-income classes. In order to counter these trends, I have argued that it is necessary to re-infuse in contemporary housing policy a recognition of the "political", a definition of the "housing question" that goes beyond mere improvements of market channels to address the fundamental questions of entitlement. More generally, it will be important to rethink the organization of contemporary cities and the delegation of much of their structures to market institutions in ways that recognize the political potential of the city.

Notes

1 The World Bank "Migration and Remittances Report 2011" lists Lebanon as the twelfth recipient in terms of total amount of remittances and indicates that 70 percent of those remittances are invested in real estate. Note Lebanon's population (four million) in relation to its position on the global scale (see World Bank 2011).

2 During the first decade after the civil war, families receiving compensations against displacement from the various quarters of the municipal city where they had squatted throughout the years of Lebanese civil war were the most coveted clients of these developers, because they had secured a comfortable first down payment through these displacement compensations. For more on these post-war displacements, see Bou Akar (2005).

References

Azuela de la Cueva, A. (1987) "Low Income Settlements and the Law in Mexico City", *International Journal of Urban & Regional Research*, 11 (4): 522–42.

Benjamin, S. and Raman, B. (2011) "Illegible Claims, Legal Titles, and the Worlding of Bangalore", *Revue Tiers Monde*, 206 (Avril/Juin): 37–54.

Benton, L. (1994) "Beyond Legal Pluralism: Towards a New Approach to Law in the Informal Sector", *Social and Legal Studies*, 3: 223–42.

Berner, E. and Phillips, B. (2003) "Left to Their Own Devices? Community Self-Help Between Alternative Development and Neoliberalism", paper presented at the N-AERUS Annual Seminar: Beyond Neo-liberal Consensus on Urban Development: Other Voices From Europe and the South, Paris, 15–17 May.

Berry-Chikhanoui, I. and Deboulet, A. (eds) (2000) *Les compétences des citadins dans le monde arabe: penser, faire et transformer la ville*, Paris: Karthala.

Bey, H. (1991) *TAZ: Temporary Automatic Zones*, New York: Autonomedia.

Blomley, N. (2004) *Unsettling the City*, London: Routledge.

Bogaert, K. (2011) "The Problem of Slums: Shifting Methods of Neoliberal Urban Government in Morocco", *Development and Change*, 42 (3): 709–31.

Bou Akar, H. (2010) "The Spatial Construction of the Religious Other: the "Making of" Sahra Choueifat, Lebanon", paper presented at City Debates 2010: Security of/in the City, American University of Beirut, Beirut, Lebanon.

Bou Akar, H. (2005) "Displacement, Politics, and Governance: Access to Low-Income Housing in a Beirut Suburb", unpublished MCP thesis, Department of Urban Studies and Planning, MIT.

Bourgey, A. and Pharès, J. (1973) "Les Bidonvilles de l'Agglomération de Beyrouth", *Revue de Géographie de Lyon*, XLVIII (2): 107–39.

Brenner, N. and Theodore, N. (2002) "Cities and the Geographies of 'Actually Existing Neoliberalism'", *Antipode*, 34 (3): 349–79.

Clerc-Huybrechts, V. (2008) *Les quartiers irreguliers de Beyrouth*, Paris: IFPO-Karthala.

Davis, M. (2007) *Planet of Slums*, London: Verso.

de Soto, H. (2000) *The Mystery of Capital: Why Capitalism Triumphs in the West and Fails Everywhere Else*, New York: Basic Books.

Deboulet, A. and Fawaz, M. (2011) "Contesting the Legitimacy of Urban Restructuring and Highways in Beirut's Irregular Settlements", in D. Davis and N. Libertun de Durén (eds) *Cities and Sovereignty: Nationalist Conflicts in the Urban Realm*, Bloomington, IN: Indiana University Press.

Denis, E. (2011) "La financiarisation du foncier observée à partir des métropoles egyptiennes et indiennes", *Revue Tiers Monde*, 206 (Avril/Juin): 139–59.

Dikeç, M. (2007) *Badlands of the Republic*, London: Blackwell.

Dikeç, M. (2006) "Two Decades of French Urban Policy", *Antipode*, 38 (1): 59–81.

Dikeç, M. (2001) "Justice and the Spatial Imagination", *Environment and Planning A*, 33: 1785–805.

Durand-Lasserve, A. (2006) "Market-Driven Evictions and Displacements", in M. Huzchemeye and A. Karam (eds) *Informal Settlements. A Perpetual Challenge?*, Cape Town: University of Cape Town Press, pp. 207–27.

Faetani, M. (ed.) (2011) *Urban Policies and the Right to the City in India: Rights, Responsibilities, and Citizenship*, New Delhi: UN-Educational, Scientific, and Cultural Organization.

Fawaz, M. (2009) "Neo-liberal Urbanity: a View from Beirut's Periphery", *Development and Change*, 40 (5): 827–52.

Fawaz, M. (2008) "An Unusual Clique of City-Makers", *International Journal of Urban and Regional Research*, 32 (3).

Fawaz, M. (2004) *Strategizing for Housing: an Investigation of the Production and Regulation of Low-Income Housing in the Suburbs of Beirut*, unpublished PhD dissertation, Cambridge, MA: Massachusetts Institute of Technology.

Fawaz, M. and Peillen, I. (2002) *The Slums of Beirut: History and Development 1930–2002*, paper prepared for United Nations Center for Human Settlement.

Fernandes, E. (2007) "Constructing the 'Right to the City' in Brazil", *Social & Legal Studies*, 16 (2): 201–19.

Fernandes, E. and Varley, A. (eds) (1998) *Illegal Cities: Law and Urban Change in Developing Countries*, London: Zed Books Ltd.

Friedman, J. (1988) "The Right to the City", in R. M. Morse and J. E. Hardoy (eds) *Rethinking the Latin American City*, Washington, DC: The Woodrow Wilson Center Press, pp. 99–109.

Harb, M. (2011) *Le Hezbollah à Beyrouth: de la banlieue à la ville*, Paris: IFPO-Karthala.

Harvey, D. (2008) "The Right to the City", *New Left Review*, 53: 23–40.

Harvey, D. (2007) "Neoliberalism as Creative Destruction", *The ANNALS of the American Academy of Political and Social Science*, 610 (21): 22–44.

Huzchemeyer, M. (2008) "Slum Upgrading in Nairobi within the Housing and Basic Services Market: a Housing Rights Concern", *Journal of Asian and African Studies*, 43: 19–39.

Isin, E. (2000) *Democracy, Citizenship and the Global City*, New York: Routledge.

Kassir, S. (2003) *Histoires de Beyrouth*, Paris: Fayard.

Lefebvre, H. (1974) *La Production de l'espace*, Paris: Antropos.

Lefebvre, H. (1968) *Le Droit à la ville*, Paris: Anthropos.

Leontidou, L. (2010) "Urban Social Movements in 'Weak' Civil Societies: the Right to the City and Cosmopolitan Activism in Southern Europe", *Urban Studies*, 47 (6): 1179–203.

Logan, J. and Molotch, H. (1987) *Urban Fortunes*, Berkeley: University of California Press.

Lopez de Souza, M. (2001) "The Brazilian Way of Conquering the 'Right to the City' ", *DISP*, 147: 25–31.

Makdisi, S. (1996) "Laying Claim to Beirut: Urban Narrative and Spatial Identity in the Age of Solidère", *Critical Inquiry*, 23: 661–705.

Martin, D., McCann, E. and Purcell, M. (2003) "Space, Scale, Governance, and Representation: Contemporary Geographical Perspectives on Urban Politics and Policy", *Journal of Urban Affairs*, 25 (2): 113–21.

McCann, E. (2003) "Framing Space and Time in the City: Urban Policy and the Politics of Spatial and Temporal Scale", *Journal of Urban Affairs*, 25 (2): 113–21.

McCann, E. (2002) "Space, Citizenship, and the Right to the City: a Brief Overview", *Geojournal*, 58: 77–9.

Michel, A., Denis, E. and Soares-Gonçalves, R. (2011) "Introduction: les enjeux du foncier urbain pour le développement", *Revue Tiers Monde*, 206 (April/June): 7–14.

Miraftab, M. and Wills, S. (2005) "Insurgency and Spaces of Active Citizenship: the Story of Western Cape Anti-Eviction Campaign in South Africa", *Journal of Planning Education and Research*, 25: 200–17.

Mitchell, T. (2007) "Dreamland", in M. Davis and D. B. Monk (eds) *Evil Paradises, Dreamworlds of Neoliberalism*, New York and London: The New Press.

Mitchell, D. (2003) *The Right to the City: Social Justice and the Fight for Public Space*, New York and London: Guilford Press.

Pamuk, A. (2000) "Informal Institutional Arrangements in Credit, Land Markets and Infrastructure Delivery in Trinidad", *International Journal of Urban and Regional Research*, 24 (2): 379–496.

Parker, C. (2009) "Tunnel-Bypasses and Minarets of Capitalism: Amman as Neoliberal Assemblage", *Political Geography*, 28 (2): 110–20.

Payne, G., Durand-Lasserve, A. and Rakodi, C. (2009) "The Limits of Land Titling and Home Ownership", *Environment and Urbanization*, 21 (2): 443–62.

Purcell, M. (2003) "Citizenship and the Right to the City: Re-Imagining the Capitalist World Order", *International Journal of Urban and Regional Research*, 27 (3): 564–90.

Purcell, M. (2002) "Excavating Lefebvre: the Right to the City and its Urban Politics of the Inhabitant", *Geojournal*, 58: 99–108.

Razzaz, O. (1998) "Land Disputes in the Absence of Ownership Rights: Insights from Jordan", in E. Fernandes and A. Valey (eds) *Illegal Cities: Law and Urban Change in Developing Countries*, London and New York: Zed Books, pp. xx.

Razzaz, O. (1994) "Contestation and Mutual Adjustment: the Process of Controlling Land in Yajouz, Jordan", *Law & Society Review*, 28 (1): 7–39.

Robinson, J. (2002) "Global and World Cities: a View from Off the Map", *International Journal of Urban and Regional Research*, 26 (3): 531–54.

Routledge, P. (2010) "Introduction: Cities, Justice, and Conflict", *Urban Studies*, 47 (6): 1165–77.

Samara, T. R. (2010) "Policing Development: Urban Renewal as Neo-liberal Security Strategy", *Urban Studies*, 47 (1): 197–214.

Sassen, S. (2000) *Cities in a World Economy*, Michigan: Pine Forge Press.

Shatkin, G. (2004) "Planning to Forget: Informal Settlements as 'Forgotten Places' in Globalising Metro-Manila", *Urban Studies*, 41 (12): 2469–84.

Shields, R. (1998) *Lefebvre, Love and Struggle*, London: Routledge.

Simone, A. M. (2005) "Uncertain Rights to the City", *Africa e Mediterraneo*, 50: 37–41.

Smith, N. (2002) "New Globalism, New Urbanism: Gentrification as Global Urban Strategy", *Antipode*, 34 (3): 434–57.

Traboulsi, F. (2007) *A History of Modern Lebanon*, London: Pluto Press.

Turner, J. F. C. (1972) "Housing as a Verb", in J. F. C. Turner and R. Fitcher (eds) *Freedom to Build: Dweller Control of the Housing Process*, New York: The MacMillan Company.

Turner, J. F. C. and Fitcher, R. (1972) *Freedom to Build: Dweller Control of the Housing Process*, New York: The MacMillan Company.

UN-Habitat (2003) *Global Slum Report*, New York: The United Nations.

Varley, A. (2002) "Private or Public: Debating the Meaning of Tenure Legalization", *International Journal of Urban and Regional Research*, 26 (3): 449–61.

Wacquant, L. (2008) "The Militarization of Urban Marginality: Lessons from the Brazilian Metropolis", *International Political Sociology*, 2: 56–74.

Ward, P. (1982) *Self-Help Housing: a Critique*, London: Mansell Publishing Ltd.

Ward, P. and Macoloo, C. (1992) "Articulation Theory and Self-Help Housing Practice in the 1990s", *International Journal of Urban and Regional Research*, 16 (1): 60–80.

World Bank (2011) "Migration and Remittances Factbook 2011", 2nd edn, Washington, DC: World Bank. Online: http://issuu.com/world.bank.publications/docs/9780821382189.

2 Cities without slums in Morocco?

New modalities of urban government and the *bidonville* as a neoliberal assemblage

Koenraad Bogaert[1]

In 2004, the Moroccan government launched the ambitious "Villes sans Bidonvilles" (Cities Without Slums) program (VSBP) following the dramatic events of 16 May 2003, when fourteen suicide bombers struck the city center of Casablanca, killing more than thirty people. The perpetrators all came from two large *bidonvilles* (slums) in Sidi Moumen, a district in the prefecture of Sidi Bernoussi on the eastern periphery of Casablanca. The events had raised awareness about the problem of slums as a threat to the security and political stability of the Moroccan city and strengthened the stigmatization of the slums as a breeding ground for radical Islamists (Zaki 2005; on the specific connection between the bombings, slum upgrading and social control, see Bogaert 2011). Poverty and the degraded living conditions in the slums were considered to be the main seedbeds for frustration and eventually radicalization. VSBP aims to eradicate all slums in Moroccan cities by 2013 (the first deadline was 2010 but this has been postponed due to several delays). This most recent slum-upgrading initiative of the Moroccan government was embedded within a discourse displaying a strong commitment towards poverty alleviation, good governance and participatory development, thereby reflecting more global shifts in neoliberal governmentality promoted by the World Bank and inscribed in the Millennium Development Goals.

I argue that Morocco's commitment to eradicate all slums in its cities is a salient example of this global shift in neoliberal governmentality. The semantic linking between government and mentality emphasizes the inseparable interrelation between modes of government and the different (political) rationalities that underpin them (Lemke 2001). Characteristic of neoliberal reform is that political and economic life is resituated in contexts designed to reflect "market" incentives and demands. As I will demonstrate, through new methods of social engineering, new forms of taxation, public-private partnerships, and financial innovations, the slum population is being required to participate in the making of a specific political world in which the ability to claim and articulate political rights are not only circumscribed by "the Arab regime", but also by the sanctions and incentives of "the free market" (Parker 2009). The Moroccan political commitments to fight poverty and uneven development form part of a neoliberal political project and demonstrate how local politics are not just imposed by neoliberal globalization but are – to follow up on Doreen Massey – actively involved in the making of the neoliberal capitalist global (Massey 2005). David Harvey considers neoliberal politics above all as class strategies that privilege the creation of new opportunities for capital accumulation over the social well being of (urban) citizens (Harvey 2006). Notwithstanding the fundamental changes in developmental policies over the last two decades, with the shift from the Washington Consensus to the Post-Washington Consensus, the very nature and objectives of these policies have not changed. The national slum program is one example of how new frontiers of capital accumulation are explored and exploited by the setting up of specific governmental arrangements and modalities of state intervention.

More specifically, this article aims to analyze the technologies of power invested in new methods of slum upgrading and participatory development, and consider what these reveal about the particular ways in which poor people are governed and how they are integrated in the "formal" city and the urban capitalist economy. This research is based on ten months of fieldwork between 2007 and 2010. I conducted over sixty interviews related to the VSBP with a wide range of actors such as, amongst others, public officials, NGO workers, private consultants, academics, journalists and members of international development and donor agencies. Furthermore, I made several visits to slums and social housing projects in cities like Casablanca, Rabat, Temara and Larache.

A new international context: the Post-Washington Consensus and the emphasis on participation.

At the turn of the century the level of urbanization in Morocco was 56 percent, and 33 percent of the urban population lived in slums (UN-Habitat estimates 2001, online at www.unhabitat.org, accessed 2 November 2010). Since their first appearance at the beginning of the twentieth century, slums in Morocco have been largely tolerated by public authorities. Yet, despite this tolerance, they always maintained a temporary and precarious status due to the illegal occupation of the terrain. Public authorities opted for a form of "management by absence": they refused to provide the *bidonvilles* with collective services such as water and electricity and restricted inhabitants' possibilities for organizing and improving their living conditions (e.g., the construction of houses out of bricks was prohibited) (Zaki 2005). By maintaining the inhabitants on the legal margin the state created a latent insecurity, which sustained the shantytown dwellers' fragility and limited the assertion of collective demands (Zaki 2008). This slum policy largely entailed the physical and symbolic exclusion of the slum dweller from the formal, the "true" city. This tension between formal and informal, this particular social and spatial imaginary, not only essentialized the slum as a homogeneous and coherent unity creeping upon the city and its public life, it also constituted the slum as a space to be treated separately from the spatial problematic of the city as a whole, thereby obfuscating its social complexity. The formal–informal divide is in essence only a perception – albeit a powerful one – and neglects the many ways in which both parts of the city are connected socially, culturally, politically and economically. Nevertheless, this perception has marked and influenced the construction of a "politics of truth" within Moroccan neoliberal governmentality and produced new forms of knowledge, new notions and concepts that contributed to the "government" of the slums and the development of new techniques of intervention and regulation (Lemke 2002). The perception of an insufficient urban integration is predominant in this politics of truth. For example, on the website of the Moroccan Ministry of Housing and Urbanism (MHU) the *bidonville* was defined as "an anarchic ensemble of houses, built with anomalistic materials, deprived of hygiene and collective services and populated by indigent people or people badly integrated in the urban fabric".[2] The question is thus *how* slum inhabitants are being integrated and *how* the public authorities facilitate this integration?

Internationally, towards the end of the 1990s, there was a general shift from repressive approaches aiming to eradicate slums to an assimilating view on the slum population (Bolay 2006). The crisis of the 1980s and the severe protest actions that arose against structural adjustment policies around the world set into motion a new political and academic discourse of "market failure" (Roy 2010). Following the so-called "lost decades", a major shift occurred in development policies (Easterly 2001). Although structural adjustment as a development practice became institutionalized as the so-called "Washington Consensus" in the early 1990s, the promotion of *good governance*, from the mid-1990s onwards, entailed a reconsideration of the Washington Consensus by the international donor community (Santiso 2001). In the negative aftermath of the era of structural adjustment, it was now clear that mere market deregulation and price incentives were not enough to spur economic growth in the Global South. Under impulse of the then Chief Economist of the World Bank, Joseph Stiglitz, the idea emerged that supporting a free market economy required state capacity (Kapoor 2008: 29). As a result, the World Bank actually distanced itself from some of the rigid assumptions of the Washington Consensus. The state, first conceived as part of the problem, came to be considered part of the solution (see World Bank 1997). This shift gave rise to the "Post-Washington Consensus" and recognized the important role of governments in regulatory economic policy. While the so-called "first generation" of neoliberal reforms aimed to stabilize and liberalize the national economies of developing countries (e.g., by rolling back state engagements in import substituting policies, food subsidies and public sector expansion), the "second generation" of neoliberal reforms turned away from the idea of autonomous self-regulating markets and aimed at reforming and strengthening governing institutions. This transition did not just entail that neoliberalism accepted the comeback of the state as such, it also entailed that neoliberalism would seriously influence *how* state intervention was to be deployed (Peck and Tickell 2002; on Morocco, see Zemni and Bogaert 2009). Past ideas on state intervention (e.g., developmentalism, protectionism, import substitution, etc.)

were replaced by more market-oriented perspectives on public policy (e.g., new public management, good governance, etc.).

As such, the euphoria concerning the so-called "end of history" and the infallibility of free market capitalism made place for a new engagement to "end poverty" (Roy 2010). Former World Bank president James Wolfensohn, expressed in a key speech in 1998, entitled "The Other Crisis", his concerns about a looming crisis which the world had failed to notice: the crisis of poverty (ibid.: 16, 74–83). With regard to slum upgrading or resettlement operations, the World Bank abandoned its views from the 1970s and the 1980s, which had been considerably influenced by the writing of the anarchist architect John F.C. Turner, who wanted to minimize the role of the state. Since the end of the 1990s, the World Bank emphasized the role of a powerful bureaucracy as a key for success (Werlin 1999). At the following Millennium Summit in 2000, all 192 member states of the United Nations agreed upon eight general development goals to halve poverty in the world by 2015. More specifically, goal seven, target eleven aspired "to have achieved a significant improvement in the lives of at least 100 million slum dwellers" by 2020 (UN-Habitat 2003: 7). This target could only be achieved through strong cooperation between the local and national governments and their development partners and international donor organizations such as the World Bank and UN Habitat. A few years later, the Global Report on Human Settlements of UN Habitat confirmed – along the lines of the post-Washington consensus – that the slums were not going to be eradicated and integrated based on neoliberal recipes of minimal government and structural adjustment. The slums, according to the report, were the result of failing global, national and local policies (ibid.: 5–6).

Besides a redefinition of the role of state institutions, the principles of good governance also opened the door for a more participatory perspective on development policies. UN-Habitat emphasized that the active involvement of slum dwellers themselves was the most important precondition in tackling the underlying causes of their living condtions (Hernandez and Kellett 2010). This emphasis on participation was another significant break with earlier methods:

> In earlier years, many internationally driven housing and slum reform projects that simply concentrated on engineering and construction solutions failed because they were not sustainable or appropriate in developing country environments. They failed to consult with and involve the people for whom they were intended. They did not work with their organizations and meet their cultural requirements. In addition, they failed to take sufficient account of good governance issues and political will, without which little can be achieved and nothing sustained once the foreign experts have gone home. (…) Incorporating the poor within the design and implementation of development projects not only helps to produce more appropriate projects, but also ensures that they are better targeted to reach those with the greatest needs.
>
> (UN-Habitat 2003: 71–2)

This increased attention to a participatory approach and the active inclusion of the population into the development schemes signified a clear break with former developmental state policies. It ostensibly discarded earlier centralized decision-making processes in favor of "bottom-up" politics, participatory development and local empowerment (Kapoor 2008). The focus on "participation" had become development's new orthodoxy (ibid.: 60). Evidently, this new global development paradigm trickled out to the individual countries and affected local development schemes, especially where they were supported or monitored by the international institutions. However, this participation paradigm is not always as neutral and "bottom up" as it appears to be. More specifically, this article wants to highlight the technologies of power invested in these new "participatory" development schemes once they are deployed in particular local settings.

The introduction of good governance and participatory development in Morocco

The structural conditions at the beginning of the 1980s marked an end to the benign neglect of the Moroccan urban peripheries and its slums. The first "bread riots" of 1981 put pressure on the approach of "management by absence", leading to a more active and systemic government intervention. Yet, during the 1980s, interventions in the *bidonvilles* were above all dictated by security motives and not so much by the poor living conditions of a large part of the urban population (Naciri 1987; Rachik 2002; Navez-Bouchanine 2002a; Bogaert 2011). This changed at the end of the 1990s. The new political context of the Moroccan *alternance*[3] – with the appointment of a socialist prime minister in 1998 and the accession to the throne of King Mohammed VI in 1999 – initiated a reflection on the social dimension of slum clearance and generated fundamental changes in the public discourse (Navez-Bouchanine 2002b). It entailed a rupture with earlier conceptions of slum clearance and started a whole new dynamic in which the fight against poverty became a central concern. Several initiatives and workshops were organized between 1999 and 2001, directed by the then Ministry of Housing, to develop a critical approach on public interventions and discuss the social impact of all the dimensions and aspects of slum clearance and resettlement (Navez-Bouchanine 2004; Zaki 2005: 90–1). This resulted in the elaboration of a new concept, that of the "*maîtrise d'ouvrage social*" (MOS),[4] which constituted a policy guideline for future slum interventions. MOS conceptualized an integrated approach in which both the technical *and* the social dimension of slum resettlement projects are equally valued. Furthermore it emphasized the participation of the local population as a key for success (Navez-Bouchanine 2002b).

The debates on MOS and the social dimension of slum resettlement implied a clear rupture with the "old" methods of public intervention (i.e., the more repressive methods, see Bogaert 2011). One of the key achievements was the recognition of the central role of intermediation between the public authorities and the local population (Navez-Bouchanine 2004). This required of course the formation of new agencies, competences and methods within the public administration (e.g., with the creation of the Social Development Agency (ADS) in 1999). It also implied cooperation with new actors on the ground, such as NGOs, local associations, foreign development agencies (e.g., USAID) and private companies (both foreign and domestic). It was clear that the conceptualization of MOS, the emphasis on the social dimension and the creation of new state agencies such as ADS reflected particular shifts at the international level. For example, Hicham Berra from ADS emphasized that this public agency was founded because:

> the World Bank and the UN talked a lot about the importance of the participation of the communities that were to be developed, about their participation in order to identify their needs. We started from this point, emphasizing proximity: we wanted to go to the people and not wait until they came to Rabat. We wanted to demonstrate that we were able to make, to propose the idea of a partnership with the population. It came down to the fact that even if we represented the state, even if we were the state, the state negotiated, the state contracted with the civil society as peers, equal to each other.
>
> (interview by Lamia Zaki, Rabat, 21 September 2010)

Following this political shift, both at the national as well as the international level, the Moroccan government launched several initiatives to fight poverty and degraded housing (for a more detailed account, see Iraki and Rachik 2005; Le Tellier 2009a). But it was not until the events of 2003 that the Moroccan government really intensified its social development initiatives and put the slums at the top of the policy agenda. The bombings were an immediate starting shot for the launching of the Villes Sans Bidonvilles program (Bogaert 2011).

As such, the VSBP could be considered as a symbolic shift towards an *active management* of the slum population in an attempt to include them in formal city life. In contrast to the more specific and isolated operations of the past, VSBP was the first nationwide development program that tried to deal with slums within a much broader urban perspective. Moreover, VSBP commits itself explicitly to the Millennium Goals (Al Omrane 2010). The program seeks the integration of all *bidonvilles* in Morocco and prioritizes the relocation of its inhabitants. It targets 348,000 households, i.e., 1.6 million inhabitants, spread over 1,000 *bidonvilles* in eighty-five Moroccan cities. Moreover, VSBP involves sixty-four city contracts, which engage the central authorities (the Ministry of Interior and the MHU), the local authorities and the public operators in charge of the specific missions. VSBP is funded to a great extent by the state via the Solidarity Fund for Housing (FSH) and the mobilization of public land.[5] The VSBP also receives support (through gifts or loans) from various international agencies, such as, amongst others, USAID, the European Investment Bank, the French Development Agency and the World Bank. According to the website of the MHU, by the end of March 2011, forty-three cities were declared slum free and 177,416 households were moved (www.mhu.gov.ma, accessed 3 May 2012). They were relocated to social housing sites, mostly situated at the peripheries of the Moroccan cities. As such, the government claims that a little more than 50 percent of the program is already realized. It is estimated that by 2013–15 large cities such as Casablanca, Rabat and Marrakech will also be declared slum free. Following these efforts, Morocco received the 2010 UN-Habitat Award.

The systematic engagement of private actors distinguishes VSBP from former initiatives. By encouraging public-private partnerships the public authorities wished to intensify the construction of low-cost housing units and the restructuring of existing neighborhoods. Construction enterprises such as Addoha and Chaabi Lil Iskane benefit from important tax reliefs. Houses built at prices in the range of US$16,800–30,000 are tax-free (Al Omrane 2010). Following the launching of VSBP, a new financially autonomous state agency was created under the supervision of the MHU: Al Omrane. This semi-public agency, as an Al Omrane official explained, is "the armed force of the Ministry" in the VSBP (interview with Al Omrane official, Rabat, 4 June 2010). It is responsible for the coordination of more than 80 percent of the slum resettlement projects. The rest of the projects are handled by other agencies. For example, due to historical reasons, Casablanca, which accommodates more than 30 percent of the entire slum population, remains a separate case within the national slum program. Despite the fact that Al Omrane is supervising large resettlement operations in the Casablanca district of Sidi Moumen, the city has not signed a city contract within the framework of the VSBP. Most of the operations in Casablanca – targeting approximately 34,500 households – are supervised by another public agency, Idmaj Sakan. This state agency was created in 2005 and is supervised by the Governor of Greater Casablanca and the Ministry of Interior.[6] Finally, in a few cases, slum resettlement operations are handled by a third semi-public operator, Dyar Al Mansour. This operator is a subsidiary of the public investment bank Caisse de Dépôt et de Gestion (CDG).[7] However, both Idmaj Sakan and Dyar Al Mansour apply the same methods of participatory development as Al Omrane in the framework of VSBP.

Citizen participation: a new method of urban government

Due to the social complexity of resettlement projects and the often fragile and mistrustful relationship between inhabitants and public authorities, new governmental methods have been deployed within the framework of the VSBP to intervene more at the individual level, trying to facilitate people's cooperation and adherence to the project. These new methods illustrate the shift towards an active management of the slum population. First of all, there was the methodology of *accompagnement social* (social accompaniment, AS). This method resulted from the debates on the social impact of slum clearance in the early 2000s and the subsequent conceptualization of MOS (Navez-Bouchanine 2002b). AS is a specific methodology to enhance the "active participation" of the inhabitants and facilitate the more technical and physical aspects of resettlement, such as construction and spatial planning (Le Tellier 2009b). As such, AS is a set of methodological guidelines to accompany (literally) the slum dweller through the whole process of moving. Between 2002 and 2009, AS was implemented in twenty-three resettlement programs (20 percent of the VSBP), targeting approximately 50,000 households (Le Tellier 2009a; Toutain 2009a). In the *bidonvilles*, AS is carried out by *cellules d'accompagnement sociale* (CAS). The CAS are teams of 3–5 persons that are set up in the *bidonville*. They mediate between the population and the public operator. Their job is to assist the slum dweller in all the administrative steps that precede his move.

It is important not to confuse AS with MOS. The latter is an integral conceptualization of the social dimension of resettlement operations. It takes into account the social, economic, urban, fiscal and juridical aspects of both the operation and the precarious living conditions of the populations. Despite the fact that MOS has been conceptualized and extensively documented, it has never been implemented on the ground. AS, on the other hand, is far less ambitious. According to the methodological guide published by the Ministry of Housing in 2004, AS refers to a social action that is "subordinated" to the technical aspect of the resettlement operation (MHU 2004: 5). The transfer of the population is the priority. AS is "meant to accelerate the rhythm of social housing [...], to put in place a measure to proceed more rapidly [...], to put in place an arrangement to facilitate people to move" (interview with Al Omrane official, Rabat, 8 June 2010). As a result, it mostly comes down to the fact that the social operator within the AS program has to persuade each slum dweller to move and accept the fact that he or she has to leave the *bidonville* and destroy his or her shack (Navez-Bouchanine and ADS 2005; Le Tellier 2009b). This contrasts with the objectives of MOS, which aims at an equal valuation of both social and technical aspects of slum clearance. The idea of "participation" in AS programs is reduced to its mere technical dimension. AS has to facilitate the technical operation and recover the costs that are imposed upon the slum dwellers. As such, although the "mediator" (e.g., the CAS) may portray itself as a neutral and fair arbiter, it is important to emphasize that the whole process of "participatory development" is managed top down, and it is not the slum dweller who decides upon the proceedings at every step of the way (Kapoor 2008: 64). Most importantly, questions of why, when and where the "beneficiaries" are to be resettled, and who is to be included and excluded from these state-subsidized operations, are all determined above the heads of the slum population. The actual process of relocation and the very issue of being relocated exist outside the participatory development framework. In reality, the intention of AS projects is not so much to improve the participation of slum dwellers but facilitate the process of relocation, which then can be referred to as "participatory".

There are three sorts of social operators implicated in AS projects: the Social Development Agency (ADS), NGOs and private consultants. The private consultants took on 45 percent of all households in AS programs but spent only 28 percent of the total amount dedicated to AS. In contrast, ADS spent 60 percent of the total AS budget while they are only responsible for 33 percent of households (Le Tellier 2009a). It was ADS that launched the first AS operations, hoping to bring forth and set the standards for a comprehensive methodology before other actors intervened in this field. However, Al Omrane quickly made an appeal to private operators by putting out public contracts based on quantifiable targets. This not only caused the emergence of new consultancies trying to compete for these contracts, but also brought about the "commercialization" of the AS procedure. Under these conditions, it is not a surprise that, depending on the type of social operator, the interpretation of the objective of AS and the specific approaches on the ground differed. Private consultants consider AS to be an act of sensibilization; their goal is to inform people. Their vision of "participation" is limited to a passive conception of the process and their objective is resettlement. As Olivier Toutain, a French consultant for the VSBP, stressed: "They [Al Omrane] need to make numbers (…) and private consultants are paid by the numbers" (interview, Rabat, 15 April 2009). Consequently, Al Omrane tends to favor contracts with private consultants in AS programs because they do it "faster and cheaper", and they can be held responsible for results (Le Tellier 2009a: 205). Yet, several cases have already demonstrated the counter-productive effect of these partnerships with private operators. In order to speed up the operation, private consultants applied the principle "first come, first served" in particular cases and exhausted the available land reserve even before all beneficiaries were resettled. Exemplary is the comparison made by a staff member of Team Maroc (one of the private consultants involved) between his role and that of ADS:

> We are a private operator, we answer to calls for public contracts [*appels d'offres*]. We have to keep ourselves to the public plan [*cahier de charge*]. (…) We don't do MOS. (…) And actually we don't do AS either. We rather do administrative accompaniment. (…) I talk in terms of results. We arrive at a number of 300 destroyed shacks per month, while public actors such as ADS arrive at an average of 10 shacks per month. We achieve five times the number of ADS. In four years time we have eradicated more than 4,000 shacks, while ADS only arrived at approximately 1,000 over the same period. (…) We are happy with Al Omrane and Al Omrane is happy with us. (…) We are bound to a *cahier de charge*. We are obliged to deliver results.
>
> (interview Rabat, 5 July 2010)

According to the logic of the private sector, the market and not the public interest is the main motive for Team Maroc's involvement. A colleague of his, working as a CAS member for Team Maroc, was a bit more cynical about his role in the VSBP: "what other role do we have than orientate people towards credit institutions [for their new homes]?" (interview, Casablanca, 7 July 2010).

In contrast, the approach and vision of social operators such as ADS and NGOs like ENDA Maghreb approximates more to the objectives of MOS. Nevertheless, one of the crucial differences between the conception of MOS and their operations within the framework of AS, according to these actors, is that MOS has to be applied *prior to* a resettlement operation, while they are contracted *during* operations. That means that to a great extent, the most important decisions (e.g., where people are to be transferred to) are already made. Additionally, social operators also complain about being constrained by the space for maneuvering that they get from semi-public operators such as Al Omrane. One staff member of ENDA Maghreb explained this as follows:

> We care about the people. Al Omrane only cares about the administrative support [we deliver]. That is why they work more with the private sector now. ENDA pays attention to women's rights. We work to find sustainable solutions. We respect the human rights and also help people in generating new revenues, etc. (…) The private sector only does what Al Omrane asks from them. (…) They have only a servicing role [*un rôle uniquement de prestataire de service*]. They carry out their research, they make an inventory [*recensement*], they send in their poll-takers [*enquêteurs*] and focus primarily on quantitative data. (…) When they finish their job, they are gone.
>
> (interview, Rabat, 2 July 2010)

In general, the more political questions related to issues of social welfare and human development are generally pushed to the background (due to a lack of interest by the private consultants and a lack of means by the social operators). This has already provoked tensions between Al Omrane and ADS (the latter has even taken the decision to disengage itself from AS projects in the future). As a result, despite the new discourse that accompanied the VSBP, the current engagements of the Moroccan authorities related to their fight against poverty in the slums are reduced to a housing question for which technical solutions are supposed to suffice. At the Ministry of Housing they are quite open about this. A staff member of the cabinet of the Minister argued that:

> [the Ministry of Housing] is a technical ministry, we deliver houses, (…). All other social aspects should be provided by other ministries. We don't do education, health care, etc. (…) we limit ourselves to houses [*Nous on fait que le logement*].
>
> (interview, Rabat, 9 June 2010)

Today, the actual elaboration and results of the VSB seem miles away from the original objectives of social and urban integration that were pushed forward so promisingly at the beginning of the century with the conceptualization of MOS. The resettlement operations elaborated in the city of Agadir are emblematic for the strictly quantitative approach. One of the destinations for the relocation of slum dwellers, Taddert, an empty site located at approximately 20 km from Agadir, experienced strong growth and a concentration of social apartments. This was due to the fact that the original convention, signed to resettle 3,200 households, had to be readjusted to resettle 5,800. Additionally, the public operator not only ignored the recommendations for social mixture, but also neglected to put in place the necessary public services and infrastructures (schools, transport, hospitals, etc.) during the time people were transferred due to the fact that the installment of some of these services depended on competences that are not included in the VSB (the Ministry of Housing only delivers houses) but depend on decisions taken in other ministries (e.g., Ministry of Education). Moreover, the pace of construction could not be sustained and some beneficiaries had to wait up to three and a half years before they could move into their new apartments. Meanwhile the public authorities had compelled the beneficiaries to destroy their former shacks, so they were obliged to find temporary accommodation. For those who could neither rely on family nor afford temporary accommodation, a return to precarious and informal housing was the only solution left. With respect to the promotion of citizen participation, the CAS team working on the Taddert resettlement operation only worked there for nine months, leaving the work largely unfinished.[8]

Poverty capital: new frontiers of accumulation

A second important aspect of resettlement operations, besides the methodology of AS and citizen participation, was the development of new mechanisms to improve the slum dwellers' access to finances and pump up, artificially, their solvability. The intention, after all, is not only to transfer people to a new living environment, but also to turn these illegal occupants into legal homeowners. However, one of the main problems is that most of the people who are being resettled possess very little means to finance this move. Additionally, for obvious reasons, there is little trust and familiarity between the poor and commercial credit providers. As such, in the wake of the VSBP, new mechanisms were created to solve that problem. One of those mechanisms is the FOGARIM convention,[9] established in 2004, which permitted slum dwellers to obtain a bank loan with a government guarantee up to a maximum amount of 200,000 Moroccan dirhams (Dh) at an interest rate of 6 percent (€1 is approximately Dh11). Intended to promote low-cost and social housing, FOGARIM was created specifically for those populations "with modest or irregular revenues". This guarantee fund permits the spread of risk between the commercial credit institutions and the state, as it guarantees a recovery of 70 percent of commercially invested capital (World Bank 2006: 16). In the following five years more than 50,000 loans, worth approximately Dh7.3 billion (almost €664 million), were granted. FOGARIM, however, is not limited to the population of the *bidonvilles*. In fact, the number of *bidonville* households that profited from FOGARIM remains relatively limited (10 percent) compared to other populations with limited revenues. Nevertheless, their number is on the rise (Toutain 2009b).[10] Yet, one of the problems remains that instruments like FOGARIM are still little known among the poorest populations. In fact, the participatory development method AS is exactly one of the ways in which the slum population is informed.

In addition to FOGARIM, Moroccan legislation has, since 2004, permitted the intervention of Moroccan microfinance institutions in resettlement and restructuring operations (Le Tellier 2009b). Microfinance has been a booming business in Morocco since the 1990s, but it was only with the launching of the VSBP that it could also be used for housing. Four microfinance institutions (MFIs) control 90 percent of the microcredit market: Al Amana, Zakoura, FONDEP and the Banque Populaire Foundation for Microcredit (FBPMC). Yet, microcredits for housing are still a minority of the portfolio of the Moroccan MFIs: in October 2010 Al Amana counted 49,000 microcredits for housing on a total of 360,000 loans (www.alamana.org.ma/images/Tab-bord-oct2010.pdf, accessed 7 November 2010). FONDEP's share is 7 percent (www.fondep.com, accessed 7 November 2010). Furthermore, the microcredits for housing are limited to a maximum amount of Dh50,000. Nevertheless, the inhabitants of the *bidonvilles* are a potentially valuable population for the MFIs because only 25 percent of them have bank accounts (Le Tellier 2009b). Still, due to the limited amount that can be borrowed these credits are foremost used for renovations and not so much for the purchase of an apartment. Also, due to the relatively high interest rate (12 percent), only those people who have a certain creditworthiness qualify for microfinance. As such, the success of these financial mechanisms is still limited and often too expensive for the poorest among the beneficiaries (Le Tellier 2009a: 206–10).

Despite their limited reach at the moment, the development of mechanisms such as FOGARIM and microcredits for housing are crucial elements in what Ananya Roy calls "the alliance that lies at the very heart of 'poverty capital' ". This alliance brings together those who "control access to the poorest" (e.g., microfinance providers, the CAS teams, the public operator, etc.) and those who "control access to capital" (e.g., the commercial banks) (Roy 2010: 31). The purpose of these financial mechanisms is to explore and exploit "the new frontier of capital accumulation" through the inclusion of the urban poor in Morocco in the urban economy, i.e., those people who hadn't been served by financial systems and markets before (ibid.: 53). The complex but crucial question at hand, according to Roy, is whether poverty capital will ensure the financial inclusion of the poor on fair and just terms, or whether these financial innovations are new ways to exploit them. More generally, is the support of the state in the VSBP – via subsidies, tax cuts and financial guarantees – a way to give the urban poor a new hope for the future, or is state support a rather convenient (class) instrument to privatize benefits and socialize risks and losses? For example, a FOGARIM loan can be obtained without any proof of a stable income or guarantee of solvability; in fact, the borrower only has to deliver a declaration with his word of honor where he confirms his average revenue and the nature of his economic activities (Toutain 2009b). If a loan cannot be repaid, the state largely guarantees the recuperation of private bank losses.

Due to the global financial crisis, Morocco's construction sector experienced a slowdown in recent years after several years of rapid growth. As a result, the Oxford Business Group (OBG), a global publishing and consultancy company focusing on emerging markets in the Arab region and Asia, indicated that both state officials and private developers are looking to offset the drop in demand for high-end development projects by stepping up support for social housing projects (OBG newsletter on Morocco, 25 August 2009). The constant high demand for affordable housing offers a reliable alternative to the business in high-end tourist and residential projects. To address the demand and support the private developers, the Moroccan government has strengthened its commitment to provide low-cost housing, hoping to promote the building of 150,000 new units a year through 2012. In February 2009, the government signed two agreements, allocating a total of Dh52 billion (€4.7 billion) for social housing projects and liberating 3,853 ha of public land for the building of 200,000 units in thirty-two different cities across the country (ibid.). According to the OBG, finance programs such as FOGARIM and other private initiatives are essential to enable Moroccans to purchase homes and "take advantage of the potential in the lower end of the market" (ibid.). Still, the OBG emphasized that some Moroccan officials have argued that more should be done to encourage investors in order for development really to take off:

> Deputy and parliamentary committee finance member Abdelhamid Mernissi told local media that the state should reduce land prices and increase bonuses from Dh30,000 (€2,665) to Dh40,000 (€3,553) for those who target social housing. Others have called on the government to ease its quantitative measurements for affordable units. Currently, it is defined as properties of 100 sq metres at a price of Dh200,000 (€18,000), while low total property value housing has a price of under Dh140,000 (€12,600) and an area of less than 60 sq metres. With space scarce in major cities, the area could be reduced to 40 sq metres, while the price could be raised to Dh300,000 (€27,000) to increase developers' profits.
>
> (OBG newsletter on Morocco, 25 August 2009)

The OBG's advice clearly emphasizes the opportunities of poverty capital. The newsletter continues:

> Still, even without these added incentives, investing in the affordable segment seems like a solid option, despite lacking the same potential for high returns as more lucrative, high-end projects. *Until the economy improves*, and big risks once again offer the prospect of great rewards, social housing, and Morocco's underprivileged citizens, stand to benefit.
>
> (Ibid., italics added)

Poverty capital delineates and opens up new spaces of investment in order to solve future "capital surplus absorption problems" (Harvey 2010). During an interview in 2007, a staff member of Al Omrane explained that the semi-public company soon wanted to enter the stock market. When asked whether this ambition would not jeopardize the social dimension of their work – after all, the stock market is driven exclusively by the profit motive – he answered, a bit excited: "no, on the contrary (…) this social dimension is our competitive advantage (…). The social dimension [*le social*] is very profitable" (interview, Rabat, 27

August 2007). The creation of poverty capital is established by methods of political intervention that attempt to bring all human action into the domain of the market (Roy 2010: 32). As such, poverty capital fits in a wider trend of the "marketization of the social" (Ismail 2006: 67). This entailed the relocation of the center of gravity of welfare and social development to the private sphere and private actors. The role of the state has shifted from providing social assistance (the developmental state) to creating "a good business environment" (the neoliberal state).

The new modalities of slum intervention in Morocco clearly show that social development and poverty alleviation are circumscribed by the market and the salutary effect of market integration. For years, even decades, many slum dwellers in Morocco have lived outside the formal market economy. Many of them had no property rights, no regular job and no legal water and electricity connection. But if the poor were not finding their way to the market, the new methods of social engineering – although still not widely applied – are now helping to bring the market *to* the poor and, if necessary, adapt the products to their possibilities (and not necessarily their needs). Due to cost efficiency calculations, the delivered "product" often shows serious faults. When visiting a social housing project like Annasr in the city of Temara, one quickly notices that these houses were constructed at an absolute minimum cost. After only two years in use, nearly all apartments already showed humidity stains and numerous cracks in the walls. In this particular case, the homeowners often do the best they can to repair the damage themselves. Moreover, the social housing apartments are frequently too small for the larger families. People are often relocated far from their former residence and detached from their original urban communities, especially the slum dwellers within the large city centers, who are usually moved towards the urban peripheries. This has new social and economic implications because many are dependent on their original living place and the more touristic city center for petty trade activities and work in the informal sector.

Furthermore, in a country where the majority of the poor population still has no bank account, concepts like debt and credit are introduced among the lower classes with instruments such as FOGARIM and microcredits for housing. The question is, what will happen when they are no longer able to pay their loans? After all, a new apartment implies a lot of extra costs: property taxes, water and electricity, public transport, etc. Today, the number of people who are unable to repay their debts remains very low (2 percent in 2009) (Toutain 2009b). Nevertheless, when I visited the Annasr project, I was told that a lot of beneficiaries had stopped paying their loans. Coincidently, the Annasr project provided no less than 17 percent of all FOGARIM loans that were granted in 2008 (i.e., 2,700 loans).

Insofar as the *bidonville* is seen to represent a threat to the city (a breeding ground for extremists), the problem is presented as one of shepherding its inhabitants towards market integration. To put it in the words of the technical director of Dyar Al Mansour: "we have to teach them a new way of life, first of all by learning them to live vertically instead of horizontally" (interview, Rabat, 1 July

2010). A staff member of Lydec, a subsidiary of the French multinational Suez in Casablanca, explained that delivering the slum population water and electricity also requires putting a lot of effort into education, teaching them how to use their new assets and how to pay for them (interview, Casablanca, 13 May 2008).[11] As such, the market becomes the norm by which good citizenship is measured. It fits within the neoliberal dogma of "helping the poor help themselves", creating an illusion of self-help (Davis 2006: 72). The VSBP aims to include the slums into the formal urban space where state intervention can define, defend and regulate the norms and boundaries for social life. This kind of approach creates new forms of exclusion. It transforms the slum dwellers "from citizens to consumers by replacing the notion of absolute (human) rights with that of a right to services" (Zaki 2008: 134). Consequently, this right is an atomized one: one that tends to generate conflicting interests between those who are able to afford market inclusion and those who are simply too poor, and thus remain excluded.

Conclusion: slum dwellers and their right to the city

This article showed that intrinsically valuable notions such as participatory development and urban integration are debased and reduced to serve top-down neoliberal rationalities. Efforts to make the slum space more governable and the focus on citizen participation were dictated by the exploitation of poverty capital and the continuation of market-oriented imaginations of development. New frontiers of capital accumulation were explored with the development of new methods of social engineering such as AS and financial mechanisms like FOGARIM. The intention of VSBP was to integrate the slum dweller into the formal market, facilitate his or her access to credit, accompany him or her during all the steps of this integration and, finally, responsibilize the individual to become a good citizen in order to be able to profit from the market opportunities and manage his or her own social development. It remains to be seen whether this will be a sustainable solution.

In his work on cities and processes of urbanization, the French philosopher Henri Lefebvre proposed the notion of "the right to the city" as an alternative to capitalist aspirations of urban development and the imagination of the city from above. As a political ideal, it implicates a collective right: i.e., the collective power of all urban residents to design, reshape and participate in the processes of urbanization and spatial (re)production (Harvey 2008). It is the right to appropriate urban space and think of the city in terms of its *use value* and not its *exchange value* (Lefebvre 1996). But the notion of the right of the city stretches beyond mere physical and official city borders. Lefebvre wanted to rethink the city as a political project in relation to society as a whole, in order to open the door for a radical re-imagination of society as well. As "a right to urban life" (Lefebvre 1996: 158), the notion involves a "claim upon society rather than a simple territorial affiliation", which includes a critique of "the social system of production" (Dikeç and Gilbert 2002: 65).

Urban poverty (and poverty in general), from this perspective, implies more than mere technocratic solutions. Besides, it can also not just be reduced to an individualized matter or responsibility. Neither should it be considered a market opportunity. Instead, poverty becomes a political problem intimately related to and even reproduced by contemporary hegemonic modalities of "development". Along the same lines, Ray Bush argues that poverty has little to do with a lack of integration. Poverty, he argues, does not emerge from the poor man's exclusion but from his particular inclusion in the neoliberalized local economies. It concerns a "differential incorporation", and – as such – Bush directs our focus to "the crucial issue of how poverty is *created* and *reproduced*" (Bush 2004: 674).

Engaging with the right to the city in regard to the Moroccan case therefore implies a radical critique of current state practices and the implementation of market-oriented reforms that tend to prioritize market opportunities over and above the lives and futures of the (urban) poor. It also implies a thorough understanding of the rationalities we see used in practice, through methods of citizen participation and market integration, and more specifically what they – at first sight – conceal: particular strategies and ideology. Within the framework of the Post-Washington Consensus, social development initiatives in Morocco have redressed the importance of state intervention and emphasized the central role of citizen participation and urban integration. Yet, already in the late 1960s, Lefebvre denounced what he called the "obsession" with integration and participation (Lefebvre 1996: 144–6). Neither integration nor participation could "be stimulated either by authoritarian means or by administrative prescription, or by the intervention of specialists" (ibid.: 146). The right to the city is a praxis that cannot be superimposed upon people or residents, certainly not under conditions that serve those in economic and political power. The new modalities of urban government discussed in this article reveal that in the case of Morocco the slum dwellers' right to the city is suppressed by other interests and strategies. After all, with the mere resettlement of slum dwellers the root causes for the perceived political instability after the 2003 events – social inequality, human deprivation, marginalization – have just been displaced.

Notes

1 I would especially like to thank Lamia Zaki for her insightful comments on the paper. I also thank Tony Roshan Samara and anonymous reviewers for their constructive critiques and suggestions. This research was made possible thanks to a scholarship funded by the Research Foundation Flanders (FWO). Of course, the author remains solely responsible for the arguments presented here and any remaining errors of facts or reasoning.

2 Original in French: "ensemble anarchique d'habitations sommaires, édifiées à l'aide de matériaux hétéroclites, privées d'hygiène et d'équipements collectifs, où vivent des populations démunies ou mal intégrées dans le tissu urbain". Online: www.mhu.gov.ma/NR/exeres/7B0C459B-A424-463D-A2BB-98D44FFA5816.htm#B (accessed 2 November 2010).

3 The term refers to a political reform process at the end of the 1990s, and more specifically it refers to the government participation of those historical opposition parties

who played an important role in the struggle for independence (in particular, the Istiqlal party and the Union Socialiste des Forces Populaire (USFP)). Often presented by the majority of the political class in Morocco as a crucial development in the democratic transition of the country, the *alternance* also permitted the monarchy to co-opt important political adversaries and strenghten its grip over the political scene.

4 Can be translated as "social management".

5 The FSH contributes Dh10 billion (US$1.176 billion). The resources of the FSH are generated by a tax of US$12 per tonne of cement sold in Morocco.

6 Ever since the first "bread riots" of 1981 in Casablanca, political control over the urban territory has been centralized into the hands of the Governor (wali) of Greater Casablanca. He is directly appointed by the King and accountable to the Ministry of Interior. The creation of Idmaj Sakan was intended to keep political control over the resettlement operations largely in the hands of the Ministry of Interior, which is *the* power base of the Moroccan monarchy. In Morocco's other cities it is the MHU, via Al Omrane, which controls slum operations.

7 Dyar el Mansour currently has three projects under its supervision: one in Rabat (Douar Al Kora) and two in Temara (the Annasr project and the Mers El Kheir project). Its first project in Rabat followed an impasse of the resettlement operation at the time. It was said that the King personally ordered the CDG to take over the project in 2002 in order to force a breakthrough.

8 I want to thank Lamia Zaki for providing me with information on Taddert.

9 FOGARIM stands for *fonds de garantie des prêts au logement en faveur des populations á revenus modestes et/ou non réguliers*.

10 The average monthly payment is Dh950–1,100. The minimum wage in Morocco was fixed in 2008 at Dh9.66 per hour, or approximately Dh2,000 per month (Toutain 2009b).

11 Suez is supported by state development programs to deliver water and electricity to informal neighborhoods.

References

Al Omrane (2010a) *Al Omrane. Leading Actor for Settlements Upgrading*, Rabat: Kingdom of Morocco. Online: www.unhabitat.org/downloads/docs/Al_OmraneGroup_Morocco.pdf (accessed 22 November 2010).

Bogaert, K. (2011) "The Problem of Slums: Shifting Methods of Neoliberal Urban Government in Morocco", *Development and Change*, 42 (3): 709–31.

Bolay, J.-C. (2006) "Slums and Urban Development: Questions on Society and Globalization", *European Journal of Development Research*, 18 (2): 284–98.

Bush, R. (2004) "Poverty and Neo-Liberal Bias in the Middle East and North Africa", *Development and Change*, 35 (4): 673–95.

Davis, M. (2006) *Planet of Slums*, London: Verso.

Dikeç, M. and Gilbert, L. (2002) "Right to the City: Homage or a New Societal Ethics?", *Capitalism, Nature, Socialism*, 13 (2): 58–74.

Easterly, W. (2001) "The Lost Decades: Developing Countries' Stagnation in Spite of Policy Reform 1980–1998", *Journal of Economic Growth*, 6: 135–57.

Harvey, D. (2010) *The Enigma of Capital. And the Crises of Capitalism*, London: Profile Books.

Harvey, D. (2008) "The Right to the City", *New Left Review*, 53: 23–40.

Harvey, D. (2006) *Spaces of Global Capitalism. Towards a Theory of Uneven Geographical Development*, London: Verso.

Hernández, F. and Kellett, P. (2010) "Introduction: Reimagining the Informal in Latin

America", in F. Hernández, P. Kellett and L. K. Allen (eds) *Rethinking the Informal City. Critical Perspectives from Latin America*, Oxford: Berghahn Books.

Iraki, A. and Rachik, A. (2005) "Quartiers non-réglementaires: intermédiation, élite de proximité et restructuration urbaine", in C. de Miras (ed.) *Intégration à la ville et services urbains au Maroc*, Rabat: INAU, pp. 97–144.

Ismail, S. (2006) *Political Life in Cairo's New Quarters. Encountering the Everyday State*, Minneapolis: University of Minnesota Press.

Kapoor, I. (2008) *The Postcolonial Politics of Development*, London: Routledge.

Lefebvre, H. (1996) *Writings on Cities*, Oxford: Blackwell Publishing.

Lemke, T. (2002) "Foucault, Governmentality, and Critique", *Rethinking Marxism*, 14 (3): 49–64.

Lemke, T. (2001) " 'The Birth of Bio-Politics': Michel Foucault's Lecture at the Collège de France on Neo-Liberal Governmentality", *Economy and Society*, 30 (2): 190–207.

Le Tellier, J. (2009a) "Programme villes sans bidonvilles et ingénierie sociale urbaine au Maroc", in J. Le Tellier and A. Iraki (eds) *Politiques d'habitat social au Maghreb et au Sénégal. Gouvernance urbaine et participation en questions*, Paris: L'Harmattan, pp. 193–212.

Le Tellier, J. (2009b) "Accompagnement social, microcrédit logement et résorption des bidonvilles au Maroc. Une étude de cas: l'opération de recasement Karyan el Oued à Salé", *Les Cahiers d'EMAM*, 17: 55–70.

Massey, D. (2005) *For Space*, London: Sage.

Ministère Délégué Chargé de l'Habitat et de L'Urbanisme (MHU) (2004) *Accompagnement social des opérations de lutte contre l'habitat insalubre. Guide méthodologique*, Rabat: Kingdom of Morocco.

Naciri, M. (1987) "L'aménagement des villes et ses enjeux", *Maghreb-Machrek*, 118: 46–70.

Navez-Bouchanine, F. (2004) "La maitrise d'ouvrage sociale (MOS) au Maroc. Une état des lieux", *Les Cahiers d'Al Omrane*, 19/20: 8–14.

Navez-Bouchanine, F. (2002a) *Les interventions en bidonville au Maroc. Une évaluation sociale*, Rabat: ANHI.

Navez-Bouchanine, F. (2002b) "Evolution de la prise en compte de la dimension sociale dans les interventions en bidonville sous le gouvernement d'alternance", *Critique économique*, 8: 285–301.

Navez-Bouchanine, F. and ADS (2005) "Organisation de la maitrise d'ouvrage social. Stratégies, objectives et taches", unpublished report, Agence de Développement Social.

Parker, C. (2009) "Tunnel-Bypasses and Minarets of Capitalism: Amman as a Neoliberal Assemblage", *Political Geography*, 28 (2): 110–20.

Peck, J. and Tickell, A. (2002) "Neoliberalizing Space", *Antipode*, 34 (3): 380–404.

Rachik, A. (2002) *Casablanca: l'urbanisme de l'urgence*. Casablanca: La Fondation Konrad Adenauer.

Roy, A. (2010) *Poverty Capital. Microfinance and the Making of Development*, New York: Routledge.

Santiso, C. (2001) "Good Governance and Aid Effectiveness: the World Bank and Conditionality", *Georgetown Public Policy Review*, 7 (1): 1–22.

Toutain, O. (2009a) "Bilan, enjeux et perspectives des dispositifs d'ingénierie sociale: le programme villes sans bidonvilles au Maroc", in J. Le Tellier and A. Iraki (eds) *Politiques d'habitat social au Maghreb et au Sénégal. Gouvernance urbaine et participation en questions*, Paris: L'Harmattan, pp. 213–24.

Toutain, O. (2009b) "Bilan evaluation du FOGARIM", unpublished manuscript.

UN-Habitat (2003) *The Challenge of Slums: Global Report on Human Settlements*, London: Earthscan Publications.

Werlin, H. (1999) "The Slum Upgrading Myth", *Urban Studies*, 36 (9): 1523–34.

World Bank (2006) *Maroc: programme Villes Sans Bidonvilles. Rapport finale: analyse d'impact social et sur la pauvreté*, Washington, DC: World Bank.

World Bank (1997) *World Development Report 1997. The State in a Changing World*, Washington, DC: World Bank.

Zaki, L. (2008) "Transforming the City from Below. Shantytown Dwellers and the Fight for Electricity in Casablanca", in S. Cronin (ed.) *Subalterns and Social Protest: History from Below in the Middle East and North Africa*, New York: Routledge, pp. 116–37.

Zaki, L. (2005) "Pratiques Politques au Bidonville, Casablanca (2000–2005)", doctoral dissertation, Institut d'Etudes Politiques, Paris.

Zemni, S. and Bogaert, K. (2009) "Trade, Security and Neoliberal Politics: Wither Arab Reform? Evidence from the Moroccan Case", *Journal of North African Studies*, 14 (1): 91–107.

3 The divisive nature of neoliberal urban renewal in Ouagadougou, Burkina Faso

Wouter Bervoets and Maarten Loopmans

> One should resist the easy temptation to elevate and idealize slum-dwellers into a new revolutionary class. It is nonetheless surprising how far they conform to the old Marxist definition of the proletarian revolutionary subject: they are "free" in the double meaning of the word, even more than the classical proletariat ("free" from all substantial ties; dwelling in a free space, outside the regulation of the state); they are a large collective, forcibly thrown into a situation where they have to invent some mode of being-together, and simultaneously deprived of support for their traditional ways of life.
>
> (Žižek 2004: 12–13)

Introduction

This chapter analyzes the extent to which neoliberal urban development, in particular slum redevelopment in Ouagadougou, capital of Burkina Faso, has the potential to support or undermine strong urban social struggles for the "right to the city". Slum redevelopment in Burkina Faso is a widespread and large-scale phenomenon, which affects a large number of residents. Large-scale slum redevelopment was introduced by the revolutionary government of Thomas Sankara in 1984 as a decisive attempt to modernize cities and achieve a more equal redistribution of resources and access to facilities. When the Sankara regime was overthrown in 1987 and a structural adjustment program (SAP) was imposed on the country, slum redevelopment continued, but gradually lost its egalitarian principles, thereby severely affecting the livelihoods of slum dwellers in a negative way. Given the scale of slum redevelopment in Ouagadougou and evidence of its negative effects, we are puzzled by the relative lack of protest and movements for improved and more egalitarian redevelopment projects. Burkina Faso has a strong and active civil society (as evidenced by the 2011 "Burkinabe Spring", Chouli 2011), but nonetheless slum redevelopment seems to figure only rarely as a reason for mobilization (Harsch 2009).

We argue that neoliberal slum redevelopment, notwithstanding its negative effects, has thus far succeeded in curbing mobilization in two ways: first, maintaining the form (yet not the content) of slum redevelopment of the Sankara years, the practice has retained a strong legitimacy among the wider populace;

second, as slum redevelopment becomes interwoven with traditional power relations and governance structures, it divides residents in oppositional interest groups differently affected by the policy, thereby prohibiting the formulation of collective claims and the development of a unified solidarity urban movement.

The article starts with an overview of our theoretical framework on urban renewal and the "right to the city" in the Global South. The next section introduces the case study of Ouagadougou and its slum redevelopment approach and gives an introduction to our fieldwork. In the third section we discuss how redevelopment results in much discontent among the local population, revealing the unrealized potential for contestation and mobilization. In a final section, we reveal how in different phases of the process, various groups and individuals are affected in a different way, leading to varying levels of support and contestation of the project. Driving wedges between residents in the same neighborhood, this variegated impact undermines the potential for large-scale mobilization against neoliberal slum redevelopment in the longer term. We conclude with perspectives and lessons for broader right-to-the-city mobilizations in Ouagadougou and other cities in the Global South.

Neoliberal urban renewal and the "right to the city" in the Global South

When Manuel Castells in 1983 published his *City and the Grassroots*, it was received as the death knell of a widespread belief in Henri Lefebvre's (1996) "Right to the City" as the catalyst of a global anti-capitalist struggle. Lefebvre's call to reinstall the right to the city's use value prior to exchange value (Lefebvre 1996; McCann 1999) had inspired a flurry of urban struggles which the author regarded as complementary to class struggle. In his seminal work on urban social movements however, Castells explains how cities are not the forefront of a wider social movement, but rather its graveyard. Castells reveals how urban social movements tend to be fragmented and factional, as the motives for urban mobilizations can be highly diverse and pit various groups against each other (see also Saunders 1980). Other prominent scholars identified the ways in which local state officials fragment urban movements. Ira Katznelson's (1981) *City Trenches* explored how the New York state apparatus created strong divides between ethnic groups in the city, and between mobilizations arising in work and residential areas, in order to weaken their political power. Piven and Cloward (1977) suggest that movements may be repressed by detaching their elites from rank-and-file members and bringing them closer to the polity through institutionalization and co-optation (see also Castells 1977: 208–9, quoted in Pruijt 2003).

However, over the past decade we have witnessed a revived optimism about "rights to the city" urban movements. Inspired by Lefebvre's earlier critique, a literature emerged which maintains that recent "neoliberal urbanization" has inspired renewed and more unified struggles for a better and more just city (Harvey 2003; Nicholls and Beaumont 2004; Purcell 2005, 2008; Brenner *et al.* 2009; Marcuse 2009). This literature asserts that neoliberal urbanization has provided urban movements with a common enemy and focus, however variegated their grievances, claims and immediate political adversaries may be. Urbanization has brought the end of history to an end. A growing urban population experiences the collateral damage of neoliberal urban development projects and recognizes the need to react as a collective force. Drawing upon a rights-to-the-city discourse provides new contentious urban actors with a stronger common identity and helps them to bridge different instances and locations of struggle. The strength of Lefebvre's right-to-the-city concept is that it is both capacious and grounded in concrete needs and experiences (McCann 2002). As Mitchell and Heynen argue, it "allows for solidarity across political struggles while at the same time focusing attention on the most basic conditions of survivability, the possibility to inhabit, to live" (Mitchell and Heynen 2009: 616). Simultaneously, recent contributions to the right-to-the-city literature have introduced alternative, more socially and ecologically just visions for the city of the future, which could provide a common purpose to variegated struggles (Marcuse *et al.* 2009).

Urban dwellers in the Global South in particular have experienced strong effects of neoliberal economic reform, as neoliberalism has both directly and indirectly intervened in the production of the city. In an indirect way, neoliberal structural adjustment programs (SAP) have played a particularly important role in changing the structural context of urbanization. Riddell (1997) argues that SAPs have first of all undermined the primacy of larger cities in national politics. Prioritizing the private sector to the detriment of the public sector, they weakened the redistributive function of the administrative city. On top of that, designating raw materials and (export-oriented) agriculture as the main comparative advantages of African countries, the primary sector was stimulated to the detriment of industries. The effects in cities have been a changing urban economy, with a rise in the informal economy and low-wage jobs to the detriment of salaried jobs in industries and state bureaucracies. In combination with reduced public services and an impoverished food supply to cities, life has become more harsh and unequal in cities in the Global South (Riddell 1997; Briggs and Yeboah 2001).

In a more direct way, neoliberalism has reframed the way policymakers consider and address cities. As in cities in the Global North, cities in the South have lost their role as redistributive nodes, but are now a major focus for attracting foreign direct investment (Harvey 1989; Myers 2011). City center redevelopment is considered a catalyst for national and regional development, as it is framed in a specific set of discourses and practices focused on private-sector-led urban regeneration, large-scale commercial urban projects and the privatization of a whole range of services from security to collective transport (Gruffydd Jones 2009). The attractiveness of the inner city becomes a major stake and is at odds with the presence of poor dwellers and informal activities. Hence the "cleansing" of city centers from "non-cosmopolitan activities" is high on the agenda, both in the Global North and South, leading to a flurry of privatized "secured spaces" and gated communities. Simultaneously, Simone (2004: 8) argues: "many states no longer even make symbolic efforts to demonstrate concern with welfare of their populations, and discourses of participatory governance or local entrepreneurship largely become performances deployed to attract donor interest." Exactly at a moment when declining conditions in the countryside pour masses of rural migrants into the cities of the Global South, large-scale attempts to improve the quality of services, including housing, electricity, water and sanitation have been practically cut short (Simone 2004; Ranganathan 2006).

Nijman (2008: 74) distinguishes three ways in which slum rehabilitation has changed in the neoliberal environment: "first, the shift from government intervention to reliance on the free market; second, the shift of responsibility from government to civil society; and, third, the rescaling of government from central to local level." The shift from state planning to market forces in slum redevelopment projects led to the microcredit revolution and land-titling programs to create urban land and real estate markets. In reality, decreasing state engagement in housing resulted in increasing informal housing for the majority of urban dwellers, with only a small minority benefiting from the increased formal housing market activity (Jenkins *et al.* 2007: 229). Today, Sub-Saharan African cities still have high degrees of informality in land occupation, and pre-capitalist systems of land rights continue to exist in parallel with "modern" systems developed by and for capitalism. The shift from government planning to the involvement of the civil society is confronted with the often weak nature of civil society in Sub-Saharan Africa. This is partly due to historic legacies of control by colonial and post-independent governments, but also abrupt social change through urbanization (Jenkins *et al.* 2007: 218). The shift of state responsibilities from central governments to local governments often resulted in a sharing of responsibility, but without powers and access to resources for the cities. As central governments often withdraw their financial support, many urban areas are becoming de facto poorer, leading to short-term horizons for decision-making (Jenkins *et al.* 2007).

With a vein of dramatism, Davis (2006) has called urban slums the dumping ground for a "surplus of humanity". Although considering slum dwellers as a redundant surplus is a deviation from reality, as most of them are strongly integrated in the global economy as informal wage earners or self-employed entrepreneurs, they are no longer considered as citizens with a formal right to adequate living conditions, healthcare or social security provision. Not only their economic integration but also their civil integration has become informalized and dependent upon a variety of livelihood strategies and tactics (Simone 2001; Myers 2011). Resistance to a reduction of citizenship rights to the city is both widespread and overt, in the form of neighborhood or urban protest movements (Verschuur 2008; Harsch 2009), and covert, in the form of non-compliance and subaltern activities (Gruffydd Jones 2009). It is this recognition of a large mass of people well aware of their deprivation of fundamental "rights to the city", or rather "right to life itself", that makes Žižek call the rapid growth of slums in the Global South "perhaps the crucial geopolitical event of our times" and the cradle of a wider anti-capitalist movement.

However, some pertinent critiques have warned against too much revolutionary optimism (Nicholls *et al.* 2012). Purcell (2006) warns that the "local trap" of particularistic local struggles as identified by Castells (1983) remains a threat to right-to-the-city movements. Moreover, neoliberal urban development has already proven its capacity to disempower counter-struggles. Mayer (2006) argues that neoliberal urban governments in Western Europe have succeeded in co-opting large segments of radical activists in the 1980s and 1990s through "partnerships" and participatory schemes. Loopmans (2008) reveals how neoliberal urban development practices and discourses can sometimes effectively secure strong legitimacy and obtain hegemonic status. Discursive struggles are waged to deflect and undermine counter-hegemonic discourses (Loopmans *et al.* 2010). These warnings from the Global North are echoed by Harrison's (2010) exposition of the depoliticizing nature of neoliberalism in Africa. Similarly, Lopes de Souza (2010) warns against the reformist reading of Lefebvre's "right to the city" concept in the circles of development NGOs and international institutions, which might lead to its trivialization and corruption. In many cases, where the "right to the city" is invoked it does not entail, as with Lefebvre, a fundamental critique of capitalism, but merely demands a more "human" life within a "reformed" capitalist city (Lopes de Souza 2010). Such an approach, Lopes de Souza argues, might severely weaken the transformative potential of any urban movement. In this article, we attempt to take this debate further with an empirical analysis of neoliberal urban redevelopment projects in Ouagadougou and the way they affect "the right to the city" of slum dwellers.

Ouagadougou = Geneva? Neoliberal visions for an urban future

Land-locked and deprived of significant natural resources, Burkina Faso considers diplomatic skills to be its most important raw material. In search for its place in the world economy, it tries to position itself as the Switzerland of West Africa (Taliercio 2008). Hosting the headquarters of the West African Economic and Monetary Union, Ouagadougou serves to attract additional service companies, international conferences and headquarters of non-governmental and supranational organizations. Apart from diplomacy, Ouagadougou positions itself as a cultural center for West Africa by events such as the Pan-African Film Festival (FESPACO). The redevelopment of Ouagadougou's city center has been instrumental to this goal, with a number of prestigious large-scale development projects under construction (Taliercio 2008). Simultaneously, neoliberal policies have affected the way Ouagadougou slums are dealt with.

Slum redevelopment was initiated in Ouagadougou in the 1970s. After independence the colonial grip on population movement loosened and rural-to-urban migration was initiated. Urban planning was not a government priority, and new urbanites largely settled in informal settlements at the city's outskirts. Informal settlements made up 60 percent of the city surface in 1980, with half of the city population living in these areas (Marie 1989; Jaglin 1993; Prat 1996). The first attempts to coordinate urban development date back to the late 1970s, when, with the help of Dutch planners, the local government developed a master plan and initiated the "gradual development method", a step-by-step approach for the redevelopment of informal settlements according to a grid plan (Beeker and Guièbo 1994; Beeker 2001). Redevelopment focused on allotment and land titling first, and only later did investments in basic infrastructure follow. Residents themselves were responsible for displacing houses and liberating space for streets and public infrastructure, thus limiting government investment to a minimum (Marie 1989; Jaglin 1993, 1994; Beeker and Guièbo 1994; Beeker 2001).

Yet slum redevelopment remained relatively low on the policy agenda until the socialist revolution of Thomas Sankara. On 4 August 1983 Thomas Sankara came to power after a *coup d'état*. For the revolutionary, Marxist-inspired government, the issue of informal settlements became a priority and a major source of popular support (Marie 1989; Jaglin 1993, 1994; Beeker and Guièbo 1994; Beeker 2001). A radical agrarian and land reform platform nationalized the land; private property was limited to buildings, with citizens having only usufruct over land. The revolutionary government initiated large-scale redevelopment projects in informal settlements. Between 1985 and 1987 60,000 lots were created (Marie 1989; Beeker and Guièbo 1994; Prat 1996). The goals of the redevelopment program were diverse (Marie 1989): to secure decent housing for every citizen, to stop property speculation through the nationalization of land and to break feudal power structures and the monopoly of the rich on land property. "Revolutionary committees" were elected to mobilize citizens to improve their neighborhoods collectively, and redevelopment operations were seen as a pedagogic tool for the transmission of a new morality. The urban project of the revolution appeared as a "clever mix between the consistency of revolutionary principles, demagogic temptations and a tenacious dream of technocratic modernity" (Le Bris 1991: 78, own translation).

Slum redevelopment turned neoliberal after the killing of Sankara in 1987. Ending the Burkinabe revolution, Sankara's right hand and current President Blaise Compaoré took over power and subjected the country to neoliberal rule. From 1991 on, the World Bank and IMF implemented a structural adjustment program in Burkina Faso, with economical liberalization and political decentralization as key elements (Harsch 1998; Beeker 2001; Biehler 2002). On the demand of international financial agencies, Compaoré re-introduced private property rights: the 1991 revision of property legislation allows for the sale of built-up lots; the 1996 revision introduced property rights for vacant lots as well (Cathélineau 2006). This re-introduced land speculation and undermined Sankara's collective urban project, yet large-scale, grid-based redevelopments continue unabated today (Harsch 2009).

Drawing upon a case study approach, four city fragments were selected in the eastern part of Ouagadougou (Figure 3.1), where mapping, observation and interviews took place between July and September 2008. To represent existing diversity in the city, each fragment represents a different phase in the redevelopment

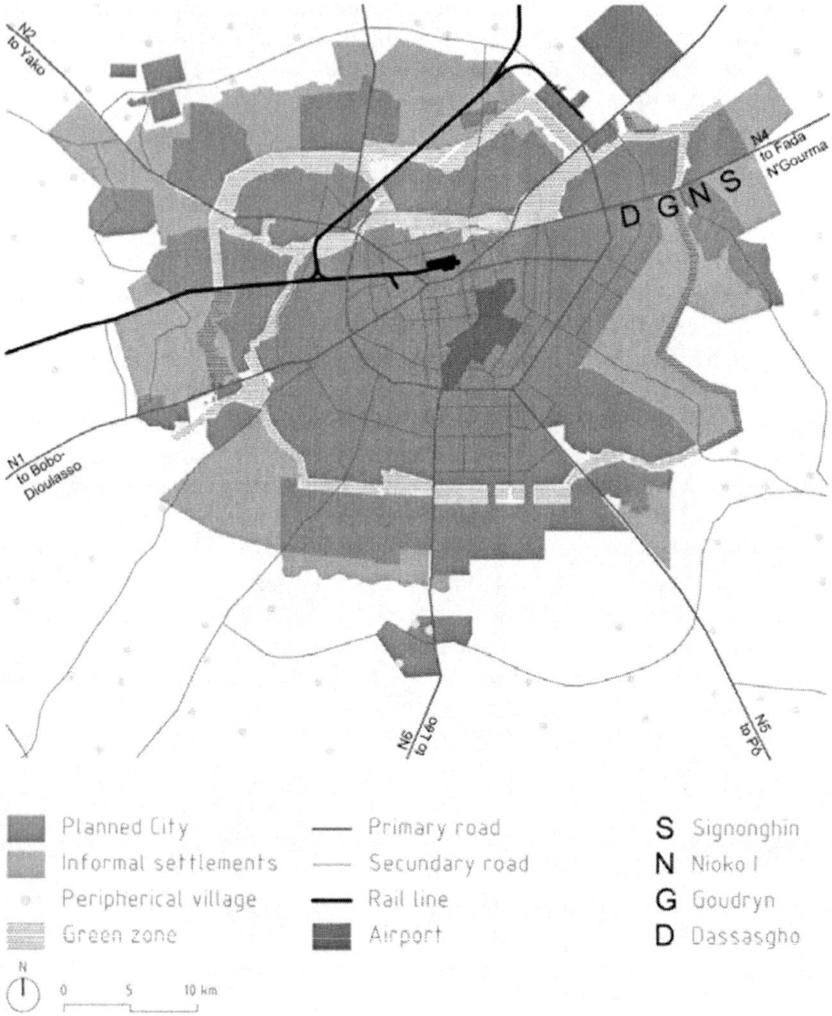

Figure 3.1 Map of Ouagadougou (source: adapted from MHU (2008)).

process, from informal settlement to planned neighborhood (see Table 3.1). The first, Dassasgho, was restructured in the mid-1980s during Sankara's revolution; Nioko I was restructured in the mid-2000s after the economic liberalization and political decentralization of the country; Goudryn was an informal settlement with a redevelopment project in preparation; and, finally, Signonghin was an informal settlement for which no redevelopment project was planned. The four neighborhoods are fundamentally different in their physical appearance and in the composition of their populations (Figure 3.2).

Figure 3.2 Selected case study areas (source: based on Google Earth, as consulted in July 2009).

Twenty years after its redevelopment, Dassasgho was completely built, and was characterized by extreme contrasts: luxurious villas next to modest huts. Dassasgho is, due to its more central location, popular with the members of the middle class, who buy up the lots of the original owners. Nioko I, only five years after redevelopment started, was still in transition. Original inhabitants lived scattered over the neighborhood, and used the many empty lots for small-scale agricultural activities. Empty lots were gradually built up with villas and

Table 3.1 The different phases of a slum redevelopment project

Phase	*Description of the redevelopment process*	*Case study neighborhood*
Phase 1: growth of the informal settlement	Informal settlements grow at the outskirts of Ouagadougou around pre-existing villages, as new rural-to-urban migrants arrive and acquire the right to live on village land from the local "chef-de-terre".	Signonghin
Phase 2: preparation of the slum redevelopment plan	Local government decides to redevelop an informal settlement and demands the authorization of the competent ministries. As news spreads, land speculation sets in and more people acquire a plot or build a house in the neighborhood. Local government prepares a population census of the informal settlement and therefore composes a commission. The commission visits the settlement, numbers all the houses and lists all the inhabitants, subdividing them in three categories: • RR (résidents réels): people really living in the settlement; • PNR (propriétaires non-résidents): people who own a house in the settlement, but do not live in it; • NCH (nouvelles constructions habitées): people who only recently built a house in the neighborhood, after the announcement of the redevelopment project. Public services prepare a master plan for the new neighborhood and get it approved by the competent civil services. Land surveyors start to install the boundary stones of future streets and lots.	Goudryn
Phase 3: attribution of the parcels	The commission attributes lots to the people enlisted during the population census. The inhabitants then need to demolish and reconstruct their houses on the attributed lots. This process can take years; meanwhile empty plots are used for agriculture. With the number of lots often being lower than the number of households, many people are forced to move out of the settlement. Some people are not able to reconstruct a house in due time and are forced to leave their lot too.	Nioko I
Phase 4: slum redevelopment is finished, lots and houses are on the market	As all lots are attributed and occupied, private property rights are clearly settled. A real estate market emerges, and as prices rise poorer households move out to new informal settlements.	Dassasgho

Sources: based on Prat (1996) and Cathélineau (2006).

apartments, bringing wealthier people to the neighborhood. Goudryn was a densely populated informal settlement, enclosed by Dassasgho and Nioko I, which had already been redeveloped. More than fifteen years after the first migrants arrived, only a few elements remained of the traditional rural village it once was. In 2006, all the inhabitants were listed and the houses numbered, but at the time of our fieldwork in 2008 people were still unsure of whether they would obtain their own lot after redevelopment. Finally, Signonghin lies further away from the city center. In this informal settlement, a transition from rural village to densely built informal settlement was still in progress: open spaces between the traditional courtyards were gradually taken up by the modest huts of rural and urban migrants in search of a place to live.

The unvoiced discontents of neoliberal slum redevelopment

The neoliberal slum redevelopment approach reveals a highly uneven division of redevelopment costs. As Marie (1989) reveals, since the introduction of large-scale redevelopment in the 1980s, most of the redevelopment costs were loaded on the shoulders of (future) residents instead of the state. Residents face a number of costs along the redevelopment process: first, they have to pay for the attribution of their lots and a tax for the usufruct, but at the same time they lose their earlier investments in the purchase of the lot on the informal market. Furthermore, residents lose a large part of their earlier investments in the informal settlement due to the enforced demolition and reconstruction of houses. This micro-scale demolition of investments also creates a conundrum at the scale of the neighborhood: the accumulated loss of individual investments decreases the affordability of basic infrastructure, which is, in the end, paid for by the residents themselves.

During Sankara's time, infrastructural costs were shared collectively. Revolutionary committees in each of the thirty Ouagadougou districts were supported by national and provincial government agencies, in the form of technical aid and partial funding. Research by Jaglin (1994) shows that the collective effort of the committees substantially improved public facilities, like school buildings or drinking water supplies. Yet Compaoré did not continue Sankara's attempt to involve inhabitants collectively in the provision of infrastructure or facilities; on the contrary, communal or collective self-organizations are intentionally undermined for the benefit of larger private or public-private companies (Verschuur 2008). Public-private companies provide basic infrastructure, like electricity or tap water, on a pay-as-you-go basis (Beeker 2001). As a consequence, neoliberal slum redevelopment in Ouagadougou consists of little more than the attribution of lots in a grid structure, with little or no provision of infrastructure and facilities. Taxes on usufruct, meant for the financing of infrastructure works, are only paid by a low number of lot owners, which reduces the profitability of a universal provision of services (Cathélineau 2006). Consequently, the under-provision of services affects the poor and rich alike in redeveloped neighborhoods, as is illustrated by the complaint of a middle-class resident in Dassasgho:

> The streets aren't good, the main roads aren't paved, the market hasn't been built yet, the indigenous people still bury their family members in the neighborhood, there is no sewage system, the electricity is often disconnected, the streets aren't lighted [...] it is like being in the bush [...]. There isn't a lot to do in the neighborhood either, we always have to go to the center for a night out.
>
> (Dassasgho, interview 20)

Furthermore, redevelopment not only increases living costs without a significant improvement in quality of life, it also has a severe impact on the income opportunities of residents. A considerable number of the inhabitants in informal settlements are active in agriculture, and apart from the loss of their earlier investments in housing, with redevelopment they also lose their fields and thus their major source of income (see also Luyten 2012):

> The redevelopment has changed our life a lot. It didn't improve anything, on the contrary the situation got worse. We don't have no farming land anymore, agriculture was the only thing we had to live from.
>
> (Dassasgho, interview 6)

With the onset of redevelopment, informal economic activities, such as laundry services, restaurants or shops, have to be restarted from scratch. Shop owners centrally located in the informal settlement might be relocated to a remote corner of the new neighborhood. Redevelopment also sweeps away the economic, social and environmental role of natural marshlands and streams. They tend to be used for urban agriculture, laundry, sand and clay extraction. In redeveloped neighborhoods streams are often straightened out and embedded in concrete, rendering these activities impossible.

As redevelopment, and in particular its neoliberal form, negatively affects the residents' ability to survive, or "right to inhabit", one would expect strong local resistance against such projects. The government is well aware of this risk, as is evident from a number of recent measures. Ouagadougou mayor Simon Compaoré organized public hearings to allow people to air their complaints (Harsch 2009). After the 2011 "Burkinabe Spring" protest in the wake of the Arab revolutions, the regime notably took on urban redevelopment in an attempt to redress popular support. In April 2011, the newly appointed Prime Minister Luc Adolphe Tiao announced the suspension of redevelopment projects for a period of twelve months to audit the problems and eventually adapt the procedures (Sawadogo 2011; Sidzabda 2011; Zouré and Patoin 2011). In a symbolic move, the mayor of the district of Boulmiougou was suspended because of irregularities in local slum redevelopment projects (Raoul 2011), while mayor Simon Compaoré suggested that local governments should no longer be responsible for slum redevelopment and that this competence should return to a national ministry (Ouédraogo 2011). Yet redevelopment projects did not figure prominently in the 2011 uprisings, and even before then slum redevelopment was only a minor reason for urban protest in Burkina Faso. Harsch (2009), reviewing urban protest between 1995 and 2007, noticed that only twenty-five out of 207 urban mobilizations were related to redevelopment projects. Civil society is not well represented in the urban policy debate, with the possible exception of traditional authorities, which exercise considerable influence in local politics (Gouëset 2009).

We claim that, although neoliberal urban redevelopment negatively affects a lot of people in Ouagadougou, it fails to stimulate strong and unified mobilization for various reasons, which will be further explored in the following sections. First, urban redevelopment has a long history in Ouagadougou, and current urban redevelopment practices maintain a strong legitimacy because of their discursive and organizational connections to prior periods. Second, neoliberal urban redevelopment interconnects with traditional and Sankarist governance structures to create cleavages and divisions within the local populace, thereby undermining unified struggle by neighborhood dwellers.

Historical legitimation of the grid-based redevelopment approach

Slum redevelopment along a grid model has a long history in Ouagadougou and has historically obtained a hegemonic status in the collective imagination of the population. The "gradual development method" model was chosen in the 1970s, against the then UN-backed small-scale "slum upgrading" approach, which preserved most of the buildings and organic street patterns (Beeker and Guièbo 1994). Land division on a gridiron projection, superimposed on the informal settlements, was perceived as more modern, echoing the layout of the "European quarters" of the colonial city and reflecting a "progressive urbanity" as opposed to the "barbarism of the bush" (Déverin 1992, quoted in Biehler 2002: 99). Furthermore, a survey among the population at the time of introduction made clear that the provision of identical lots was perceived as the fairest solution by most respondents (Beeker and Guièbo 1994). "Becoming part of the city" is still an important reason for people to support redevelopment of their informal settlements, and respondents from Goudryn and Signonghin were in general looking forward to redevelopment and its promise of an improved quality of life:

> The redevelopment will bring us tap water, electricity, paved roads and sanitary facilities. Because roads will be constructed, it will be easier for us to reach the city center by taxi and the neighborhood will be accessible for ambulances.
>
> (Goudryn, interview 2)

People were also looking forward to their official lot with tenure security:

> My life will be much better. I will finally be at ease and be secured, then I will be really at home [...] being at home in your own place is the best thing out of everything.
>
> (Goudryn, interview 14)

Since Sankara's massive urban redevelopment program, slum redevelopment is also regarded as a remnant of the revolution and purveys a message of social justice and equality that still legitimizes the program itself, as is exemplified by the comment of a Signonghin respondent:

> Today the problem of the redevelopment projects is not at the level of the authorities [...] but rather at our level. The interests have changed. At the time of the revolution and today, redevelopment projects are executed in the same way, but the people have changed. The actual problem of the redevelopment projects is currently at the level of the residents of the informal settlements.
>
> (Signonghin, interview 4)

For this respondent, the problems with slum redevelopment are caused by the ignorance of slum dwellers instead:

> I wish the population is mobilized to generate an awareness on the merits of solidarity, on the need to unite and to truly understand the redevelopment projects. Because until now there are people that do not really understand these redevelopment projects.
>
> (Signonghin, interview 4)

When people protest against slum redevelopment projects, they criticize the actors involved instead of the project itself. This has fragmented the protest against redevelopment. Neoliberal reform decentralized urban redevelopment, thereby diverting protest to the widespread corruption of specific local authorities and commissions (RENLAC 2003: 45–6):

> The commissions are fraudulent, they are the ones that make us suffer. The administrative authorities should have a look at these commissions which lie and sell lots. The commissions should be composed of competent persons, so that every resident could obtain a lot.
>
> (Signonghin, interview 10)

Dividing neighborhoods

Secondly, slum redevelopment has strongly uneven effects within neighborhoods, both in the period before and during the redevelopment project. For Sankara, slum redevelopment was an instrument to wipe out traditional power structures and privileges, but failed to do so completely. Even today, traditional structures interact with the neoliberal logic of the slum redevelopment process to deliver different outcomes for different residents, depriving slum dwellers of a strong collective interest and undermining collective mobilization. Throughout the process, tensions emerge between different groups and individuals (MHU 2008: 176–7). The commodification of land and the increase of land speculation, especially after Thomas Sankara's downfall, turned the periphery of Ouagadougou into a "battleground for land property" (Jaglin 1993; Prat 1996).

Elder vs younger generations of autochthonous villagers

First of all, tensions arise amongst autochthonous residents of the original peripheral villages themselves (MHU 2008: 176). Confronted with a growing demand for lots by migrants, the autochthonous start selling lots to rural and urban migrants, as we observed in the neighborhood of Signonghin. According to customary right, the village chief is the only party responsible for the allocation of land to newcomers (Gouëset 2009: 148). Sometimes the village chief abuses this right to make profit at the expense of other autochthonous residents, by selling their farmland without any compensation. But it also happens that the younger generation sells family land without consultating the village chief. This causes conflicts between different generations in one family, but also between different families because sometimes even the land of other families is being sold.

The village chief might also abuse its connections with the local administration or the registration commission for personal benefit. In Nioko I, the autochthonous population accused the old village chief of having obtained more than 800 lots for himself and his close family, while some other autochthonous residents failed to obtain even one single lot and find themselves without farmland or other resources (see also Gouëset 2009).

Autochtonous villagers vs immigrants

Second, conflicts arise between the autochthonous population and rural and urban immigrants to the area (MHU 2008: 177). These conflicts are diverse, but are in essence related to the ambiguous rules for the attribution of lots. With the reform of property legislation during the revolutionary period, the feudal land ownership system and customary laws were officially halted. But the traditional system still affects redevelopment as autochthonous residents were offered some privileges to secure their support. The minimum age for obtaining a lot is fifteen years for autochthones, instead of eighteen years, and the financial contributions required for purchasing a lot are lower for autochthones. But the most important privilege is the priority in the attribution of lots and the right to obtain multiple lots as compensation for the loss of agricultural land. (Cathélineau 2006) On the eve of the redevelopment project the autochthones insist on their privileges:

> I was born here, I grew up here and I will die here. In principle they should start attributing lots to the autochthones and only afterwards to those people that came to install themselves between us.
>
> (Signonghin, interview 21)

The privileges of the autochthonous residents lead to tensions with migrants, who often live in the peripheral villages for a long time. These migrants reject the distinction between autochthonous and non-autochthonous and claim the same rights.

> In case of a redevelopment we should for sure get a lot. We live here that long, that I can say we became part of the village, we became children of this place.
>
> (Signonghin, interview 7)

The attribution of the lots in the redevelopment project is the responsibility of a communal commission composed of the local mayor, representatives of the public services for urban planning and the land register and representatives of the local inhabitants (Cathélineau 2006). Rural and urban migrants question the unfair amount of autochthonous representation in the commission, which disadvantages migrants in the attribution of lots.

> There are conflicts between immigrants and autochthones, people that don't like each other manage that others do not get a lot. The autochthonous inhabitants have too much power over the immigrants in the attribution of lots. Look at Nioko I, today [after the redevelopment] there are autochthonous that used to have only a bicycle, but now they have already built a villa. So there is also a lot of corruption involved. We need an independent commission in which also newcomers are represented.
>
> (Signonghin, interview 3)

As a reaction to these irregularities, the rural and urban migrants of Signonghin organized themselves into a neighborhood committee that lists all the inhabitants and their date of arrival. They hoped that this list would help them to obtain a lot the day the redevelopment project started.

Migrants vs speculators

On the eve of a redevelopment project important conflicts also arise between migrants and speculators, who are settling in the informal settlement with the sole purpose of obtaining the maximum number of lots after redevelopment. Being registered during the population census at the beginning of the redevelopment process is crucial. Speculators construct small, uninhabited houses, called "matchboxes" or "alibi houses", to obtain the right to a lot. Although legislation stipulates that every citizen has the right to only one lot, speculators with several "alibi houses" borrow names from family members living in rural areas to register more than once (see also Marie 1989). As the number of lots after redevelopment is mostly lower than the number of households living in the informal settlements, speculation decreases the chances of real residents to obtain a lot. Particularly when speculators have more financial resources to bribe the allocation committee, tensions increase in the neighborhood. In the informal settlement of Goudryn, where the local government was supposed to attribute the lots in the near future, the inhabitants were anxious about what the planned redevelopment project would bring to them.

> We are waiting since the population census, but we still don't know where our lot will be situated [...]. The only thing we know is that there won't be enough lots for everybody [...] maybe they will give our lot to somebody else and then they will arrive here on a certain day with bulldozers to chase us away.
>
> (Goudryn, interview 20)

Rich vs poor

Whereas slum redevelopment promises a higher quality of life for the residents, in reality the social and economic consequences of redevelopment projects are rather disastrous for the most vulnerable inhabitants (see also Marie 1989). Tenants and people living for free in the informal settlement are the most vulnerable; in general they are not allowed an official lot and are forced to leave the neighborhood after redevelopment. But a considerable number of families that have been registered during the population census as lot owners are often forced to leave the neighborhood after the redevelopment, as the number of lots is often lower than the number of households in the informal settlements. Money acts as a lubricant, where a "financial gift" helps to secure a lot, but not everyone has the financial means to bribe commissioners or state officials. Poorer families are excluded, often on unclear grounds:

> We are living here since 1997, but we didn't obtain a lot at the time of the redevelopment although we were already living here at that time. The local authority told us that our demand was addressed wrongly. We now try to obtain more information, but with no result until now [...] if we don't obtain a lot, we will be obliged to move further.
>
> (Nioko I, interview 17)

But families that do obtain a lot might also be forced to leave the neighborhood after the redevelopment. Many of these families lack the financial means to build their lot in less than five years as the law prescribes, and are thus forced to sell their lot and move again to an informal settlement. Other families are forced to sell their lot because of financial problems, for example to pay for medical treatment for a family member. These lots are mostly bought by more affluent middle-class families for the construction of a villa, or by investors for the construction of rental housing. The social exclusion of original inhabitants and the influx of more affluent newcomers transform the original informal settlement into a middle-class villa neighborhood. This social exclusion generates tensions between different groups, as one autochthonous resident testified:

> The redevelopment changed our life a lot [...] social life was turned upside down. Today everybody is living on his fenced off lot. There are also some

> troubles between us, autochthonous people and outsiders that came here to take advantage of the lots.
>
> (Nioko I, interview 3)

Also social ties and networks seem to suffer from redevelopment operations, making the daily struggle to survive even more difficult for the most vulnerable groups.

> Many people have left the neighborhood because they didn't obtain a lot or because they sold it because of lack of funding. In the past we used to know everybody here, but not any longer [...]. Life became more difficult; when in the past I urgently needed money for my children, I could always pass along my neighbors, but they don't live any longer in the neighborhood.
>
> (Nioko I, interview 7)

When basic infrastructure is finally provided in the neighborhood, high connection fees keep drinking water or electricity out of reach of the most vulnerable. Despite the redevelopment project, living conditions remain hard for the urban poor, in sharp contrast with their more affluent neighbors:

> We live here on our lot in a very simple house with no connection to water or electricity. But our neighbors are living in big villas with multiple floors and all comfort. Our toilet and shower are open air, so our neighbors can see everything out of their villa.
>
> (Dassasgho, interview 6)

Conclusion

The present analysis has questioned whether neoliberal urban development projects could stimulate "right to the city"-based urban movements in the Global South. Drawing upon a case study of Ouagadougou, Burkina Faso, it takes issues with the fact that few collective mobilizations have taken place about slum redevelopment, even though Ouagadougou is unique in terms of the scale and number of such slum redevelopment projects initiated. Through indepth interviews and observations in four neighborhoods in different phases of redevelopment, we have reviewed the effects of slum redevelopment upon its original dwellers and analyzed the way these affected the potential for mobilization.

Our findings nuance the all too general and optimistic claims of the coming of urban revolutions inspired by the "right to the city", and a mass of slum dwellers taking their aversion for the neoliberal transformation of their living environment as a source of inspiration for a wider anti-capitalist struggle. Even though the masses in the slums of the Global South are rightly considered as destitute and thrown together in an enraging situation, we cannot conclude that they form a uniform collective or that neoliberal urban projects bear the potential to forge them into a revolutionary collective. On the contrary, neoliberal urban projects enrage indeed, but do so in a very selective and divisive way, and do more to undermine, rather than reinforce, a collective reaction of "the slum multitude".

Neoliberal urban projects develop in particular contexts and engage with local histories in a way that neoliberal competition and profit-driven urban development combine with locally particular non-market institutions and traditions. Such local mediation of neoliberal urbanization renders it specific and increases its local legitimacy, even among slum dwellers. Moreover, it undermines the emergence of a collective rights identity around a notion of inhabitance of the city, within as well as across cities, as it recombines with local social structures to reinforce particular identities (migrant, native, elder, youth,...).

In Ouagadougou, slum redevelopment was introduced on a large scale by the revolutionary government in 1983, with the purpose of modernizing the city and securing a decent home for everyone, while simultaneously breaking traditional power structures. Yet the ensuing counter-revolution imposed neoliberal reform on the country. It reintroduced land tenure and speculation, but maintained the spirit of modernization and the grid-based "one household, one lot" land titling of the Sankarist revolution. As a result, urban redevelopment still resonates of "social justice" and of "progress", and continues to enjoy popular support.

Concrete projects do stir discontent, but in highly divisive ways, as different people are affected differently. The case of Ouagadougou shows that individual land titling benefits the wealthier slum dwellers, but is disastrous for poorer inhabitants who simply cannot afford incorporation into the fully commodified formal economy. Yet the poor are not homogeneous and their subdivision rests on the recombination of neoliberal market pressures with local governance structures, such as customary law and local state institutions. Tying land titling to customary primacy for autochthones, autochthonous residents are pitted against immigrants as oppositional identities; local communities are disrupted as village chiefs abuse their customary right to land allocation to reap off land rent to the detriment of younger villagers; and the corruption of state officials involved in the allotment procedures pits speculators against actual residents.

The local mediation of neoliberalism is something often overlooked by its opponents, yet the recombination of neoliberal markets with local non-market institutions is exactly what undermines their resistance; any rights-to-the-city movement will need to confront this local mediation of neoliberal forces so as to avoid the "local trap" of becoming locked in particularistic struggles that undermine its unity and strength.

Note

This chapter is partly based on the Masters thesis project by Wouter Bervoets, "Towards a Sustainable Urbanization Model for Ouagadougou – Burkina Faso", supervised by Professor Maarten Loopmans and Professor Rob Geys at the Erasmus University College Brussels. Fieldwork in Ouagadougou between July and September 2008 was made possible with a VLIR-UOS travel grant.

Bibliography

Beeker, C. (2001) "Pitfalls in Planning: a Look at Ouagadougou in Burkina Faso", in L. J. de Haan, I. Baud, T. Dietz and J. Post (eds) *Re-aligning Government, Civil Society and the Market: New Challenges in Urban and Regional Development*, Amsterdam: AGIDS.

Beeker, C. and Guièbo, J. (1994) "Plotting the Urban Field of Ouagadougou", *Third World Planning Review*, 16 (3): 309–24.

Biehler, A. (2002) "Enjeux et modes de constitution des espaces publics", unpublished thesis, Université de Paris I, Sorbonne.

Brenner, N., Marcuse, P. and Mayer, M. (2009) "Cities for People, Not for Profit: an Introduction", *CITY: Analysis of Urban Trends, Culture, Theory, Policy, Action*, 13 (2–3): 173–81.

Briggs, J. and Yeboah, I. E. A. (2001) "Structural Adjustment and the Contemporary Sub-Saharan African City", *Area*, 33 (1): 18–26.

Castells, M. (1983) *The City and the Grassroots: a Cross-Cultural Theory of Urban Social Movements*, Berkeley: University of California Press.

Castells, M. (1977) *The Urban Question*, Londen: Arnold.

Cathélineau, O. (2006) "L'étalement urbain excessif à Ouagadougou – Burkina Faso: Vers quels moyens d'action?", unpublished thesis, Institut d'Urbanisme de Paris, Université Paris XII – Val de Marne.

Chouli, L. (2011) "People's Revolt in Burkina Faso", in F. Manzi and S. Ekine (eds) *African Awakening: the Emerging Revolutions*, Oxford: Pambazuka Press.

Davis, M. (2006) *Planet of Slums*, New York: Verso.

Déverin, Y. (1992) "Le corps de la terre. Moose de la région de Ouagadougou: Représentations et gestion de l'environnement", unpublished thesis, Université de Paris I – Sorbonne.

Gouëset, V. (2009) "Développement urbain et mobilités à Ouagadougou: le rôle des politiques publiques", in F. Boyer and D. Delaunay (eds) *Peuplement de Ouagadougou et développement urbain: rapport provisoire*, Paris: IRD – Université Paris I.

Gruffydd Jones, B. (2009) " 'Cities Without Slums'? Global Architectures of Power and the African City', paper presented at African Perspectives conference – The African Inner City: [Re]sourced, Pretoria, September.

Harrison, G. (2010) *Neoliberal Africa: the Impact of Global Social Engineering*, London: Zed Books.

Harsch, E. (2009) "Urban Protest in Burkina Faso", *African Affairs*, 108 (431): 263–88.

Harsch, E. (1998) "Burkina Faso in the Winds of Liberalisation", *Review of African Political Economy*, 25 (78): 625–41.

Harvey, D. (2003) "The Right to the City", *International Journal of Urban and Regional Research*, 27 (4): 939–41.

Harvey, D. (1989) *The Condition of Postmodernity*, Oxford: Blackwell.

Jaglin, S. (1994) "Why Mobilize Town Dwellers; Joint Management in Ouagadougou, 1983–1990", *Environment and Urbanization*, 6 (2): 111–32.

Jaglin, S. (1993) 'Les banlieues de Ouagadougou', *Afrique Contemporaine*, 168: 54–66.

Jenkins, P., Smith, H. and Wang, Y. P. (2007) *Planning and Housing in the Rapidly Urbanizing World*, Routledge: New York.

Katznelson, I. (1981) *City Trenches: Urban Politics and the Patterning of Class in the United States*, Chicago: University of Chicago Press.

Le Bris, E. (1991) "Crise urbaine et effets urbains de la crise: le cas de l'Afrique Noire", *Espaces et Sociétés*, 65: 61–81.

Lefebvre, H. (1996) *Writing on Cities*, Oxford: Blackwell Publishers.

Loopmans, M. (2008) "Relevance, Gentrification and the Development of a New Hegemony on Urban Policies in Antwerp, Belgium", *Urban Studies*, 45 (12): 2499–519.

Loopmans, M., De Decker, P. and Kesteloot, C. (2010) "Social Mix and Passive Revolution. A Neo-Gramscian Analysis of the Social Mix Rhetoric in Flanders, Belgium", *Housing Studies*, 25 (2): 181–200.

Lopes de Souza, M. (2010) "Which Right to Which City?", *Interface*, 2 (1): 315–33.

Luyten, S. (2012) "De regulatie van de West-Afrikaanse stadstuinbouw. Diversiteit in de kleinschalige groente- en fruitteelt in Ouagadougou en Dakar", unpublished thesis, University of Leuven.

Marcuse, P. (2009) "From Critical Urban Theory to the Right to the City", *CITY: Analysis of Urban Trends, Culture, Theory, Policy, Action*, 13 (2–3): 185–97.

Marcuse, P., Connolly, J., Novy, J., Olivo, I., Potter, C. and Steil, J. (eds) (2009) *Searching for the Just City: Debates in Urban Theory and Practice*, New York: Routledge.

Marie, A. (1989) "Politique urbaine: une révolution au service de l'état", *Politique Africaine*, 33: 27–38.

Mayer, M. (2006) "Manuel Castells' The *City and the Grassroots*", *International Journal of Urban and Regional Research*, 30 (1): 202–6.

McCann, E. J. (1999) "Race, Protest, and Public Space: Contextualizing Lefebvre in the U.S. City", *Antipode*, 31 (2): 163–84.

McCann, E. J. (2002) "Space, Citizenship and the Right to the City", *GeoJournal*, 58 (2–3): 77–9.

MHU – Ministère de l'habitat et de l'urbanisme (2008) "Schéma Directeur d'Aménagement du Grand Ouaga – Horizon 2025".

Mitchell, D. and Heynen, N. (2009) "The Geography of Survival and the Right to the City: Speculations on Surveillance, Legal Innovation, and the Criminalization of Intervention', *Urban Geography*, 30 (6): 611–32.

Myers, G. (2011) *African Cities: Alternative Visions of Urban Theory and Practice*, London: Zed Books.

Nicholls, W. and Beaumont, J. R. (2004) "The Urbanisation of Justice Movements? Possibilities and Constraints for the City as a Space of Contentious Struggle", *Space and Polity*, 8 (2): 119–35.

Nicholls, W., Uitermark, J. and Loopmans, M. (2012) "Cities and Social Movements: Theorizing Beyond the Right to the City", *Environment and Planning A*, forthcoming.

Nijman, J. (2008) "Against the Odds: Slum Rehabilitation in Neoliberal Mumbai", *Cities*, 25 (2): 73–85.

Ouédraogo, D. E. (2011) "Lotissements à Ouagadougou: Simon conseille le retrait de la gestion aux maires", *L'Observateur Paalga*, 20 June. Online: www.lobservateur.bf/index.php?option=com_content&view=article&id=600:lotissements-a-ouagadougou-simon-conseille-le-retrait-de-la-gestion-aux-maires&catid=23:chez-nous&Itemid=30 (accessed 22 May 2012).

Piven, F. F. and Cloward, R. A. (1977) *Poor People's Movements*, New York: Pantheon.

Prat, A. (1996) "Ouagadougou, capitale Sahélienne: croissance urbaine et enjeu foncier", *Mappemonde*, 1: 18–24.

Pruijt, H. (2003) "Is the Institutionalization of Urban Movements Inevitable? A Comparison of the Opportunities for Sustained Squatting in New York City and Amsterdam", *International Journal of Urban and Regional Research*, 27: 133–57.

Purcell, M. (2008) *Recapturing Democracy: Neoliberalization and the Struggle for Alternative Urban Futures*, London: Routledge.

Purcell, M. (2006) "Urban Democracy and the Local Trap", *Urban Studies*, 43 (11): 1921–41.

Purcell, M. (2005) "Review of the Right to the Redevelopment", *Antipode*, 34 (3): 519–40.

Ranganathan, M. (2006) "Understanding the Link between Tenure and Services for the Peri-Urban Poor: Case Studies from Senegal and India", paper presented at the Breslauer Graduate Student Symposium: The Right to the City and the Politics of Space, Berkeley, April.

Raoul, B. (2011) "Lotissement à Boulmiougou. En attendant la révocation de Séraphine Ouédraogo", *L'Indépendant*, 8 November. Online: www.independant.bf/article.php3?id_article=207?&sq=arti (accessed 22 May 2012).

RENLAC (2003) "Rapport 2003, état de la corruption au Burkina Faso", Online: www.renlac.org/ (accessed 6 June 2012).

Riddell, B. (1997) "Structural Adjustment Programmes and the City in Tropical Africa", *Urban Studies*, 34 (8): 1297–307.

Saunders, P. (1980) *Urban Politics: a Sociological Interpretation*, London: Penguin.

Sawadogo, A. K. (2011) "Maires et lotissements: Il faut boire la coupe jusqu'à la lie", *L'Observateur Paalga*, 21 June. Online: www.lobservateur.bf/index.php?option=com_content&view=article&id=605:mairma-et-lotissements-buvez-la-coupe-jusqua-la-lie-&catid=9:grille-de-lecture&Itemid=32 (accessed 22 May 2012).

Sidzabda (2011) "Problèmes de lotissement: une Séraphine peut en cacher bien d'autres", Le Faso, 10 May. Online: www.lefaso.net/spip.php?article41957 (accessed 22 May 2012).

Simone, A. (2004) *For the City Yet to Come: Changing African Life in Four Cities*, Durham and London: Duke University Press.

Simone, A. (2001) "Straddling the Divides: Remaking Associational Life in the Informal African City", *International Journal of Urban and Regional Research*, 25: 102–17.

Taliercio, P. (2008) "Un déguerpissement exemplaire à Ouaga (Burkina Faso)", *Revue Agone*, 38/39: 89–107.

Verschuur, C. (2008) "Neighbourhood Movements, Gender and Social Justice: the Cultural Reinvention of Politics by Women", *International Social Science Review*, 193: 411–22.

Žižek, S. (2004) "Knee-Deep", *London Review of Books*, 26 (17): 12–13.

Zouré, A. and Patoin, J. (2011) "Avec la suspension des lotissements, on saura quelle est le mal. Interview avec Dieudonne Tapsoba, DG de l'Urbanisme", *Le Pays*, 31 May. Online: www.lepays.bf/?DIEUDONNE-TAPSOBA-DG-DE-L (accessed 22 May 2012).

4 Greening dispossession

Environmental governance and socio-spatial transformation in Yixing, China

Jia-Ching Chen

> [In] that picture that lacks all spatial coherence, is a precise region whose name alone constitutes for the West a vast reservoir of utopias. In our dreamworld, is not China precisely this privileged site of space? In our traditional imagery, the Chinese culture is the most meticulous, the most rigidly ordered, the one most deaf to temporal events, most attached to the pure delineation of space; we think of it as a civilization of dikes and dams beneath the eternal face of the sky; we see it, spread and frozen, over the entire surface of a continent surrounded by walls.
>
> (Foucault 1994: xix)

> China is changing from the factory of the world to the clean-tech laboratory of the world. It has the unique ability to pit low-cost capital with large-scale experiments to find models that work. China has designated and invested in pilot cities for electric vehicles, smart grids, LED lighting, rural biomass and low-carbon communities. They're able to quickly throw spaghetti on the wall to see what clean-tech models stick, and then have the political will to scale them quickly across the country. This allows China to create jobs and learn quickly.
>
> (Peggy Liu, Chairperson of the Joint US–China Collaboration on Clean Energy, quoted in Friedman 2010)

This chapter examines a master-planned eco-city centered on renewable energy industries in the city of Yixing, Jiangsu Province, arguing that it illuminates the quintessential strategy, ideology and social–environmental contradictions of what might be called China's "Green Leap Forward". By now, many readers familiar with the problem of global climate change mitigation have heard a series of facts. First, since 2007, China has been the leading national emitter of greenhouse gases. Second, China's economic growth is spurred by rapid expansion in both production and consumption, and these lead to further increases in energy demand, carbon dioxide and other greenhouse gas emissions. Third, China's cities are growing at an unprecedented rate: the nation's urban population surpassed its rural counterpart in 2011, and is projected to exceed 1 billion by 2030. Many observers conclude that, in terms of addressing global climate change, "everything is won or lost in China" (Lovins 2008).

This problematization of Chinese urbanization and development is attended by a rapidly expanding field of transnational green development expertise, state and corporate activity. Increasingly, experts and boosters point to China's combined economic power and political system as the largest – and perhaps most important – venue for the development of "a clean energy future" (Finamore 2011; Friedman 2010). As in Foucault's description of Orientalist imaginaries, this future "dreamworld" is predicated on what makes China wholly other – its authoritarian-modernist state power and its faceless masses of laboring bodies. This perception is bolstered by China's rapid ascendance as the world's leading producer of solar photovoltaics and wind turbines, as well as internationally prominent examples of "eco-city" and "eco-industry" planning and construction.[1] Overhyped initiatives serving as city marketing and controversial failures in implementation notwithstanding, these efforts of "green development" are being backed by central government policies including new systems of standards, piloting programs, legislation and the official development policies set forth by the Eleventh and Twelfth Five-Year Plans (covering 2006–10 and 2011–15, and commonly referred to as the "11–5" and "12–5" Plans).

As in previous moments of Chinese state-led developmental nationalism, this Green Leap Forward explicitly targets rural space and society as primary sites of transformation. Whereas land grabs have sparked massive rural protests in the past few years and prompted the central government to enact regulations on arable land preservation, green development projects are able to justify processes of rural dispossession as environmentally rational and socially progressive. Contrasting earlier rural development patterns based on rural township and village enterprises, rapid growth in green industries such as solar photovoltaics has been planned top-down, with significant state support in technology and business incubation in special economic zones (SEZs). These projects entail large-scale land enclosure and displacement of agricultural villages, and thus a refiguring of tenure rights and livelihoods as land is designated as "urban" and enclosed under direct state control.

In order to examine how green development unfolds across national and local scales of intervention, this chapter utilizes an empirical study of environmental and urban planning processes, and ethnography of rural transformation and contestation in Yixing, where an SEZ focused on solar photovoltaics has enclosed 106 sq. km of rural land since 2006, displacing approximately 50,000 residents from over 200 villages. I argue that as a result of this *green dispossession*, villagers are socially and politically marginalized, and previous patterns of urban–rural inequality are entrenched in new spaces of peri-urban segregation. Furthermore, dispossessed villagers have their assets commodified and transferred into urban development projects. I find that these projects reveal a politics of aesthetics and expertise that construct local land as a national environmental resource, while deeming rural people and livelihoods as environmentally irrational. I argue that these dynamics reveal conflicts between the different geographic scales of sustainability objectives, and contradictions between the global green economy and local social and environmental outcomes. This chapter analyzes these tensions and their implications for current understandings of sustainable development.

After a brief overview of Yixing's green development history in the context of national policy mandates, I will focus on the planning and implementation of a master-planned eco-city project within the Yixing SEZ. This will be followed by analysis of rural land conversion and dispossession under green development. In conclusion, the chapter will explore how these phenomena open a field of politics that presents new opportunities for linking transnational struggles against neoliberal environmentalization and the false solutions to social–environmental problems presented by capitalist models of "low carbon development" and the like.

Urban environmentalization in the Chinese countryside: the Yixing case in national context

Contemporary Yixing grew from an ancient water town on the western shore of Lake Taihu (see Figures 4.1, 4.2). Its picturesque rural villages are still laced by small streams, ponds and irrigation canals that feed some of China's most productive farmland. The urban core, now centered on a shopping and leisure development district, fills an area between two lakes and is traversed by a grid of canals that are now used to move industrial freight and the raw materials feeding the local construction boom. Yixing is a county-level city in the Wuxi prefecture of Jiangsu, consisting of an urban center of 66 sq. km within a total administrative area of over 2,000 sq. km (Figure 4.2).

Long before the founding of the Yixing Economic Development Zone ("the Zone") and its eco-city project, Yixing's claim as China's "hometown of environmental protection" was bolstered by its history in the field of wastewater and air pollution control. A first wave of green development established manufacturing industries in pollution control equipment in the early 1990s. As one of the

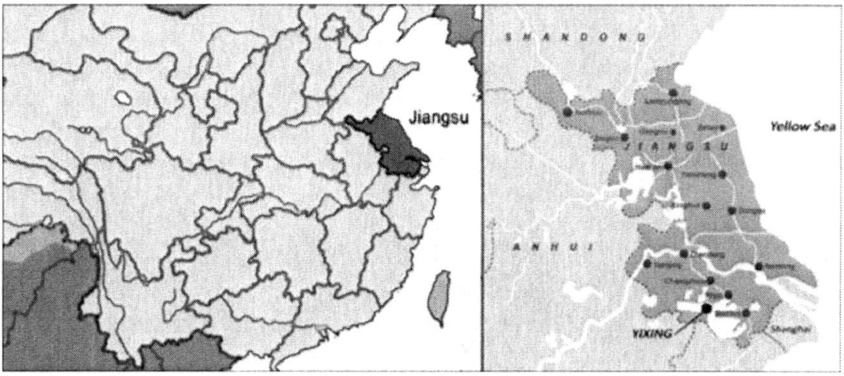

Figure 4.1 Location of Jiangsu Province and Yixing City (source: the author)

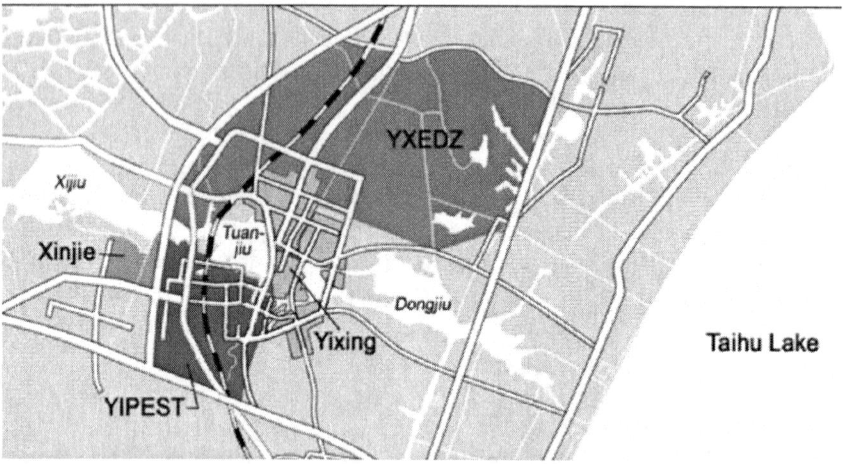

Figure 4.2 The Yixing city region, Two green development zones are shown in grey: YXEDZ; and the YIPEST, the first national designated zone for environmental protection (source: the author).

most industrialized rural regions of China at the time (Bramall 2007), the Taihu basin had hit its local environmental limits for the absorption of untreated pollutants. Foreign-educated engineers in Yixing seized upon manufacturing equipment for mitigating water pollution as an economic opportunity. Local officials supported the efforts with township joint ventures, supplying land and political capital. As a base industry, pollution control had a strong industrial clustering effect, requiring the adaptation of machine, pipe, filter, pump and other manufacturing industries. This led to the establishment of the Yixing Industrial Park for Environmental Science and Technology (YIPEST) as a pillar of China's Climate Protection Program under the UN Rio Declaration in 1992. By the time China signed onto the Kyoto Protocol in 1998, Yixing generated 18 percent of the national total value added in the environmental industry (Zhang 2002). Based on the subsequent success of the Zone in constructing a national base for solar photovoltaics, and the "prominent enhancement" of "environmental carrying capacity" and "people-oriented sustainable development" through its eco-city planning, Yixing was designated a national sustainable development experimental community in 2009 by the Ministry of Science and Technology (Xu 2009), and a national ecological city under the Ministry of Environmental Protection (Yu 2009).

These processes are examples of China's ongoing environmentalization, a term environmental sociologist Fred Buttel defines as, "the concrete processes by which green concerns and environmental considerations are brought to bear in political and economic decisions … [and] in institutional practices" (1992). Yixing's distinctive role in this history makes it an excellent case for examining the interaction of global markets, national industrial policy and the politics of local implementation and transformation.

Central government support for Yixing's model of green development is further evident in the reorientation of discourse and policy on national development. The Eleventh and Twelfth Five-Year Plans (covering 2006–15) designate eco-industrial zones and eco-cities as primary strategies for accelerating the transformation of the prevailing model of economic development, and for attaining the goal of an "environmentally, economically and socially harmonious society" (NPC 2005, 2011). The 11–5 Plan set targets for energy intensity (the amount of energy used per unit of GDP) and renewable energy generation. The 12–5 Plan includes new benchmarks for reducing the carbon intensity of GDP and introduces "low-carbon" (*di tan*) and "green development" (*lüse fazhan*) as fundamental concepts in official development discourse (NPC 2011). Moreover, the plans introduced a new vocabulary of environmental governance into China's official development lexicon.

The core approach to development and environment revolves around the Scientific Development Concept, which is a "summation of the 'comprehensive, coordinated, and sustainable development' " as the means to achieve a Harmonious Society (Fewsmith 2004: 1). This central pillar of "Hu Jintao thought" purports to address the main contradictions in China's development path at the current historical conjuncture through the implementation of the "five balances" (*wu ge tongchou*) (Fang 2003). This overarching vision links "sustainable all-around development" to the "comprehensive" and "people-centered" resolution of urban–rural, uneven regional, socioeconomic, environmental and geopolitical contradictions. This reorientation is observable in the rise of new characterizations of socialist construction, particularly in the terms "harmony" and "harmonious society" (*hexie shehui*) (Fan 2006), which make explicit reference to ameliorating the uneven social, economic and environmental outcomes of the past decades of growth. A prominent aspect of this new ideology of development is the manner in which it explicitly targets rural China as a site of transformation.

From its roots as a peasant revolution through the Great Leap Forward, rural society has continuously been a key site of socialist construction in China. It is with the advent of China's urban revolution that the rural has once again emerged as a fundamental problem of development and social mobilization. However, in the conjuncture of late socialism – characterized by the entrenchment of the market economy and the retrenchment of socialist entitlements – the rural question appears as the negative space from which China's urban success has risen. Rural society is frequently characterized as "backward" (*luohou*), and urban China as inherently "modern" (*xiandai*) (see Zhang 2006). For example, the influential modernization theorist He Chuanqi draws heavily upon Walt Rostow's *The Stages of Economic Growth* in his hierarchical conception of a linear progression of Chinese development through stages of "primitive", "agrarian", "industrial" and "knowledge"-based societies (He 2007a, 2005). The process of modernization is viewed as inherently "progressive," and rural transformation –

including processes of rural dispossession and the destruction of livelihoods and tenure rights – is assumed to produce greater equity through increased economic growth.

Yixing's master-planned New City, centered on environmental industries, represents the quintessential strategy of contemporary green development and modernization. The model exemplifies the 12–5 Plan's call for for the construction of "innovation-oriented cities ... to enhance sustainable development" as the platform for regional development, with particular reference to the roll of Jiangsu Province (NPC 2011). By effecting a simultaneous transformation of rural space, culture and economy, such a strategy promotes a "canal" approach to bypassing intermediary rungs of development to attain ecological modernization (He 2007a, 2007b). According to the China Center for Modernization Research at the Chinese Academy of Sciences, such strategies exemplify an integrated approach to creating "ecological balance" in a positive-sum relationship with development (He 2007b). This pathway of "integrated ecological modernization" is theoretically comprised of coordinated advances in "green urbanization" and "green industrialization" (He 2007b: 87, 215).[2]

Master-planning ecological value in the Yixing economic development zone

As pursuit of environmentalization strategies become explicitly linked to urbanization, the discursive and technical practices of planning and design attempt simultaneously to address the various aspects of greening society. In its pursuit of such urban–ecological greening strategies, Yixing has deployed the disciplines of environmental, industrial and economic development planning for the design and construction of a master-planned eco-city project within the Zone, with the ambition of constructing a "high-tech, low-carbon model community" (YXEDZ Committee 2010). Dubbed the "Scientific Innovation New City" (*ke chuang xin cheng*), the project occupies a 22 sq. km swath of village land and wetlands in the eastern district of the Zone (Figure 4.3).[3] Though formal annexation did not take place until 2009, conceptual planning began in 2008. The initial phase of the project occupies 10.3 sq. km, with a planned 450,000 sq. m of newly constructed residential floor area accompanying 480,000 sq. m of new business and commercial space. By 2020, the New City will serve as Yixing's urban center, with a planned population of 150,000. New urban residents will occupy housing distinctly separate from the relocation settlements provided for the 50,000 villagers and migrant workers displaced by the development.

Early in the conceptual and master-planning process, the lead director of the zone was dissatisfied with the work of the local Yixing Institute of Planning (Chen 2010). The Zone subsequently sought out collaboration with NITA, a Dutch planning firm, and Shi Kuang, a designer famous for his master plan of the China-Singapore Suzhou Industrial Park and in his role as chief architect at CCDI, the firm behind the Beijing Olympics Water Cube. Having described the original plans as "very ugly", the management committee of the Yixing Zone

Figure 4.3 An aerial view rendering of the New City project as a "green tapestry", emphasizing an aestheticized urban–ecological landscape (source: Scientific New City Urban Design 2008, Yixing Economic Development Zone Administration).

placed a high degree of importance on aesthetic markers that distinguished the new plan from the industrial districts that preceded the Zone. A "comprehensive urban layout" (*zonghe chengshi buju*), with a heavy emphasis on urban design including greenbelts, landscaping, an incorporation of water landscape features and consideration of views became as important as the allocation of lots for industrial construction (Chen 2010) (Figure 4.4). The early prioritization of what was explicitly conceived of as urban–spatial over industrial development demonstrates the degree to which urbanization has superseded industrialization as a policy priority (see Hsing 2010).

This subsumption of industrialization into green urban modernization is a matter of policy in Yixing. The head of the Zone's investment bureau told me,

> People think our purpose is to build new energy industries. In fact, it is not. New energy is simply a rapidly rising sector at this time. It has a good future potential, but we don't know for how long. The only way to ensure sustainable development is to build a modern innovation city.
>
> (Chen 2010)

As a mode of development, this "greening of urbanization" seeks to weld the functions, culture and environment of the city to the internalized market and material metabolisms of an eco-industrial economy.

That greening is embedded within this ideological conception of urbanization is exhibited in statements by the management committee describing its master-plan goals to develop the Zone by drawing upon the natural and historical heritage of Yixing – to cultivate a place-specific urbanism and culture using " 'born and growing from nature' as the planning ideal … to construct a green tapestry with the mountains and waters, a beautiful and elegant new city" (YXEDZ Committee 2010). Dense with environmental imagery and allusions to national cultural values and local heritage, statements like these construct a discursive aggregate of nature and development. The statement conflates an idealized nature "out there" with an aesthetic image. In the master-planning process, these aesthetic representations became decisive, recasting the urban as a harmonious balance with nature and in juxtaposition with the industrial. Jacques Rancière (2004) conceptualizes aesthetics as "the distribution of the sensible", meaning a sort of common sense that shapes what can be perceived and understood as correct in a social system of valuation and ethics. In the common sense of green development, this nature was seamlessly woven into the eco-city project as a new form of urbanism through its representation in plans and discourse, and its mobilization in the planners' aesthetics of expertise.

In his genealogy of planning, Söderström (1996) argues that the visual forms of representation central to expert practices of zoning and master-planning are tools to abstract social space and construct the urban as an object of political intervention. In the New City planning process, this abstraction began from initial surveys of villages and their environmental conditions. In order to construct their model of green urbanism, the planners shaped much of the conceptual process around environmentalized physical features observed during cursory site visits. From the designer's viewpoint, the site was interpreted according to observations of physical "present conditions" (*xiankuang*) (Figure 4.4). Space was reified from social contexts and analyzed as single-factor elements: for example, as a particular "land use" or type of surface. In representing areas without existing construction as "open spaces" and "green spaces", the planners constructed the countryside as a *tabula rasa*, belying the great extent to which the land itself has been built out of the marshes over hundreds of years. Instead, these wetlands are now referenced as "ecological resources" in need of protection, and as a nature discernible from human causes that can yet be integrated with the new sustainable development aspirations of eco-city construction. Planners claim that this "ecological framework" (*shengtai goujia*) has been incorporated into the eco-city design in order to maintain the overall environmental integrity of the site (Figures 4.6, 4.4): for example, by serving as a network of contiguous habitats (Figure 4.8). However, in the urban design, the key physical features of this framework have been paradoxically transformed into a space of recreation and commercial activity (Figure 4.8). Within globally circulating concepts of new urbanism, this type of urban greening is also broadly associated

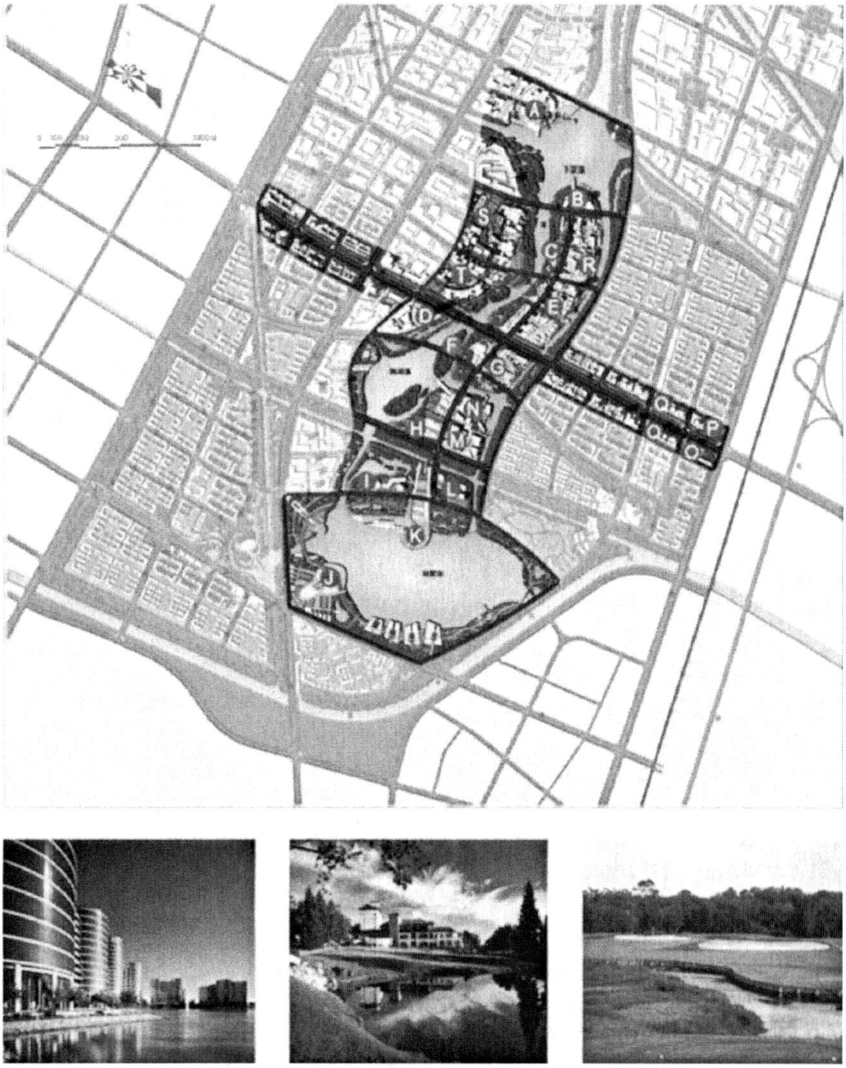

Figure 4.4 The emphasis on comprehensive urban design, e.g. in this diagram of "public cultural services distribution", overlays the "cruciform ecological structure" (shown in Figure 4.6) with waterside greenways linking business service centers and shopping distrcits, and de-emphasizes the industrial location (source: Scientific New City Urban Design 2008, Yixing Economic Development Zone Administration).

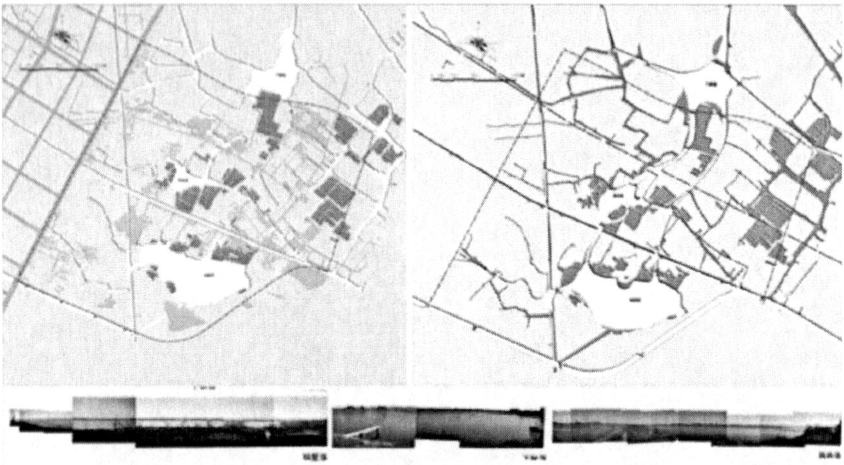

Figure 4.5 The map of "land use present conditions" (left) diagrams rural uses as spatially discrete forms of land cover. Aquaculture is shown in dark grey, totally distinct from "natural water systems", irrigation and navigable canals, which are categorized together in white. The "surface water conditions analysis" map and illustrations (right and bottom) emphasize an artificial category of "natural landscape conditions" and ignore actual hydrology. Water features to be filled for construction land are shown in dark grey (source: Scientific Innovation New City Control Plan 2008, Yixing Economic Development Zone Administration).

with the concept of "quality of life," and ecological progress is increasingly interpreted through physical, aesthetic signifiers such as parks, clean public space and newly constructed housing (see Hoffman 2011; Zhang 2006).

During the master-planning process, the "ecological value" of the entire site was defined and assessed according to an environmentalized aesthetic that justified rural dispossession. Despite references to ecological service functions, the planning agencies did not employ wildlife biologists, hydrologists, soil scientists or other ecologists. Instead, the designers analyzed the site by abstracting the landscape into a series of aerially mapped layers, separated as surface water or land use, and also describing "ecological service corridors" (Figures 4.5, 4.8). With the purported goal of identifying and preserving ecological services and value, the analysis proceeded based largely on a purely visual interpretation of land cover that conflated observable surface features with discrete land uses, and made a bright-line distinction between "natural landscape condition" (*ziran jingguan xianzhuang*) and "land use conditions" (*tudi shiyong xianzhuang*) (Figure 4.5). The resulting analysis placed primary "ecological service value" (*shengtai fuwu jiazhi*) on three bodies of water and connecting canals, and assessed the site accordingly in a gridded spectrum of preservation values and suitability for construction (Figure 4.7).

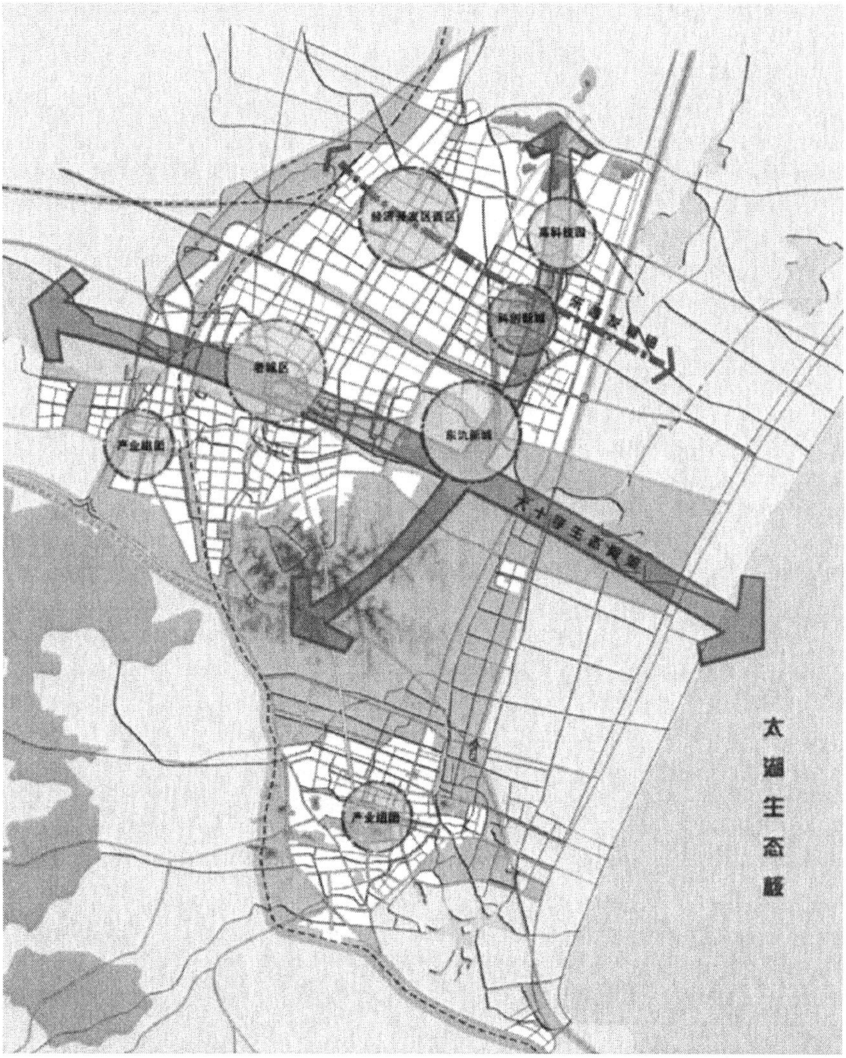

Figure 4.6 "Regional ecological and spatial structure" diagram, emphasizing the area's "cruciform ecological framework" (source: Scientific New City Urban Design 2008, Yixing Economic Development Zone Administration).

Such practices and discourses of environmentalization work to separate ecological functions from human activity. This has the consequence of obscuring the ways in which rural people actually make their livelihoods, and the complex social–environmental interactions between ostensibly distinct land uses. Although the small lakes and ponds, fringed by orchards and riparian plants, do

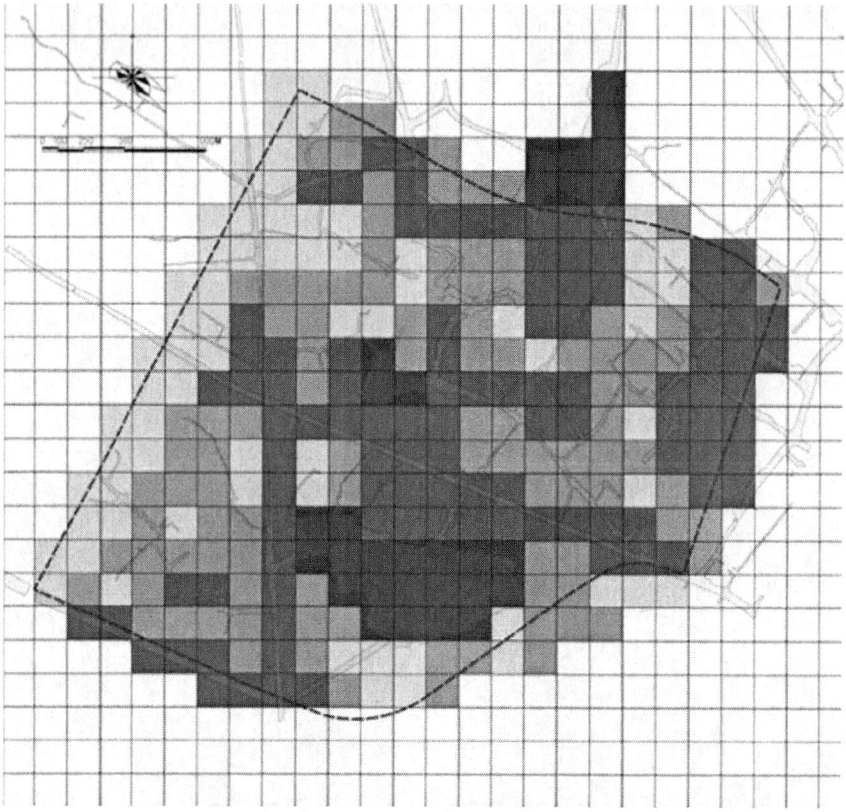

Figure 4.7 The "ecological value appraisal map" categorizes ecological values in order to determine preservation zones and suitability for construction. "Higher" values, shown in darker shades of grey, emphasize "natural" bodies of surface water (source: Scientific Innovation New City Control Plan 2008, Yixing Economic Development Zone Administration).

serve as habitat to birds and other animals, the wetlands now enclosed within the Zone have been continuously manipulated and managed for centuries. The edges of water features have been transformed through aquaculture, and reclamation for settlement and agriculture. Aquaculture stocks interact with fish and other organisms in adjacent waterways and ponds, which are also stocked for subsistence fishing. The fact of a long history of social–environmental interaction undermines any conception of an *a priori* natural environmental condition for preservation or restoration, and foregrounds the political–economic and aesthetic–cultural choices behind the construction of the New City project. This aesthetic construction of a dichotomy between human and ecological space is especially striking in a resultant hydrology plan, which highlights surface water features that will be filled in to provide construction land (Figure 4.5).

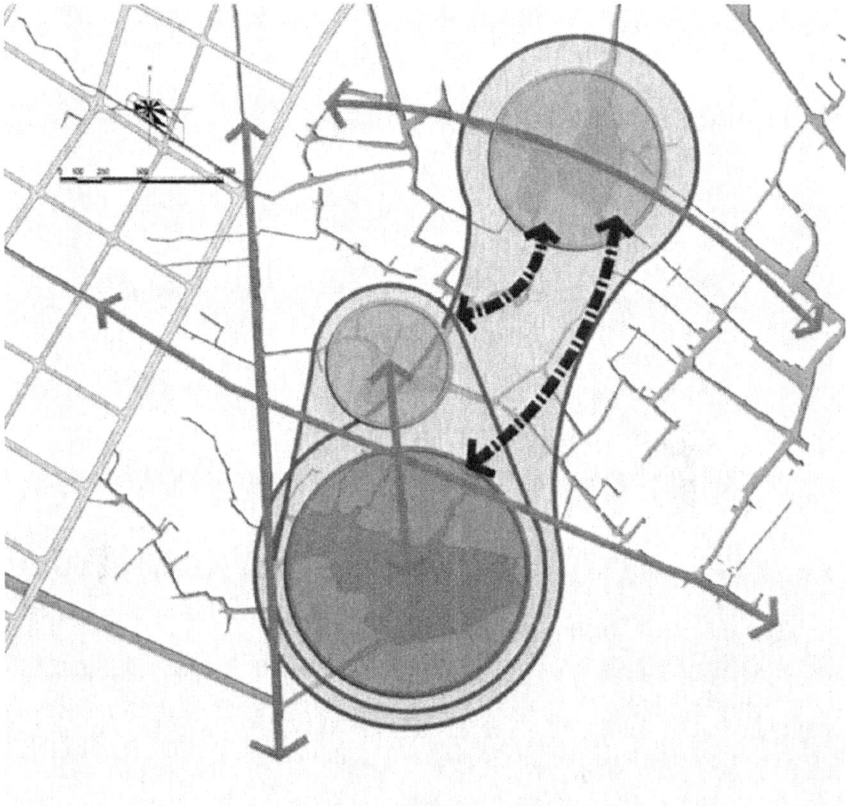

Figure 4.8 The "ecological spatial structure and conditions" diagram shows ecological "cores", "connection axes" and "corridors" as functions centered on three bodies of water. These appraisals were driven by design ideas rather than actual biological and ecological study of the region (source: Scientific Innovation New City Control Plan 2008, Yixing Economic Development Zone Administration).

This aesthetic produces ideological connections between the social–environmental transformations of urbanization, industrialization, dispossession and village demolition. The resulting "green tapestry" of the Scientific Innovation New City belies the paradox of mobilizing local cultural and environmental heritage as a motive justification for the project even as rural social space is being refigured in the name of the "new" and "modern." These discursive and technical practices demonstrate how, as agricultural land is constructed as a green development resource, it is divorced from rural society, effectively making rural people obstacles to the urban–ecological vision of sustainable development.

Constructing national land resources, greening agricultural land loss

The Yixing case demonstrates the trade-offs and conflicts between the different constructed scales of green development. Underpinned by political-economic agendas and techno-scientific expertise, participation in markets for renewable energy and carbon reduction credits construct climate change as a global "trans-boundary" issue. This conceptualization makes carbon emissions globally fungible as both market commodities and mitigation quanta. The Chinese national development agenda constructs greening as a strategy for meeting goals of economic growth while restructuring the industrial–environmental metabolism of production. In Yixing, central economic and environmental policy mandates intersect with local state-led strategies for capital accumulation and expansion of territorial authority. Yixing's model of green development presents eco-cities powered by environmental industries as an ideal form of intervention, linking across scales with a comprehensive solution of rural transformation.

Within the common sense of green development, China's declining farmland area is paradoxically a justification for rural dispossession. The Scientific Innovation New City project enclosed over thirty-six villages previously inhabited by an estimated 20,000 residents, each with 670–1,000 sq. m of farmland (Chen 2010).[4] In order to convert rural land to other non-agricultural uses, local governments must first transfer collectively owned village land to direct government control as state-owned urban land.[5] Local authorities must also clear land use changes through the Ministry of Land and Resources (MLR). In 2006, the central government set a 1.8 billion mu (120 million hectares) "redline" national minimum threshold for arable land protection. For the purposes of regulating land supply, the Land Administration Law (1998) classifies all land as agricultural land, construction land or unused land. Conversion of rural land must conform to a series of requirements, including local land use master plans and the local administration of the national land-supply regulations. To maintain a zero net loss of agricultural land in the face of rapid urban and industrial development, the MLR coordinates provincial-level quotas of land conversion and massive "land reclamation" projects to add arable land to the national balance sheets. These regulations are intended to protect agricultural land, which was enclosed at an alarming rate during the 1990s and early 2000s for speculative development by local authorities (Hsing 2010; Lin 2009).

However, in projects of green development, industrial and rural transformations have intersected to justify further land conversion. This confluence can be seen clearly in the urbanizing countryside, where the elimination of unregulated township and village enterprises and the enclosure, consolidation and privatization of agricultural land are furthered by an environmental rationality (Zhang 2003; Shi and Zhang 2006). Policies such as the National Climate Change Program (NDRC 2007) and Ministry of Land and Resources statements on land consolidation represent small-scale agriculture and small landholding as environmentally irrational (MLR 2006). In what functions as a de facto quota trading system for rural land as a national environmental "resource," local authorities are able to create net gains in land area available for development by reducing the footprint of various rural uses (such as residential construction), constructing arable land on land reserves and through reclamation from hydrologic features such as wetlands. This rationalization of land resources is underpinned by a number of environmental justifications, including food security, ecological buffers, habitat and overall efficiency.

This approach to constructing and managing rural land in aggregated metrics of national land resources explicitly places village land uses at odds with fundamental goals of urban ecological modernization. For example, in 2006 the MLR encouraged the consolidation and transfer of small land holdings to private ownership, to better manage the net quantity of agricultural land nationally (MLR 2006). This policy is part of the general process of reform that enables the "rationalization" of tenure within and around urban areas, and facilitates the privatization of agricultural land "in the service of economic development" (MLR 2006).[6] This anti-rural bias replicates the historical contradictions between rural and urban forms of development in China, and recasts them in terms of an environmental rationality that equates rural urbanization with ecological soundness (May 2008). Furthermore, it reflects a widespread assumption that "compact" urban forms are inherently more ecologically sustainable than geographically extensive settlements, an assumption predicated on a false representation of the relationships between the city and its hinterlands (see Williams 1975; cf. Neuman 2005). As a result of these policies, farmland is further marginalized geographically, contributing to a reification of urban–rural relationships in a hierarchy of green modernization.

In order to justify the conversion of collectively held croplands, orchards and horticultural land, woods, grasslands, waterways and village construction land, planners must conform village enclosures to larger environmental management goals and policies. In Yixing, the enclosure of rural land is further justified under the policy of "returning farmland to forest" (*tuigeng huanlin*). In one 2009 "grain to green" project over 530 hectares of village agricultural land, the holdings of over 2,500 households were converted as part of a 1,333-hectare greenbelt between the Zone and Lake Taihu (Liu 2010b).[7] According to government statistics, such "ecological withdrawal of agriculture" and afforestation projects, including some for certified carbon credits, account for the majority of agricultural land losses nationally (Tian 2007; UNFCCC 2012b). In the Yixing project, the reallocation of land resources and populations is linked to policies to restructure agriculture as "pollution free, organic … modern ecological agriculture", with goals to increase efficiency and generate climate mitigation outcomes (Min 2008).

Under the 11–5 and 12–5 Plans, policies addressing urban–rural uneven development have emerged under the banner of "constructing the new socialist countryside" (*jianshe shehuizhuyi xin nongcun*). New socialist countryside policies purportedly address the rural–urban income gap by modernizing agricultural practices and improving infrastructure. These policies come in response to steadily increasing social unrest among rural residents and migrant workers, and are in part responding to the critiques put forward by the Chinese New Left, particularly those associated with Wen Tiejun and the New Rural Reconstruction Movement (Day 2008). In 2006, Yixing developed its current policy of urban–rural integrated development under this general mandate to restructure agriculture as a means of social and economic development, and adopted slogans like "without industry, no wealth; without agriculture, no stability" and "industry nourishes agriculture, the city supports the countryside" (Liu 2010a).

In practice, however, the modernization of Yixing agriculture and the support of rural areas have unfolded in differentiated spaces and without clearly benefiting many rural residents. In the areas of the immediate periphery outside the urban core, agricultural land has been eliminated with the justification that providing a new industrial base in the Zone will raise rural incomes while simultaneously improving communication and transportation infrastructure for farther outlying areas, which are planned to be new sites of high-efficiency industrial agriculture run by private enterprises. Such justifications are made highly visible, in explicit terms of social–spatial transformation across pervasive public media such as billboards and poster campaigns, and in official statements of project aims and progress in television, Internet and print media. Following village evictions billboards are erected over the demolition sites with images and slogans promoting the comprehensively transformational projects. After the 2008 demolition of Siqian village for the New City project, a billboard rose over the rubble touting, "new talent, new industry, new city" on one side, and "Yixing Solar Valley: construct a new energy industry base on the west banks of Taihu" on the other.

I talked with Siqian residents before and after the demolition. Aside from only vaguely understanding the Zone's expansion, villagers did not know what specific projects had required their eviction. Regarding the billboard, I asked about the message of transformation and the villagers' ongoing role in it. Wry comments were made: "Apart from the words on the signs, this isn't our talent, industry or city. … We've only been taught what the slogans mean." The sense of exclusion grew from early on in the eviction notification process. Although their fields were not being bulldozed and they were not receiving compensation for the land itself, they were required to vacate the village before the fall harvest. Evicted residents returned on bicycles, some from an hour away. They trod new paths through the decorative landscaping fringing the rubble of their village to tend crops and harvest vegetables for their families.

Hsing (2010) describes this cascading effect of land development as evidence of urbanization becoming the "spatial fix" for Chinese capital accumulation. In its imbrication with processes of green development, it is clear that such patterns of uneven development also produce new contradictions specific to attempts to govern rural land and its embedded social–environmental relations at a national scale. Put another way, China's "urban revolution" is fundamentally a transformation of its social–environmental relations across all scales of governance and differentiation. In this context, the ideology of green urbanization can be seen as clearly presenting challenges to equitable and just rights to the city *and* countryside.

Social justice and the eco-city

Together, Yixing's various green development zones have required the enclosure of over 300 sq. km of rural land – over twice the area of Manhattan – and the forceful eviction of approximately 100,000 villagers. These enclosures, wrought partly in the name of rural development, have resulted in greater social inequality. The failures of such efforts at rural development are stark. One National Statistics Bureau survey found that nearly half of dispossessed villagers are impoverished by eviction and relocation processes (Hsing 2010: 209, n. 18). As villagers are dispossessed of their land and livelihoods, transformations to social–environmental relations, cultural values and the places people live present new terrains of politics and social division.

The proliferation of the eco-city model of green development demonstrates a dialectical reshaping of state–society relationships that can be understood in two ways. First, as Buttel (1992) argues, environmentalization proceeds in relationship to structural transitions. In the US case, the move to neoliberal social and economic policies with the decline of Fordism shaped the politics and ethical claims of scientized sustainable development discourse, which was "crucial in leading to the substitution of environmental for social justice discourse" (Buttel 1992: 16). This analysis is consonant with China's current emphasis on scientific sustainable development in the context of the gutting of rural collective property rights and social welfare entitlements. In the context of this neoliberal environmentalization, the restructuring of property extends beyond the establishment of leasehold and other private forms of holding and rent seeking. Land resources and enclosure itself are also greened in integrated schema linking carbon credit afforestation projects to greenbelt tourism parks, and new ecological industries and spaces to the embodiment of new talents and urban civilities. These practices demonstrate an emphasis on environmental rationalities that systematically produce and address rural land and people as objects and subjects of governmental action (see Scott 1998).

In Yixing's eco-urbanization, relocation communities are spatially and socially segregated from the rapidly expanding urban core and new city developments for which residents were displaced. As villages are divided for phased demolition, village committees are dissolved and their authority is subsumed under larger administrative village structures and the privately owned demolition company. Social cohesion is lost as residents scatter to find rental housing and transitional livelihoods. During this transitional period, villagers are not technically classified as urban residents and must maintain their rural household status until the separate compensation process for their land is worked out with local authorities. Depending on the rate of investment, financing and construction, this process may take years. The land enclosure boom and "development zone fever" of the 1990s left over 85 percent of seized land undeveloped (Ren 2003). With the loss of access to land and livelihood, villagers are forced into a new, more proximate but more explicitly marginalized relationship with the city. In extreme cases, dispossessed villagers are referred to as a new "underclass" with "three nothings" – no land, no work and no social benefits.

A class of "four nothings" is also emerging as some villagers lose permanent housing. Because the compensation system requires dispossessed families to pay for the difference between the "market prices" of their demolished homes and their relocation housing, many families are frequently impoverished in the process. Poor families are frequently unable to pay the fees, and they lose the "compensation" for their demolished homes in the process, as the money is tied to a compulsory mortgage system for the relocation housing. Some families are left with no option other than to attempt to sell the property through a broker. However, as no market exists for the resettlement housing apart from renting to recently displaced families, this is generally unsuccessful. Many families are forced to purchase a home further outside of the city. Others may move in with relatives as circumstances allow.

Furthermore, because they are forced into temporary housing for one to two years as they await the completion of relocation housing, all households must spend their savings and small cash incomes on rent. In my survey of one village, government subsidies covered an average of 76 percent of rent, leaving nothing for increased expenses of water, transportation and cooking fuel. Most villagers also face the very fundamental issue of finding suitable and affordable transitional housing that allows access to their fields or marginal land for subsistence livelihoods. The compensation package projects an eighteen-month transition to permanent housing and provides a monthly stipend of CNY500. However, since the current package was first implemented in 2007, rents have increased in some areas and many residents found that they could not house their full extended families for less than an average of CNY650. This shortfall meant that older-generation family members had to take up some kind of wage labor, and frequently also had to deplete their savings for rent and other new living expenses, which total an average of around CNY500 during summer months. These transitions to an urban cash economy contribute directly to the impoverishment of many villagers.

The vast majority of displaced villagers cannot find work in the new high-tech industries in the region. Largely as a result of village demolition and displacement, unemployment has increased by 9.3 percent over the past five-year planning period (YXEDZ Committee 2010). For those who no longer have access to their fields, increased food expenses can easily eat into savings as well. Farmers impoverished by displacement seek small plots of land on the margins of developments and factories to help make up for their losses. However, there is not enough space. The question of tenure is precarious as evicted villagers sometimes lease out land without any clear assurances from local officials about demolition and construction plans and schedules. The degree to which socioeconomic outcomes are differentiated and uneven is startling. Key factors are previously held assets and timing. Villagers who are better off in terms of village property and cash income tend to fare better as they are more able to purchase relocation housing. In some cases, well-off villagers are able to become relatively wealthy as they sell or rent out extra units of relocation housing. Allegations of corruption and disputes over uneven compensation are frequent.

In the political economy of displacement and land conversion for urbanization, these changes amount to a transfer of previously uncommodified rural assets into processes of development. Though the net amounts may be very small, they are significant in important ways. The cash amount of the subsidy shortfall is equal, on average, to over CNY2,700 (over eighteen months). This is a significant amount for a rural household. For retirees who depend mainly on subsistence farming, cash income may be as little as CNY60–100 per month. However, simply multiplying this amount across displaced households does not give an appropriate picture of its net economic significance. In addition to discounting (the future value of money, interest and inflation as well as opportunity costs), this shortfall also produces a "multiplier effect" in the local economy by increasing the supply of cheap and flexible labor. That said, this process is not centered on proletarianization, as in classic analyses of primitive accumulation (see Glassman 2006).

It is important to understand that this labor is both fully incorporated into the local economy at the same time that it is irregular in character. Employers hire workers from job to job, and do not cover payroll taxes and other fees. Wage rates are reflective of the rural rather than the urban economy. The ability of the urban and industrial development process to utilize such labor flows underwrites the cost of the overall transformation and externalizes these costs by placing socioeconomic burdens on individual households. These dynamics outline a circuit of accumulation through the extra-economic means of state violence (Glassman 2006; cf. Harvey 2003a). The extent to which the environmental state relies upon enclosure as a "spatial fix" to construct new territories for the production and absorption of capital surpluses reflects the primacy of land and urbanization as a source of revenue and state authority (Hsing 2010). Here, I argue that these patterns of dispossession for urban-spatial accumulation strategies cannot fully explain the forms of "circulation" of rural land examined above. Rather, such forms of accumulation (and sometimes their failure) demonstrate that local social–spatial transformation takes place as a part of broader processes of transformation mediated at national and global scales. In the case of Yixing, rural land enclosure has played a functionally multivalent and multiscalar role in producing local land rents, meeting national renewable energy targets, balancing national land resource quotas and serving the sustainable development objectives represented by Euro-American markets for solar energy and certified emissions reductions.

Linking nail households, rural reconstruction and the right to the city

The question of the right to the city emerges within the context of rural transformation and changing notions of what constitutes correct and "harmonious" social–environmental relations. If we take for granted that cities have historically been founded for different locational advantages, grow through the slow accumulation of various types of surplus and subsequently birth new forms of sociality, then these examples of master-planned green developments circumvent the processes of accretion and change for the amplification of a single agenda. In the case of green development, this is an agenda that has broad international support and a mandate underscored by the politics of crisis (cf. Swyngedouw 2007). At its largest horizons, green development qua ecologically sustainable development purports to remake the environmental bases of production and, therefore, of accumulation. The usual sites and forms of creative destruction also trace changing circuits of ecological capital, entrenching old inequalities in new sites of segregation. Village fields are plowed for solar farms and forests of carbon credits. Given this context, one might better say that the struggle for the right to the city (and the countryside) *re-emerges* from the dialectical contradictions of such an attempt to totalize a new urban–ecological sociality.

The top-down nature of master-planned urban and economic growth threatens massive disenfranchisement, producing class divides and socio-spatial segregation. Neoliberal environmentalization of land and social space continue the radical asymmetry of state–society relationships of the past decade. This model of socioeconomic development refigures social entitlements as a focus on what I refer as "internal developmentalism", as a form of disciplinary value in the production of human capital (see Anagnost 1997; Greenhalgh and Winckler 2005). With its emphasis on rural transformation, the urban–ecological approach to green development modeled in Yixing maps a clear contradiction between the purported goals of neoliberal environmentalization and the displacement of long-sustained patterns of rural livelihood and environmental interaction.

Villagers in Yixing have primarily engaged in isolated and everyday forms of resistance to these processes. Forms of resistance included "nail households" (*dingzi hu*), whose active refusal of eviction processes have been successful in some cases in winning concessions, but have also been met with violence. Everyday forms of resistance include the illegal appropriation of small plots and returns to enclosed fields for subsistence. Although levels of organized mobilization have been low in Yixing, these forms of contestation reveal an expanding domain of politics directly addressing social–environmental change in China. In the context of transnational farmers' movements, this politics may potentially provide ground for the intersection of international and national environmental and social equity mobilization agendas. Especially as China's land grabs extend beyond its own borders to locales in Southeast Asia, Africa and Latin America, these agendas will increasingly produce a political economy of dispossession for food and energy resources.

As local processes of dispossession are environmentalized at national and global scales with the legitimating force of globalized sustainable development science, it becomes increasingly necessary for local resistance to transcend local and cellularized forms. Likewise, mobilization that targets the Chinese state at various levels must move beyond the conception of rights constrained by property entitlements and "proper" procedures of compensation and relocation. Here, Chinese villagers have little in the way of hope to alter the logic of dispossession (see Hsing 2010: 201–7), nor recourse to democratic forms of planning participation and review. This field of politics calls for urban–rural solidarities that identify the uneven sustainability outcomes of green development.

The grassroots construction of alternatives has historically shown to be fruitful in China. In particular, the politics of maintaining village livelihoods can take lessons from struggles to reinvigorate rural identities and forms of organization under the New Rural Reconstruction Movement. Wen Tiejun, the most influential advocate of the NRRM, radically shifted the policy debate on rural development in the late 1990s. He reoriented the agrarian question from a focus on the relations of production and political economy to the well being of farmers in the context of rural society and agriculture (the "three agrarian problems", *san nong wenti*). Policy debates could no longer be framed in terms of agricultural economics, technological solutions or urbanization (Day 2008). Major points of contribution relevant to these contemporary struggles are the foregrounding of a material analysis of the non-commodified nature of land as a "subsistence resource" as opposed to a "production resource", and the requirement for equitable distribution that it implies (ibid.). Furthermore, the reconstruction collective forms of organization and property holding build solidarity and power to resist land seizures.

As social and environmental activists seek to reshape cities as healthier and more equitable places in the United States and elsewhere, we must recognize how our efforts are implicated in neoliberal environmentalization and false solutions to climate change. In this case, the massive social–environmental costs of green development for the production renewable energy and carbon reduction commodities cannot be ignored. If, as Harvey suggests (2003b, 2008), the city is the historic place where the world can be "re-imagined and re-made", then that imagination must encompass an understanding of how neoliberal environmentalization is linking these new urbanisms and reshaping us all in the process.

Notes

1 The most famous among these include two failed projects: Dongtan eco-city and Huangbaiyu eco-village. Both projects continue to generate positive attention for their celebrity designers, Peter Head and William McDonough, respectively. For a recent study of Huangbaiyu, see May (2011).

2 Although the work of He Chuanqi (2007b) follows the analysis of Western theorists in describing ecological modernization as an "inexorable global tide" (*bu ke nizhuan de shijie chaoliu*) and "historical necessity" (*lishi biran*), his policy prescriptions recall contemporary invocations of Mao Zedong thought in his approach to the "historical contradictions" and "opportunities" that reveal a distinctive pathway for Chinese development.

3 The project name plays on the use of the character for "new" (*xin*) in joining "creation and innovation" (*chuangxin*) to "new city" (*xin cheng*). The emphasis on newness and its paradoxical conflation with environmental protection will be elaborated below.

4 Due to administrative changes to village-level jurisdictions made to facilitate different

phases of enclosure, demolition and relocation, the exact number of natural villages (*ziran cun*) counted under the six administrative villages (*xingzheng cun*) within the New City project area varied between 2006 and 2010.

5 According to the PRC Constitution, all urban land is state owned and rural land is collectively owned.

6 Despite strong reactions against the illegal conversion of agricultural and "urban village" land to other purposes by state and capital forces (Hsing 2010), the MLR has failed to maintain its policy of "no net loss" of arable land (vs *cultivated* land), notwithstanding "land reclamation" and the transfer of topsoils from converted cultivated land to reclamation sites.

7 The Lake Ecological Zone project, supported by Premier Wen Jiabao, will ring the lake with 200 to 1,000 meters of "recovered" forests, grasslands, wetlands and lake with a stated policy goal of constructing model eco-tourism industry (Wuxi Bureau of Science and Technology 2009).

References

Anagnost, A. (1997) *National Past-Times: Narrative, Representation, and Power in Modern China*, Durham: Duke University Press.

Bramall, C. (2007) *The Industrialization of Rural China*, New York: Oxford University Press.

Buttel, F. H. (1992) "Environmentalization: Origins, Processes and Implications for Rural Change", *Rural Sociology*, 57, 1–27.

Chen, J. C. (2010) Yixing City Fieldnotes, unpublished, author's archives.

Day, A. F. (2008) "The End of the Peasant? New Rural Reconstruction in China", *boundary*, 2 (35), 49–73.

Fan, C. C. (2006) "China's Eleventh Five-Year Plan (2006–2010): From 'Getting Rich First' to 'Common Prosperity' ", *Eurasian Geography and Economics*, 47, 708–23.

Fang, C. (2003) "Grasp the 'Five Overall Plans' of the Scientific Development Concept", *Guangming Daily*, 13 November 2003 (Chinese).

Fewsmith, J. (2004) "Promoting the Scientific Development Concept", *China Leadership Monitor*, 11: 1–10.

Finamore, B. (2011) "A Clean Energy Future: A Shared Vision of Presidents Obama and Hu", NRDC. Online: http://switchboard.nrdc.org/blogs/bfinamore/a_clean_energy_future_a_shared.html (accessed 30 January 2011).

Foucault, M. (1994 [1970]) *The Order of Things*, New York: Vintage.

Friedman, T. L. (2010) "Aren't We Clever?", *New York Times*, 19 September: WK9.

Glassman, J. (2006) "Primitive Accumulation, Accumulation by Dispossession, Accumulation by 'Extra-Economic' Means", *Progress in Human Geography*, 30: 608–25.

Greenhalgh, S. and Winckler, E. A. (2005) *Governing China's Population: From Leninist to Neoliberal Biopolitics*, Stanford: Stanford University Press.

Harvey, D. (2008) "The Right to the City", *New Left Review*, 53: 23–40.

Harvey, D. (2003a) *The New Imperialism*, Oxford: Oxford University Press.

Harvey, D. (2003b) "The Right to the City", *International Journal of Urban and Regional Research*, 27: 939–41.

He, C. (ed.) (2007a) *China Modernization Report (2001–2007)*, Beijing: Beijing University Press (Chinese).

He, C. (ed.) (2007b) *China Modernization Report 2007: Ecological Modernization Research*, Beijing: Beijing University Press (Chinese).

He, C. (ed.) (2005) *China Modernization Report 2005: Economic Modernization Research*, Beijing: Beijing University Press (Chinese).

Hoffman, L. (2011) "Urban Modeling and Contemporary Technologies of City-Building in China: the Production of Regimes of Green Urbanisms", in A. Roy and A. Ong (eds) *Worlding Cities*, Malden: Wiley-Blackwell.

Hsing, Y. T. (2010) *The Great Urban Transformation: Politics of Land and Property in China*, Oxford: Oxford University Press.

Lin, G. C. S. (2009) *Developing China: Land, Politics and Social Conditions*, New York: Routledge.

Liu, P. (2010a) *Yixing Constructs Urban-Rural Integration for a New Pattern of Development*. Online: www.js.xinhuanet.com/xin_wen_zhong_xin/2010–07/06/content_20788313.htm (Chinese).

Liu, P. (2010b) "Yixing Spent 1.3 Billion on Taihu Lake Water Pollution Control in 2009", *Xinhua, Yixing*, 2 March (Chinese).

Lovins, A. B. (2008) "Natural Capitalism: the Next Industrial Revolution", public lecture, Weinstock Lecture Series, University of California, Berkeley, 28 October.

May, S. (2008) "Ecological Citizenship and a Plan for Sustainable Development", *City*, 12: 237–44.

May, S. (2011) "Ecological Urbanization: Calculating Value in an Age of Global Climate Change", in A. Roy and A. Ong (eds) *Worlding Cities: Asian Experiments and the Art of Being Global*, Malden, MA: Blackwell.

Min, D. (2008) "Yixing Comprehensively Implements Taihu Lake Belt 'Withdraw Agriculture, Withdraw Pasture, Withdraw Aquaculture' ", *Wuxi Daily*, 15 June (Chinese).

MLR (2006) "Communique on Land and Resources of China (2 November 2006)", in *Resources*, Beijing: Ministry of Land Resources.

NDRC (2007) "China's National Climate Change Program", Beijing: National Development and Reform Commission.

NPC (2011) "The Twelfth Five-Year Plan for Economic and Social Development of the People's Republic of China", Beijing: National People's Congress (Chinese).

NPC (2005) "The Eleventh Five-Year Plan for Economic and Social Development of the People's Republic of China", Beijing: National People's Congress (Chinese).

Neuman, M. (2005) "The Compact City Fallacy", *Journal of Planning Education and Research*, 25: 11–26.

Rancière, J. (2004) *The Politics of Aesthetics: the Distribution of the Sensible*, New York: Continuum.

Ren, P. (2003) "The End of the 'New Land Enclosure Movement", *Caijing*, 16: 111–16 (Chinese).

Scott, J. (1998) *Seeing Like a State: How Certain Schemes To Improve The Human Condition Have Failed*, New Haven: Yale University Press.

Shi, H. and Zhang, L. (2006) "China's Environmental Governance of Rapid Industrialisation", *Environmental Politics*, 15: 271–92.

Söderström, O. (1996) "Paper Cities: Visual Thinking in Urban Planning", *Cultural Geographies*, 3: 249–81.

Swyngedouw, E. (2007) "Impossible 'Sustainability' and the Postpolitical Condition", in R. Krueger and D. Gibbs (eds) *The Sustainable Development Paradox: Urban Political Economy in the United States and Europe*, New York: Guilford Press.

Tian, C. (2007) "A New Round of Land Use General Planning", *China Land*, 5: 10–15 (Chinese).

UNFCCC (2012b) *Issuance Certified Emission Reductions*, Bonn: UN Framework Convention on Climate Change Secretariat. Online: http://cdm.unfccc.int/Issuance/cers_iss.html (accessed 18 March 2012).

Williams, R. (1975) *The Country and the City*, London: Oxford University Press.
Wuxi Bureau of Science and Technology (ed.) (2009) "In Order to Urge and Speed the Construction of a Leading Innovation Economy City, Premier Wen Jiabao Visits Wuxi on Three Inspection Visits", 12 January.
Xu, N. (2009) "Yixing Successfully Establishes a National Sustainable Development Experimental Zone", Yixing: Jiangsu Xinhua News Service, 23 June (Chinese). Online: http://yx.js.xinhuanet.com/2009-06/23/content_16889065.htm (accessed 19 March 2011).
Yu, X. (2009) "Yixing City, Jiangsu Province: Creation of National Ecological City Receives High Marks", *People's Daily*, Beijing (Chinese). Online: http://qingyuan.people.com.cn/GB/14748/9284184.html (accessed 19 March 2011).
YXEDZ Committee (2010) "12–5 Planning Documents", Yixing: Yixing Economic Development Zone Administration (Chinese).
Zhang, L. (2006) "Contesting Spatial Modernity in Late Socialist China", *Current Anthropology*, 47: 461–84.
Zhang, L. (2003) "Ecologizing Industrialization in Chinese Small Towns", in A. Mol and J. Buuren (eds) *Greening Industrialization in Asian Transitional Economies: China and Vietnam*, Lanham: Lexington Books.
Zhang, L. (2002) "Ecologizing Industrialization in Chinese Small Towns", unpublished PhD dissertation, Wageningen University.

Part II

Governance and cosmopolitanism

Escaping the South

5 Urban governance, mega-projects and scalar transformations in China and India

Xuefei Ren and Liza Weinstein

Introduction

Recent research on urban mega-projects has revealed the increased prevalence of large, iconic and costly commercial housing and infrastructure projects underway in cities throughout the world. While this research has been conducted in diverse locations, little attention has been paid to mega-projects' uneven geography and, more specifically, to why this mode of development has become more prevalent in some locations than others. Highlighting the experience of mega-project development in Chinese and Indian cities, where local state government tends to play a key role commissioning, sponsoring and regulating urban development, this chapter argues that variations in the planning and implementation of mega-projects can be explained, in part, through the lens of state rescaling. China's non-electoral, single-party system has enabled the relatively smooth transfer of power to the metropolitan level, thus facilitating the consolidation of the financial and political resources needed to plan and implement Shanghai's large, iconic and usually disruptive mega-projects. While similar efforts have been made in India to rescale power to the metropolitan level, reforms have been hampered by inter-party and inter-scalar competition.

Shanghai and Mumbai: making Asia's global cities

In late 2003, soon after the international consulting firm McKinsey & Company released a set of recommendations for turning Mumbai into "a world-class city" in the model of Asia's other commercial capitals, Chief Minister Vilasrao Deshmukh announced his intensions to turn "Mumbai into Shanghai" (Katakam 2005). Soon this idea spread among government officials and eventually became a key promise of the national government's re-election campaign (Srivastava 2005). Throughout the next year, politicians proposed new initiatives that would presumably replicate the development successes in Shanghai. Meanwhile, most public actions – including a series of slum demolitions that rendered more than 300,000 people homeless – were justified as necessary means to achieve this end (Katakam 2005; Weinstein and Ren 2009). Yet despite the ubiquity of "Shanghai" – used most commonly as a verb – most of the politicians and civic administrators invoking this concept had never been to the Chinese megacity and had only a vague idea of what it might mean to "Shanghai" Mumbai. It was clear to these politicians, however, that in a relatively short period of time Shanghai had transformed itself into a clean, slum-free city, with state-of-the-art skyscrapers and impeccably modern infrastructures. In order to do what Shanghai had done, they claimed that Mumbai would have to undertake similar large-scale housing and infrastructure projects (Indian Express 2005; Ghoge 2007). Paying little attention to political context or institutional differences, India's politicians announced that Mumbai should simply build big.

Meanwhile, these contextual differences – including variations in governance institutions, systems of public finance and the nature of state power – have also been overlooked in recent scholarly accounts of urban mega-projects. While many of these accounts have sought to explain temporal variations, including how this current phase of mega-project development differs from earlier phases (Altshuler and Luberoff 2003; Flyvbjerg *et al.* 2003; Diaz Orueta and Fainstein 2008; Lehrer and Laidley 2008), a few more recent studies have emphasized regional variation (Fainstein 2008; Bezmez 2008). This article contributes to the growing concern with variations in urban mega-project development in geographically and institutionally diverse settings. In doing so, it argues against a crude convergence hypothesis that suggests cities are increasingly converging around a common mode of mega-project development.

In order to explain variations in the scale, pace and mode of urban development apparent in Chinese and Indian cities, this chapter compares mega-project development in Shanghai and Mumbai through the lens of state rescaling. We find that the particular distribution of power within Chinese and Indian regional states helps account for variations. The rapid transformation of Shanghai's built environment – achieved primarily through high-profile mega-projects – has been facilitated, in large part, by the Chinese central government's effort to reconfigure power at the metropolitan scale. Meanwhile, despite similar pressures in India to devolve authority and resources downward, these capabilities have been retained at the regional state level (by the State of Maharashtra, in the case of Mumbai), leaving a power vacuum at the metropolitan level. In some cases, non-state actors have sought to fill this power vacuum and have assumed greater authority over projects, resulting in a more contested and piecemeal development process in Mumbai. We argue that these arrangements of regional state power are the product of deliberate efforts – outlined in five-year plans, legal mandates and constitutional amendments – to rescale and reconfigure the power of Chinese and Indian states over the past three decades. The character and degree of state rescaling, we argue, depends upon historically structured administrative and political institutions, which shape the ability of local governments to plan and implement urban mega-projects.

After a brief discussion of state rescaling in India and China, this chapter compares the planning and implementation of two globally oriented mega-projects in Shanghai and Mumbai. The project we examine in Shanghai is the acquisition of land and relocation of residents for the World Expo 2010 (WE). In 2002, the Shanghai city government designated prime land on the riverfront as the World Expo site. In a period of six years, it has relocated thousands of residents and dozens of manufacturing facilities, warehouses and shipyards to make land available for the project. The comparable Mumbai project we discuss is the Dharavi Redevelopment Project (DRP). The DRP also seeks the dramatic transformation of a currently vibrant mixed-use district in the city center, seeking to transform the settlement of Dharavi from an industrial and residential "slum" into a middle-class residential and office district. Although these are very different projects – with the World Expo, as a mega-event, having a firm deadline and the DRP, as a slum redevelopment project, without a deadline – both are land conversion schemes designed to bolster the exchange value of centrally located properties, attract international investment and better position their respective cities as global commercial capitals. Moreover, both projects have required the consolidation of political and financial resources to bring them about. The data on the Shanghai case include official planning documents, fieldwork conducted in 2005 and online postings from relocated residents. The data on the Mumbai case include fieldwork interviews conducted in 2005 and 2006, media reports and government and planning documents.

The mega-project as a research question

Although urban mega-projects have garnered significant attention in recent years, most research on mega-project development has barely explored the different processes and outcomes of mega-projects in a comparative perspective. Contributing to understandings of this current moment of global urban renewal, in which cities throughout the world are being transformed by large-scale developments, the majority of this research has sought to explain the political causes, and the symbolic and administrative implications of this "new" mode of development. The newness of these projects, however, is referenced quite self-consciously, as an either explicit or implicit comparison is usually made between the current wave of mega-project construction and the wave that marked the Fordist cities of post-World War II North America and Western Europe (Altshuler and Luberoff 2003; Diaz Orueta and Fainstein 2008). In a recent symposium on mega-projects, the newness of these projects' aims and consequences was addressed directly. Overall, the contributors attached a neutral meaning to the term "new", asserting that its use simply acknowledges "the perception that there has been a recent wave of large schemes within major cities" (Diaz Orueta and Fainstein 2008: 760). However, these authors, with others, have pointed out that the new mega-projects do differ from their predecessors in that they seek to minimize displacements and tend to be framed in the rhetoric of diversity and inclusiveness – that is, all interests are being served, thus defusing potential opposition (Diaz Orueta and Fainstein 2008; Lehrer and Laidley 2008). This point has also been made by Altshuler and Luberoff (2003), who contrast the current era of "do no harm" against the "great mega-project" era of the 1950s and 1960s, asserting that opposition to the large-scale displacements brought about by urban renewal led governments to avoid such disruptive projects, particularly those located in residential areas. It has been suggested, however, that while this shift may be discernable in North America and Europe, rapidly developing countries like India and China are currently experiencing their own period of urban renewal and undertaking socially and environmentally disruptive projects (Diaz Orueta and Fainstein 2008; Weinstein and Ren 2009).

Beyond the question of "newness", research on mega-projects has generally focused on the political causes of the current wave and on the symbolic and administrative implications of this mode of development. With respect to the political causes, several authors identified mega-project development as part of a broader strategy of entrepreneurial urban governance (Flyvbjerg *et al.* 2003; Beazley *et al.* 1997). Noting that economic restructuring over the past several decades has placed cities in a competitive frame, they argue that urban and regional governance has become increasingly dominated by high-profile projects that rely heavily upon private financial contributions. Skeptical about the local benefits actually produced by mega-projects, the focus of this work has primarily been on the decisions of entrepreneurial politicians and local administrators to pursue such projects, often at the expense of redistributive policies. In a related area of inquiry, several authors have focused on the symbolic value of mega-projects, as both a motivation for local governments to pursue this mode of development and as a consequence of large, iconic projects. Much of this research has focused on the manner in which the architectural designs of mega-projects signify a city's or region's emergence as "global" (Marshall 2002; Olds 2002; Douglass 2000). Recognizing that projecting a global image has become almost as important as "actually" being global in the current era of capital mobility and inter-urban competition, this literature has identified the monumental or iconic architectural designs featured in mega-project development that symbolize the global character of their host city (Marshall 2002; Olds 2002; Ren 2008). And third, a significant emphasis of this research has concerned the financing and administrative arrangements between the public and private sector that underlie the planning and implementation of mega-projects (Altshuler and Luberoff 2003; Flybjerg *et al.* 2003). This research also focuses on the imperative of coordinating intergovernmental relations in order to leverage the political and financial capital required to build big.

While both Shanghai and Mumbai have adopted mega-projects as an entrepreneurial governance strategy, this has not been a defensive response to cope with deindustrialization as has been the case in many Western cities. Mega-project development in these cases, rather, reflects the deliberate regional state project to build global cities in order to tap in the international investment, facilitate capital accumulation and, in the Chinese case, also legitimate the Communist Party-state. Yet while mega-projects in Shanghai have been relatively streamlined, marked by large-scale urban renewal and residential displacements, those in Mumbai have been pursued in a more piecemeal fashion, entailing contestations and negotiations between the private sector, the public sector and local communities (Weinstein and Ren 2009). Among the possible explanations for these differences (which we acknowledge are numerous), we argue that the particular distribution of power, authority and resources among the scales of governance in India and China helps account for the variations in mega-project development in Mumbai and Shanghai. Specifically, we argue that the convergence of regional state capabilities – a process we term "metropolitanization" – helps explain the greater prevalence, larger scale and more rapid implementation of Shanghai's mega-projects. In doing so, we bring a comparative analysis of state rescaling into the study of urban mega-projects. Our case studies reveal that state rescaling – i.e., the reconfiguration of regional state power among different levels of government – can significantly alter the trajectories and outcomes of mega-project development.

State rescaling and mega-project development

As studies of urban mega-projects have focused overwhelmingly on cities in the West, inquiries into state rescaling have also neglected questions of regional variations and configurations of regional state power in non-Western contexts. Broadly speaking, the literature on state rescaling has focused on how formerly Fordist nations in North America and Europe have responded to the pressures associated with globalization by unbundling or de-territorializing power and reconfiguring it in a manner consistent with global restructuring. While state rescaling usually refers to the devolution of power from the national to the local level, it may also refer to the "scaling up" of authority and resources from municipalities to metropolitan governments, or to shifts in authority to the private sector, civil society groups or other agencies or branches within governments (Brenner 2004; Sassen 2006). Yet while the literature on state rescaling has focused overwhelmingly on North America and Europe, non-Western nations have undergone similar reconfigurations of power, although perhaps with different objectives and consequences.

Research on state rescaling reflects an effort to understand how nations have been impacted by economic and political globalization. Typically, political theorists have adopted one of the two competing conceptualizations of state power in the context of globalization. The first asserts that power has been gained on a global scale at the expense of the nation state (Ohmae 1995; Hay and Marsh 2000; Miyoshi 1993). The second argues that the nation has retained its power, given the institutionalization of national power and the questionable legitimacy of global governance (Mann 1997; Weiss 1997; Hirst and Thompson 1995). In the late 1990s, however, urban geographers and sociologists presented a third conceptualization, pushing past the zero-sum perspective of power implied by the earlier two. These theorists posited that rather than diminishing or maintaining the power of the nation state, globalization entails the territorial descaling and rescaling of power *within* the nation (see, for example, Swyngedouw 1997; Brenner 1998, 2004; Jessop 2001; MacLeod 2001).

Examining the rescaling of state power in Western Europe, Brenner (2004) asserts that urban regions have become key institutional sites of state rescaling. In response to broader structural changes since the 1970s, such as deindustrialization, the retrenchment of the welfare state and European integration, the economic contexts of cities have changed and major metropolitan regions have been forced to adopt entrepreneurial policies to attract capital investment. Although administrative boundaries remain fixed, changes in practices and regulatory policies reconstitute territorial divisions and intergovernmental relations. Instead of adopting a view of linear decentralization, Brenner (2004) emphasizes the reshuffling of state power and circulation of resources between different geopolitical scales.

Although certain neoliberal strategies and rescaling processes observed in the Western city are present in Shanghai and Mumbai, the historical and social contexts of post-independence China and India have produced different assemblages of state power. Similar to the rescaling processes in the West, China and India have undergone reconfigurations of state power over the past thirty years, including rescaling power to the local or metropolitan scales. Yet these processes have not been pursued in the same manner or with the same sense of urgency in both contexts, or for the same reasons they have in the post-Fordist nations of the West. Since the late 1970s, the Chinese nation state has increasingly shifted authority and resources to the metropolitan scale – including both cities and city regions. In doing so, it has given Shanghai the power it needs to undertake mega-project development, including the ability to displace residents, acquire land and raise necessary revenue. Meanwhile, the Indian central government has similarly attempted to devolve certain authorities and resources to the local level. But, as we discuss in more detail throughout the chapter, India's administrative framework and federalist structure – to a large degree, a legacy of the colonial period – have hindered these efforts. As conflicting interests, operating at multiple geopolitical scales, make claims on the city, the resources needed to plan and implement urban mega-projects rarely reach the metropolitan level.

In the remainder of this chapter, we highlight a reconfiguration of state power that helps explain the variations in mega-project development in Shanghai and Mumbai. This reconfiguration, which we term the "metropolitization of the state", refers to the deliberate redirection of power (i.e., the ability to realize mega-projects), authority (i.e., the ability to decide independently) and resources (i.e., fiscal independence) to the metropolitan scale in order to plan and implement large-scale development projects. We highlight two particular elements of this process. We note the centrally directed nature of state rescaling processes, directed from above by central or national governments, and the strategic importance of the metropolitan scale – including both cities and city regions. We suggest that, because the metropolitan level has been recognized to interface more smoothly with the global economy than the national level, nation states may employ the strategy of metropolitanization in order to facilitate global restructuring. Our use of the term metropolitanization – as opposed to urbanization, localization or even decentralization or devolution – reflects the importance of the metropolitan region, understood here as the geographic area under the administrative jurisdiction of the metropolitan government. Not necessarily the devolution of power to the lowest possible scale, metropolitanization may entail the "scaling up" of some capacities from the local (e.g., villages, towns and townships), as well as the "scaling down" of other capacities from the national to the metropolitan scale.

China's gradual metropolitanization

Scholarship on state rescaling in China has mainly focused on the manipulation of administrative boundaries of cities by the central government to promote urban development and spread industrialization to rural areas (Cartier 2005; Ma 2005; Shen 2005, 2007). Cartier (2005) and Ma (2005) argue that the administrative rank of a place is one of the most important factors affecting China's local space economy.

In the socialist period (1949–78), the Chinese governing regime was organized along a vertical system of government ministries and horizontal territorial authorities, such as provinces, prefectures and cities. This system, widely referred to in Chinese as "*tiaotiao-kuaikuai*" (i.e., lines and pieces), constituted the basic administrative grid that governed from the socialist period through the reform period (Lieberthal 2004). In terms of territorial authorities, China's administrative hierarchy was characterized by a two-level structure composed of central and local governments. All major national policy decisions were made by the central government based in Beijing. The "local" encompassed a wide range of administrative ranks from provinces, prefectures and cities, to counties, towns and villages. During the socialist planned economy, local governments had little real power even in the decision-making of development policies in their own territories.

Since the early 1980s, the reorganization of state power has gradually taken place along both vertical and horizontal administrative lines. The central government shifted decision-making and financial responsibilities away from central ministries to lower levels of state administration, but it still maintains control over local governments though legislative and personnel appointing systems. In the 1982 report on the sixth five-year plan (1981–5), then Premier Zhao Ziyang called for the use of large and medium-sized cities as engines for economic growth. The central government gave metropolitan governments more autonomy over urban policy in order to lease land and attract investment. The assemblage of power and authority at the metropolitan scale explains the incentive to designate new cities and expand city boundaries. Specific mechanisms used to "scale up" rural counties to cities include converting counties to cities, city-administering counties and annexation of suburban counties as urban districts (Ma 2005; Shen 2005, 2007). As a result, the total number of cities increased from 193 in 1978 to 668 in 1997, and all major urban regions have experienced rapid geographic expansion (Shen 2007).

The reorganization of state power has had significant impact on urban governance in Shanghai. By administrative rank, Shanghai is one of the four metropolitan areas at the provincial level that report directly to the central government (Figure 5.1). During the socialist period, the anti-urban national policy largely ignored investment in large cities. Although Shanghai concentrated many

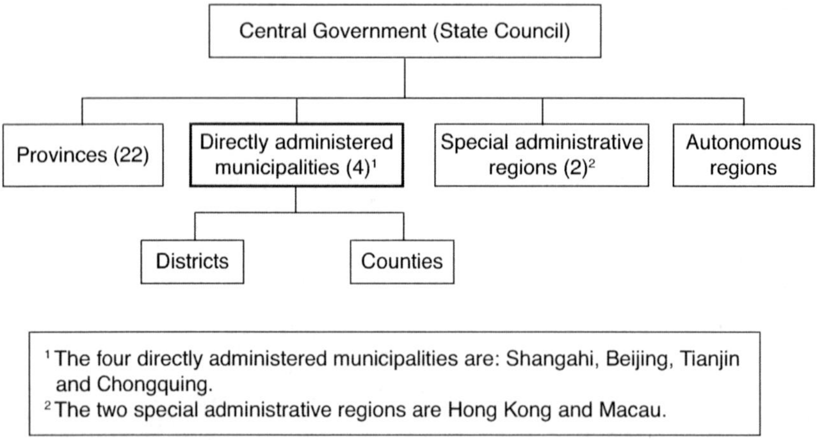

Figure 5.1 The administrative hierarchy in China.

capital-intensive industries and contributed the largest proportion of the national economy, the city's revenue was tightly controlled by the central government. Shanghai was left with little revenue to reinvest in its housing and basic infrastructure sectors. In the 1990s Shanghai's economy finally took off, benefiting from the dual processes of scaling up and scaling down of state power and resources. First, Shanghai expanded its administrative territory by annexing surrounding counties and turning them into suburban districts. The geographic area of Shanghai has expanded from 282 to more than 6,000 square kilometers since the 1980s. Shanghai has thus been transformed from a single municipality into a metropolitan area with nine central city districts, five inner suburban districts, four outer suburban districts and one rural county. The resources of former rural hinterland were scaled up to the metropolitan government of Shanghai. The metropolitan government could reorganize its economy by relocating industries to the new suburban districts, and opening up space in the central city districts for specialized service sectors. Second, the power and authority to devise policies to attract investment capital have been scaled down from central ministries to the metropolitan and district government, which can approve mega-projects without reporting to the central government. The dual processes of scaling up and scaling down have simplified bureaucracy and consolidated authority and resources in the hands of metropolitan and district governments.

India's contested devolution

Despite similar efforts made by the Indian central government to devolve power to the metropolitan scale, the authority and resources required for urban development remain at the middle rung of India's three-tiered federalist structure (see

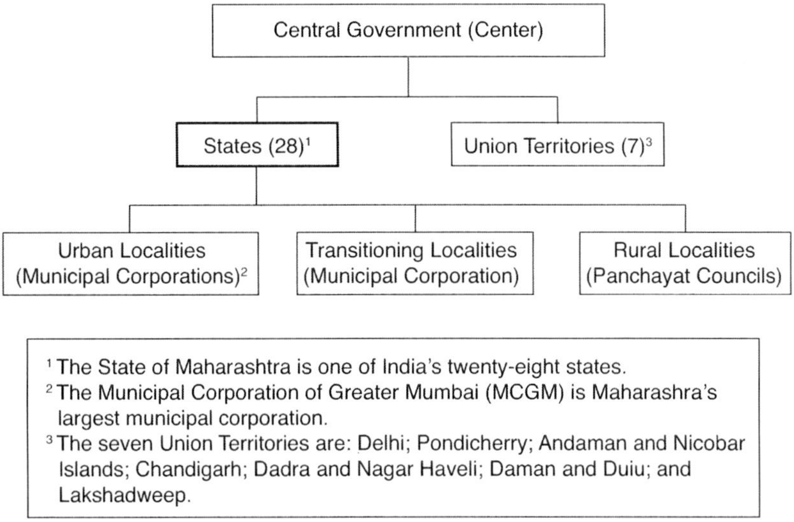

Figure 5.2 The administrative hierarchy of India.

Figure 5.2). In 1992, the central government adopted the Seventy-Fourth Amendment to the Indian Constitution, in order to devolve substantive powers and resources to municipal governments – including the "powers and responsibilities ... for the implementation of development schemes". Yet India's municipalities and metropolitan regions remain largely disempowered. Although some reforms have been made in Mumbai's local governance, most observers agree that the Seventy-Fourth Amendment has done little to alter the distribution of power between the state government of Maharashtra and Mumbai metropolitan government (Pinto and Pinto 2005).

Although the framers of the Indian Constitution identified federalism as one of the five core elements of the independent Indian nation state (along with sovereignty, democracy, socialism and secularism), the three-tier federalist structure was left incomplete, defining responsibilities only for the central and state governments (Corbrige and Harriss 2000; Tummala 1992; Verney 1995). Concerned about regional fragmentation along religious, ethnic or linguistic lines, they granted considerable power to the regional states, in part to hold the union together (Stepan 1999). One of the principle framers, Dr B. R. Ambedkar, explained that the Constitution instituted a dual polity, with the Union at the centre and the states at the periphery, each endowed with sovereign powers (*Frontline* 1998). Urban development – including housing, infrastructure, economic development and poverty alleviation – was a field to be financed and implemented by the regional states. Although regional states could choose to devolve power and resources to localities, most – including the government of Maharashtra (and its precursor, the state of Bombay) – did not. The competitive stance between regional states and municipalities arises in part from inter-party competition. While devolution in China is simplified by the unified party structure, the rescaling of power in India often entails the handing of power from one political party to another, as we will discuss in the later section on the Dharavi Redevelopment Project.

The political weakness of India's municipal governments also stems from colonial legacies and an inherent fear of the local. Prior to independence, municipalities were organized to carry out basic service-delivery functions, while the more substantive powers were retained by the colonial (presidency) government, particularly the land use powers and industrial policy, which would directly impact the colonial economy. When power was handed from the British Raj to the independent Indian nation state in 1947, the "real" authority over urban municipal bodies was transferred from the colonial presidencies to the regional state governments, while municipalities remained primarily service-delivery bodies (Pinto and Pinto 2005). The decision not to empower the local scale was also a deliberate one, rooted in an anti-local bias. Although Gandhi famously asserted that India "lives in her villages", most of the English-educated delegates to the Constitutional Assembly, including the first Prime Minister, Jawaharlal Nehru, perceived India's villages to be backward, archaic and inherently corrupt. They believed that local elites (in both villages and cities) organize power on the basis of communal sentiments, rather than on enlightened democratic principles. The less local the political system, the more democratic many believed it would be (Corbridge and Harriss 2000).

Beginning in the 1980s, however, the pressure to devolve authority and resources to the local level began to emerge from two disparate corners: pro-democracy social activists and pro-liberalization economic reformers. In the mid-1980s, development economists and social activists began making the connection between India's rampant inequalities and the lack of political representation at the local level (Drèze and Sen 1995, 2002; Plummer 2000; Heller 2005). "Democratic deepening", it was argued, would ensure political participation and more equitable resource allocation. The Sixty-Fourth and Sixty-Fifth Constitutional Amendments were proposed in 1989 to devolve political representation to rural and urban local bodies respectively, but they failed to garner the requisite support in Parliament. In the wake of India's adoption of economic liberalization reforms in 1991, however, the political movement to empower India's municipalities was reinvigorated; this time recast as a component of structural adjustment. Included in the package of reforms advocated by the IMF, political decentralization was identified as a precondition for liberalization, diminishing the authority of the central government, introducing a greater role for the private sector and enabling more openness to foreign investment. With support from these strange political bedfellows, the Seventy-Third and Seventy-Fourth Amendments, defining a wider scope of authority for rural localities and urban municipalities, were adopted in 1992. Yet with the central government merely encouraging implementation of the amendments, few regional states have chosen to enact the reforms. This is particularly true in the domain of urban development, where regional states have chosen to retain the budgets and many key powers associated with land, housing and infrastructure schemes.

The different configurations of state power produced by China's gradual metropolitanization and India's more contested devolution process can be discerned in the planning and implementation of mega-projects in the country's two largest and most global cities. In the next sections we discuss how these arrangements have affected the ability of Shanghai and Mumbai to acquire land, raise revenue and plan globally oriented mega-projects.

The Shanghai World Expo 2010

Hosting prestigious mega-events has become a widely adopted strategy by urban governments for place promotion and city regeneration (Hiller 2000; Roche 2000). Shanghai is no exception. In 2002, Shanghai won the bid to host the World Expo 2010. The Expo site is 5.28 square kilometers of land spreading onto both sides of Huangpu riverfront. The riverfront had been occupied by a dozen large-scale steel plants, shipyards and warehouses, as well as nearly 20,000 households. By hosting the Expo on the riverfront, the city government could relocate these manufacturing facilities and residents to suburbs and make the prime central city real estate available for redevelopment after the Expo. By examining state-sponsored land acquisition and relocation initiatives for the Expo, this section shows that the process of land acquisition and relocation has been facilitated by the rescaling of state power at the metropolitan level and was achieved at the expense of citizen rights.

It took only thirteen months – from April 2004 to May 2005 – to relocate nearly 20,000 households from the designated Expo site. Even in Shanghai, this is considered extremely fast and was described as "Pudong speed" in local government-controlled media outlets. Fast residential relocation would not be possible without legislative support from the central government. In 1991, the State Council issued the Urban Housing Demolition Regulation, which allowed municipal governments to issue demolition permits without seeking consents from affected residents. When disputes occur, if developers can provide "some form" of temporary rental housing and compensation, then the law allows developers to start demolition while residents appeal to local courts. Municipal governments have the authority to enforce demolition if residents refuse to be relocated without "proper reasons". Responding to growing complaints about demolition and real estate speculation, the central government later included a few items in a new 2001 regulation intended to protect residents' rights. However, there is still no mention of seeking consent from affected residents before relocation (Weinstein and Ren 2009).

The municipal and district governments had a clear division of labor for relocation. Among the 272 enterprises to be relocated, the city government was in charge of seventy-three ministry and municipal-level state-owned enterprises (SOEs). The three district governments were responsible for the remaining privately and collectively owned enterprises, as well as residents. As most SOEs have close ties with the government, there was no strong resistance to the relocation. The largest landowner on the Expo site, Baoshan Steel Plant, was offered an undisclosed compensation package to relocate to the outer suburbs.

While the relocation was moving ahead, the city government carried out a massive fundraising program to secure the estimated US$30 billion of investment for the Expo. According to the Bureau of Shanghai Expo, 40 percent of the funds were provided by central and city governments, 40 percent by state and commercial bank loans and 20 percent by corporate donation and other channels. The fundraising was carried out by a sophisticated organization composed of different layers of government. At the top level, the Expo Organizing Committee is chaired by Vice Prime Minister Wu Yi and consists of all major ministries of central government. Central government approved the request of Shanghai city government to issue Expo bonds as a way of raising funds for capital construction. RMB15 billion of bonds were issued in 2006, followed by another RMB40 billion in 2007. At the middle level, the Expo Executive Committee is chaired by Mayor Han Zheng and includes all major municipal bureaus. The Mayor has the final voice in all major decision-making. At the bottom level, Land Holding Co. Ltd and the Center of Land Reserve were established to oversee land acquisition, construction of housing projects for the relocated and infrastructure provision. Thus, the initial fund for land acquisition, relocation and construction was provided primarily by the government. The fast implementation of the WE project would not have been possible without the consolidation of these financial and political resources within the metropolitan government.

From 2004 to 2006, two large-scale housing projects, named "World Expo Home Gardens", were completed in Pudong and Minhang districts to accommodate the relocated residents. Displaced residents could purchase new apartment units in these two housing projects at prices lower than market price. Residents could also choose monetary compensation instead, but staying was not an option. In March 2006, the residents of Pujiang Expo Garden set up their own blog. Using online chat forums, residents posted hundreds of messages about the inadequate transportation infrastructure. Pujiang Expo Garden is about fifteen kilometers from central Shanghai. Although the government promised to provide a few bus routes and a subway (M8) line, currently there is only one bus route to serve more than 10,000 residents. The grievance over public transportation escalated into violence towards bus drivers. In May 2007, three bus drivers were attacked by passengers. As these bus tickets are priced higher than those for other bus routes, residents also refused to buy tickets. In the online chat rooms, residents drafted letters to the Mayor and Transportation Bureau (Box 5.1). The Internet might provide an alternative venue for residents to express criticism, but a true platform from which to leverage government has yet to be formed. The residents relocated for the Expo are among the millions of Shanghainese who have been moved from the city center to urban fringes for various urban renewal programs. As Shanghai is transforming itself into a global city, millions of Shanghainese have been accommodated in such new residential enclaves on the city's periphery. Residents' responses to the relocation illustrate that the rescaling of state power at the local metropolitan level was not accompanied by democratic citizen participation in the planning of mega-projects.

Box 5.1 Postings from residents in Pujiang Expo Garden

> The buses are overcrowded, and the operating time is too short. After 9:30 p.m., there are not even taxis in the area. If there is any medical emergency, people have to call ambulance to get to hospitals.... We feel neglected and we hope the government will not ignore us relocated residents.
>
> (xiaocong01, 19 March 2006)

> We are former residents from the inner-city Huangpu district. We were very concerned about the transportation, but the city government promised to complete transportation infrastructure quickly. The government didn't deliver its promise so far.... The bus fare was decided without consulting with residents. A ticket costs RMB4 – for most of us lower-income residents from the old inner city districts, we can't afford this.
>
> (blueguy222, 17 July 2006)

> It's been six months since I moved here. The difference between Shanghai and the countryside is huge. There're no taxis, no public phones, no newspaper stands, few banks, and even the Internet is slower here.
>
> (laolesh, 28 April 2007)

Mumbai's Dharavi Redevelopment Project

The Dharavi Redevelopment Project (DRP) is the latest in a long line of efforts to redevelop Mumbai's largest and most notorious squatter settlement and clear centrally located land for higher-income development. The project was designed and promoted by private architect and developer Mukesh Mehta, who began exploring the feasibility of Dharavi's redevelopment in 1997. In 2004, the project received backing from the state of Maharashtra, when Dharavi's redevelopment was linked to the state's broader efforts to transform Mumbai into a competitive, slum-free city, "like Shanghai." But still, more than eight years later, construction had not yet begun. After national-level endorsement was secured, a financing scheme still had to be formulated, regulations had to be altered and resident consent had to be elicited.

Dharavi is a "slum" settlement on the northern edge of Mumbai's island city, with a history predating the city's colonial period. Dharavi today is roughly three square kilometers, comprising tens of thousands of single-story hutments, hundreds of midrise buildings and countless industrial and commercial enterprises. Unlike earlier schemes, the current plans to redevelop Dharavi were not initiated by the government, or even by non-governmental organizations (NGOs). Rather, they were designed by a private architect as a for-profit land development scheme. Mukesh Mehta's initial plan proposed the removal of Dharavi's slums, relocation of residents to highrise buildings along Dharavi's perimeter and clearance of the center for luxury housing, office buildings and a golf course. Yet encountering resistance from residents and elected representatives, these plans underwent considerable changes in the nine years before they were adopted as policy (Weinstein 2009). According to the plan, finalized in November 2006, Dharavi will be divided into five sectors of roughly equal size, each containing the residents, commercial establishments and industry currently located in that sector, with new civic amenities to be constructed by developers. Current slum dwellers deemed "eligible for rehabilitation" will be given single-room tenements in midrise buildings, free of charge. However, tenants will be responsible for paying municipal taxes and maintenance fees, significantly raising the housing expenses of most Dharavi residents.

After relocating to Mumbai from New York in 1997, Mukesh Mehta spent the next decade designing and promoting the DRP. Most of his efforts were expended on building a political consensus and garnering government support for the project. He received initial support from the right-wing Shiv Sena Party in the late 1990s, then ruling in a coalition at the regional state level. But when a Congress Party-led coalition took control of the regional government in October 1999, the earlier support was lost and Mehta began working again to build political support. These efforts focused entirely on the state government, as Mehta recalls expending no effort to win favor with the politically anemic municipal government. His efforts paid off in 2002, when his project captured the attention of the Urban Development Minister. Although general support was given, competing priorities within the government of Maharashtra prevented the leadership from fully backing the project at the time. In January 2004, the project won an important victory when the central government – ruled by a right-wing coalition that included the Shiv Sena Party – committed five billion rupees for Dharavi's redevelopment. It was remarkable that the central government would offer finance, even though the project had not yet received backing from the state government of Maharashtra, thus subverting the regional state government's authority over urban development. The intervention of the central government into Mumbai's urban affairs was viewed as a direct affront to the Congress-led regional government, reflecting both inter-party and intergovernmental conflicts. In response, the state government of Maharashtra announced the following week that it would partner with Mehta on the project and provide financial backing. But later that year, the ruling national government lost to a Congress-led coalition; and with the Congress Party ruling at both the central and regional level, the inter-party conflicts were resolved. The central governmental never released the promised funds and the project fell back into the domain of the regional state government.

After the announcement of backing from the regional state government, a financing scheme still had to be devised and regulations had to be changed. Although much of the funding could come from private sources, as per Maharashtra's existing slum housing scheme, the extensive infrastructure and public amenities proposed in the plan would require additional resources. With the promised central government funds never released and the Municipal Corporation of Greater Mumbai (MCGM) unable to secure revenue from independent sources, these resources would have to come from state government. But with competing demands emanating from Maharashtra's rural and urban districts, the regional government could not justify devoting public resources to an urban development project of this magnitude. Eventually a scheme was devised whereby private developers would finance the construction of all infrastructure, amenities and housing for slum dwellers in exchange for additional land to develop and sell at market rates.

The efforts to garner support from Dharavi's residents have also been marked by contention. According to the housing policy, slum dwellers could not be rehoused without at least 70 percent consenting to the finalized plans. With ambiguous eligibility rules and the project in a continuous state of revision, residents were growing increasingly concerned about their residential tenure. After a failed attempt to acquire written consents from residents in 2004, opponents came together in a series of direct actions. In response, Mehta and a cadre of Community Development Officers from the state government held town hall-style forums, as well as smaller meetings with mobilized resident groups and industrial associations (Weinstein 2009). Meanwhile, bureaucrats in the regional state government were quietly working to remove the consent clause from the housing policy. While the movement was quieted for some time, the removal of the consent clause, along with other developments in the planning process, resulted in periodic protests and direct actions by project opponents.

While mobilized opposition had arisen, in part, out of anxieties produced by project delays, these fears were partly a consequence of the power vacuum at the local level, which had been heightened by the inter-scalar and inter-party conflicts. Despite the efforts taken over the past decade to devolve authority for development planning to the municipal level, local leaders and municipal representatives had not been empowered to participate in the project's planning process. With their local representatives largely excluded from the project's planning process and the project being directed by regional-level bureaucrats and a private consultant, the residents had little assurance that their interests would be represented in the project. In contrast to the pace of Shanghai's WE project, the DRP became entangled in the scalar conflicts and administrative incoherence that typically define Mumbai's mega-projects. The concentration of resources and public authority at the government of Maharashtra, as well as the political incoherence among the various administrative and political actors, analyzed above, has led to the contentious dynamics of mega-project implementation in Mumbai.

Conclusion

Different from the "new mega-projects" in the West that seek to minimize displacements and are framed in the rhetoric of diversity and inclusion, mega-projects in cities of the global South mostly involve large-scale land conversion and residential displacement. Studying the planning and implementation of such disruptive mega-projects can help to provide a comparative perspective on urban governance, social justice and citizenship rights conditions. This chapter has focused on the contrasting experiences of Shanghai's WE and Mumbai's DRP, examining how mega-projects are being planned, negotiated and implemented in these two globalizing cities. In spite of their differences, in both cases urban residents were clearly in disadvantaged positions as regards to the decision-making of mega-projects.

The WE and DRP are both high-profile projects designed to attract investment, increase the exchange value of centrally located land, and dramatically reshape the cities' built environments. However, the differences between the projects' planning processes are stark. In the case of the WE, with the concentration of power, authority and resources at the metropolitan level, enabled by China's one-party, non-electoral system, the Shanghai government was able quickly to remove residents and acquire land for development. Residents were excluded from decision-making and relocated to the urban periphery to make space for the city spectacle. In contrast, inter-party competition and intergovernmental conflicts have prevented the scaling down of power from the regional state government of Maharashtra to Mumbai municipal government. Even after the endorsement from the state government of Maharashtra, the DRP was delayed because of popular opposition, party turnover, fragmentation of authorities and, most importantly, because of the inability of the metropolitan government of Mumbai to act to address these issues in a direct way.

Some might argue that the different financial capacity of Shanghai and Mumbai accounts for their different development trajectories. As the sole owner of the city's land, the Shanghai government has amassed revenue by leasing land to private investors. Shanghai was able to finance the initial stages of land acquisition, residential relocation and infrastructure provision for the WE through bank loans, bonds and direct subsidies. In contrast, the city of Mumbai lacks the resources to invest in development schemes and relies upon the regional state government for funding. Although it too owns large tracts of land in the city (including much of the land in Dharavi), the MCGM lacks the authority to lease land independently to raise revenues. Meanwhile, the Chinese central government also provided political and financial support for the WE. Yet the Indian government's promise of funds for the DRP was met with suspicion by the state government of Maharashtra, due to inter-scalar competition. Although financial capacity is a key to understanding the divergence between the WE and DRP, the difference in financial capacity has to be understood in relation to the degree to which state power has been metropolitanized in the two contexts.

Meanwhile, it has been suggested that democracy has hindered Mumbai's efforts to oversee a similar transformation – that development delays are simply one price India is paying for its democracy. However, the difficulty of building a democratic consensus for the DRP is itself a symptom of Mumbai's disempowered metropolitan governance. The contentious relations between the regional state government of Maharashtra and Dharavi residents, resulting in project delays, stem in part from its political distance. Although Mumbai may be institutionally democratic, the limited direct representation at the local level (particularly over matters concerning urban development and mega-project planning) has led project opponents to employ strategies of direct action, rather than working through democratic institutions, thus slowing and complicating the project's implementation.

What has enabled Shanghai to carry out mega-projects like the WE is its access to authority and resources, scaled down from the central government and scaled up from its hinterland of the Yangtze River Delta. The Chinese central government has identified urban regions as growth engines, deregulated the land and housing sectors and metropolitanized state capabilities. In China, the redistribution of power has not been a zero-sum game, in which local governments have gained power at the expense of central government. Rather, central government has reshuffled political and financial resources amongst its geopolitical scales in order to interface with global agendas and systems more easily. In the Indian case, because regional state governments have tended to view their power as zero-sum, intergovernmental competition has resulted in the containment of power, authority and resources above the metropolitan level and created a power vacuum for urban mega-project developments. With the government of Maharashtra constrained by competing priorities, mega-projects such as the DRP have progressed more slowly, which has in turned created space for non-state actors to take the lead, such as the role played by political entrepreneur Mukesh Mehta.

Highlighting metropolitanization as a particular configuration of state power shaping mega-project development, this chapter has examined variations in the planning and implementation of relatively comparable mega-projects underway in contextually distinct locations. In doing so, we have shown that postcolonial and post-independence countries such as India and China follow a rescaling process that diverges somewhat from the Western experience, because neither country has undergone the particular transition from a Fordist to post-Fordist mode of production, an assumption upon which the rescaling theory is built. The process of state restructuring must, therefore, be understood within historically structured institutional frameworks. We have illustrated with the case studies that the degree to which nation states metropolitanize affects their ability to implement urban mega-projects. The divergence from the neoliberal model of state restructuring opens a new direction for comparative research on mega-projects, urban governance and globalization. Future studies might examine divergent processes of state rescaling and their implications for mega-project developments in other dynamic global city regions.

References

Altshuler, A. and Luberoff, D. (2003) *Mega-Projects: the Changing Politics of Urban Public Investment*, Washington, DC: Brookings Institution Press.

Beazley, M., Loftman, P. and Nevin, B. (1997) "Downtown Redevelopment and Community Resistance: an International Perspective", in N. Jewson and S. MacGregor (eds) *Transforming Cities: Contested Governance and New Spatial Divisions*, New York: Routledge, pp. 175–86.

Bezmez, D. (2008) "The Politics of Urban Waterfront Regeneration: the Case of Haliç (the Golden Horn), Istanbul", *International Journal of Urban and Regional Research*, 32 (4): 815–40.

Brenner, N. (2004) *New State Spaces: Urban Governance and the Rescaling of Statehood*, New York: Oxford University Press.

Brenner, N. (1998) "Global Cities, Glocal States: Global City Formation and State Territorial Restructuring in Contemporary Europe", *Review of International Political Economy*, 5 (1): 1–37.

Cartier, C. (2005) "City-Space: Scale Relations and China's Spatial Administrative Hierarchy", in L. J. C. Ma and F. Wu (eds) *Restructuring the Chinese City: Changing Society, Economy and Space*, New York: Routledge, pp. 21–38.

Corbridge, S. and Harriss, J. (2000) *Reinventing India*, Cambridge: Polity Press.

Diaz Orueta, F. and Fainstein, S. S. (2008) "The New Mega-projects: Genesis and Impacts", *International Journal of Urban and Regional Research*, 32 (4): 759–67.

Douglass, M. (2000) "Mega-Urban Regions and World City Formation: Globalisation, the Economic Crisis and Urban Policy Issues in Pacific Asia", *Urban Studies*, 37 (12): 2315–35.

Drèze, J. and Sen, A. K. (2002) *India: Development and Participation*, Delhi: Oxford University Press.

Drèze, J. and Sen, A. K. (1995) *India: Economic Development and Social Opportunity*, Delhi: Oxford University Press.

Fainstein, S. S. (2008) "Mega-Projects in New York, London, and Amsterdam", *International Journal of Urban and Regional Research*, 32 (4): 768–84.

Florestano, P. (1980) "Areawide Government and Multiple Jurisdictions: an Examination of Selected Counties in Maryland", *Publius*, 10 (3): 77–99.

Flyvbjerg, B., Bruzelius, N. and Rothengatter, W. (2003) *Megaprojects and Risk: an Anatomy of Ambition*, New York: Cambridge University Press.

Frontline Magazine (1998) "Article 356, Bommai and Fair Play", *Frontline Magazine*, 15 (4).

Fuguitt, G., Heaton, T. and Lichter, D. (1988) "Monitoring the Metropolitanization Process", *Demography*, 25 (1): 115–28.

Ghoge, K. (2007) "Maharashtra in for $3 Billion Facelift", *Hindustan Times*, 18 July.

Government of India (1992) The Constitution (Seventy Fourth Amendment) Act.

Hay, C. and Marsh, D. (2000) "Introduction: Demystifying Globalization", in C. Hay and D. Marsh (eds) *Demystifying Globalization*, Indianapolis, IN: Macmillan Publishers, pp. 1–16.

Heller, P. (2005) "Reinventing Public Power in the Age of Globalization: the Transformation of Movement Politics in Kerala", in R. Ray and M. Katzenstein (eds) *Social Movements in India: Poverty, Power and Politics*, New York: Rowman and Littlefield, pp. 79–106.

Hiller, H. (2000) "Mega-Events, Urban Boosterism and Growth Strategies: an Analysis of Objectives and Legitimations of the Cape Town 2004 Olympic Bid", *International Journal of Urban and Regional Research*, 24 (2): 439–58.

Hirst, P. and Thompson, G. (1995) "Globalization and the Future of the Nation State", *Economy and Society*, 24 (3): 408–42.

Indian Express (2005) "Crores to be Spent to Turn Mumbai into Shanghai", *Indian Express*, 9 February.

Jessop, B. (2001) "The Crisis of the National Spatio-Temporal Fix and the Tendential Ecological Dominance of Globalizing Capitalism", *International Journal of Urban and Regional Research*, 24 (2): 323–60.

Katakam, A. (2005) "For a New Mumbai, at Great Cost", *Frontline Magazine*, 22 (2).

Lehrer, U. and Laidley, J. (2008) "Old Mega-Projects Newly Packaged? Waterfront

Redevelopment in Toronto", *International Journal of Urban and Regional Research*, 32 (4): 786–803.

Lieberthal, K. (2004) *Governing China: from Revolution through Reform*, New York: W. W. Norton.

Ma, L. J. C. (2005) "Urban Administrative Restructuring, Changing Scale Relations and Local Economic Development in China", *Political Geography*, 24 (4): 477–97.

MacLeod, G. (2001) "New Regionalism Reconsidered: Globalization and the Remaking of Political Economic Space", *International Journal of Urban and Regional Research*, 25 (4): 804–29.

Mann, M. (1997) "Has Globalization Ended the Rise of the Nation-State?", *Review of International Political Economy*, 4 (3): 472–96.

Marshall, R. (2002) *Emerging Urbanity: Global Urban Projects in the Asia Pacific Rim*, New York: Routledge.

Miyoshi, M. (1993) "A Borderless World? From Colonialism to Transnationalism and the Decline of the Nation-State", *Critical Inquiry*, 19 (4): 726–51.

Ohmae, K. (1995) *The End of the Nation State: the Rise of Regional Economies*, New York: Free Press.

Oi, J. (1992) "Fiscal Reform and the Economic Foundations of Local State Corporatism in China", *World Politics*, 45 (1): 99–126.

Olds, K. (2002) *Globalization and Urban Change: Capital, Culture, and Pacific Rim Mega-Projects*, New York: Oxford University Press.

Pinto, M. and Pinto, D. A. (2005) *Municipal Corporation of Greater Mumbai and Ward Administration*, Delhi: Konark Publishers.

Plummer, J. (2000) *Municipalities and Community Participation: A Sourcebook for Capacity Building*, Sterling, VA: Earthscan Publications.

Ren, X. (2008) "Architecture as Branding: Mega-Project Developments in Beijing", *Built Environment*, 24 (4): 517–31.

Roche, M. (2000) *Mega-Events and Modernity: Olympics and Expos in the Growth of Global Culture*, London: Routledge.

Sassen, S. (2006) *Territory, Authority, Rights: from Medieval to Global Assemblages*, Princeton, NJ: Princeton University Press.

Shen, J. (2007) "Scale, State and the City: Urban Transformation in Post-Reform China", *Habitat International*, 31 (3–4): 303–16.

Shen, J. (2005) "Space, Scale and the State: Reorganizing Urban Space in China", in L. J. C. Ma and F. Wu (eds) *Restructuring the Chinese City: Changing Society, Economy and Space*, New York: Routledge, pp. 39–58.

Srivastava, S. (2005) "Mumbai Struggles to Catch Up with Shanghai", *Asia Times Online*, 16 March 16.

State of Maharashtra (1971) Maharashtra Slum Areas (Improvement, Clearance, and Redevelopment) Act of 1971.

Stepan, A. (1999) "Federalism and Democracy: Beyond the U.S. Model", *Journal of Democracy*, 10 (4): 19–34.

Swyngedouw, E. (1997) "Neither Global nor Local: Globalization and the Politics of Scale", in K. R. Cox (eds) *Spaces of Globalisation*, New York: The Guilford Press, pp. 137–66.

Tummala, K. K. (1992) "India's Federalism under Stress", *Asian Survey*, 32 (6): 538–53.

Shanghai City Government (1991) Urban Housing Demolition Regulation.

Verney, D. (1995) "Federalism, Federative Systems, and Federations: The United States, Canada, and India", *Publius*, 25 (2): 81–97.

Weinstein, L. (2009) "Democracy in the Globalizing Indian City: Co-optation and Participation in Mumbai's Slum Redevelopment", *Politics & Society*, 37 (3): 397–427.

Weinstein, L. and Ren, X. (2009) "The Changing Right to the City: Urban Renewal and Housing Rights in Globalizing Shanghai and Mumbai", *City & Community*, 8 (4): 407–32.

Weiss, L. (1997) "Globalization and the Myth of the Powerless State", *New Left Review*, 225: 3–27.

Wong, C. P. W., Heady, C. and Woo, W. T. (1995) *Fiscal Management and Economic Reform in the People's Republic of China*, Oxford: Oxford University Press.

Wu, F. (2003) "The (Post-) Socialist Entrepreneurial City as a State Project: Shanghai's Reglobalisation in Question", *Urban Studies*, 40 (9): 1673–98.

Wu, F. (2007) "From State-owned to City Inc.: the Case of Shanghai", in K. Segbers (ed.), *The Making of Global City Regions: Johannesburg, Mumbai/Bombay, Sao Paulo, and Shanghai*, Baltimore: Johns Hopkins University Press, pp. 207–31.

6 Bourgeois environmentalism, leftist development and neoliberal urbanism in the City of Joy

Pablo S. Bose

Introduction

More than a decade of neoliberal urban development both within Kolkata and its periphery has steadily transformed the entire region's political and urban landscape. While the state of which it is the capital (West Bengal) is perhaps best known as a stronghold of leftist and revolutionary traditions, and a fertile ground for land reforms and political decentralization, much of the post-Independence period has seen a reification of Kolkata's image as a city overwhelmed by poverty, pollution, population growth and economic decline. But the state and city have been undergoing remarkable changes in recent years. Globalization and the embrace of neoliberal economic growth strategies, both nationally and regionally, have significantly altered the landscapes of Kolkata. Luxury housing complexes, industrial parks devoted to the IT and business process outsourcing sectors, foreign-funded environmental and infrastructure projects and the construction of numerous luxury amenities for the middle classes and elites hold the promise of reintegrating Kolkata into transnational circuits of economy and culture after decades of apparent stagnation.

Yet such apparent progress has come at a heavy price: a continuous cycle of evictions, expropriations, ecological degradation and social protest speak to the weight of such costs. The future appears, therefore, as simultaneously bright, contested and uncertain. In recent years in particular, the power of leftist governments at both state and municipal levels has evaporated, due in large part to controversial plans that have included embracing the development of special economic zones, the privatization of a range of services and the conversion of rural farmland into factories. Yet while such controversies have received significant attention in both local and national public discourse, the ongoing struggles regarding urban change in Kolkata have received less notice. This paper examines some of these dynamics with special attention to the impacts of the recent changes to the urban landscape, especially in terms of various forms of displacement. At what cost is the once and future global city imagined, and on which particular bodies are these burdens borne? How is it that strategies of neoliberal urbanism have come to be embraced so wholeheartedly in what has been for decades a heartland of Marxist political power in India?

The chapter takes up this question in three parts. In the first, I focus on some of the political shifts that have taken place at multiple scales, and what these might tell us about the role of neoliberal development within West Bengal. In the second, I focus on recent urban developments in Kolkata and the ways in which narratives of the global city and resurgent economy are reshaping life and urban space. In the final section of the paper, I turn my attention to some of the struggles over Kolkata's recent transformations, conflicts based primarily on differing notions of the natural and built environments, and the use, abuse and access to each by citizens in the city.

I explore in particular some of the conflicts that have arisen between middle-class and poorer residents in various parts of the city, especially in and around upscale housing developments. What we see at play in these dynamics are struggles over the right to the city, the presence of certain bodies in certain places, and of the reshaping of public space itself. Drawing on extensive ethnographic fieldwork conducted over eight years – including interviews, archival research and participant observation – this paper argues that what we see in Kolkata today is an attempt by certain groups to reshape the landscape, to change Calcutta, City of Postcolonial Decay, to Kolkata, City of Post-Liberalization Joy.

1 Kolkata, West Bengal and the politics of neoliberalism

The past decade has witnessed a sea change in the regional and local politics of West Bengal. As recently as 2007, a coalition of Marxist and left-leaning political parties known as the Left Front were celebrating thirty years of continuous rule in West Bengal, the longest-serving democratically elected communist government in the world. Federal, state and municipal elections over the previous three years had confirmed the popularity and power of the coalition; at each level the political partners increased their share of both the popular vote and representation in legislatures, state assemblies and city halls. In 2005, the Left Front took power in fifty municipalities out of eighty statewide, including a victory in Kolkata, traditionally a stronghold of the centrist Indian National Congress (INC). Both in municipal elections and in the 2006 state elections, the Left Front solidified its power, increasing its representation in the legislature, capturing 235 out of 293 available seats and continuing the trend of extending their reach beyond traditional rural strongholds and into urban centers and municipalities (Chakrabarti 2006).

Yet fast-forward only a few years and the ground seems to have collapsed beneath the Left Front. In the 2009 general election, the Left Front saw its federal representation significantly diminished by a resurgent All-India Trinamool Congress Party (TMC, a regional breakaway from the INC). The 2010 municipal elections resulted in an even greater defeat for the Left Front, as it lost the vote in key suburbs such as Bidhannagar (Salt Lake) and won only thirty-nine out of 139 wards in Kolkata. Control of the capital city rests now with the TMC, which has emerged as the Left Front's greatest rival. In the 2011 statewide elections, the TMC ended the Left Front's thirty-four-year rule with a landslide victory, winning 227 out of 294 seats, while the Left Front was reduced to a mere sixty-two seats, a decline of some 171 from the previous election (Chief Electoral Officer 2011). What has precipitated this political earthquake? Why has the Left Front, which had seemingly solidified its popularity with the rural poor while extending its appeal to the urban middle classes, suddenly lost its way? Is the turn in its fortunes an indication that voters in Kolkata and West Bengal have rejected the forms of neoliberal developmentalism that the Left Front explicitly adopted in the late 1990s under the aegis of what Ananya Roy (2003) calls "New Communism"?

We need to pause here to consider what neoliberalism means in this specific context. Like "globalization", it is a seemingly ubiquitous term, simultaneously full and empty of meaning. Many on the left in particular have derided neoliberalism for its apparent connections to "Western/American" cultural, economic and political domination, neoconservative policies and capitalist integration. Yet as Ong (2006: 1) points out, "despite such widespread criticism, Asian governments have selectively adopted neoliberal forms in creating economic zones and imposing market criteria on citizenship."

What do neoliberalism and globalization mean in the context of Kolkata's recent past? Critics would argue that they have manifested in a series of ongoing evictions, expropriations, the privatization of city services, a lack of enforcement of labor and environmental regulations, increasing private-public partnerships in a range of arenas and the ongoing marginalization of the urban poor (Roy 2010; Sen 2011; Samaddar 2011). The CPI(M) (Communist Party of India (Marxist)) would no doubt disagree; certainly, it has not embraced the ideas of neoliberalism or globalization explicitly. Indeed, the official CPI(M) stance on both concepts is not markedly different from that of many others across the world. In the past, the party's leadership called globalization a buzzword that "offers respectability to a nakedly aggressive design of imperialism" (CPI(M) 2000) and blamed the neoliberal policies of the central government for the "deep distress" faced by farmers across India today (CPI(M) 2007). And yet, as many critics have pointed out, while the CPI(M) may continue to critique these concepts – especially at the national level – it has adopted key elements of these concepts for its own policies in the arenas where it holds power. Such ideological flexibility is perhaps easier to reconcile if we follow the lead of Tickell and Peck (2003: 165) in examining neoliberalism – and more specifically what they refer to as "neoliberalization" – as a set of processes rather than as an end result.

From this perspective we see that neoliberalism is not simply an existing condition, but rather a disciplinary apparatus through which market-based economic rationalities help to engender new forms of governance and subjectivity. If we follow the historical trajectory that Tickell and Peck use to describe the rise of neoliberalization as a dominant form, the position of the CPI(M) is less awkward than it would appear. In particular, the CPI(M) experiments would seem especially appropriate to the third stage of Tickell and Peck's formulation, where neoliberalism moves from its relatively marginal position prior to the 1980s – standing outside of Keynesian economic orthodoxy – through what they call the "roll-back neoliberalism" that saw the structural adjustment programs of the 1980s and early 1990s, to the present context of "roll-out neoliberalism". The latter focuses on the extension of neoliberal ideals throughout the state apparatus, reproduced in new contexts and in new forms, often framed by the rhetoric of taming the worst excesses of the capitalist system while simultaneously affirming its ascendance and inevitability. This analysis fits the view that scholars such as Roy have of contemporary developments in Kolkata:

> The Left Front has long been known for its agrarian reformism. But more recently it has engaged in a strategy of urban developmentalism. In the 1990s, in keeping with India's liberalization drive, it launched a New Economic Policy, partly in order to adapt to the electoral challenges of the new millennium. The policy envisaged a broad range of public-private partnerships in the provision of housing and infrastructure, thus marking a shift from prior policies of slum improvement to investments in the upper end of the housing market.
>
> (Roy 2004: 152)

It is important here once again to emphasize that the urban developmentalism of which Roy speaks – coded in popular discourse as Kolkata's "resurgence" – is not merely a product of a neoliberal moment, if there is such a thing. In India today, all change, it appears, is linked to 1991 and the arrival of the so-called liberalization era (which is only one in a series of moments in the opening of the Indian economy). But development and evictions, reclamations and resistance, planning and politics are as old as the city itself. It has been undergoing periodic bursts of "renewal" and "revitalization" almost from the moment of its inception as a British colonial site. The foundations for much of what we see as new in Kolkata today were indeed laid in the 1960s. The governments of that day brought in Ford Foundation experts to help plan a newer and better Kolkata, one in which the already felt industrial decline would be reversed. A basic development plan was created, one that envisaged an explicitly "entrepreneurial city" (CMPO 1966), Bengalis' lack of entrepreneurship being among the supposed culprits of the city's decline. Ironically, the decision on the part of the governments at the time to bring in Ford Foundation experts was at least in part to forestall the rising power of the Communists (Banerjee 2005). Following their ascension to power, however, the CPI(M) adopted plans to build the entrepreneurial city, with a second metropolitan center, new port facilities, reclaimed townships, peripheral highways, bridges and other infrastructure.

In a sense, then, we see elements of proto-neoliberalism in the early visions of the remade Kolkata. The change that is occurring in Kolkata today, therefore, is dependent both on existing social, political and economic infrastructure and tradition, and on the mobilization of contemporary neoliberal market logics. For example, government agencies may still repeat the mantras about building social housing, and they may develop projects that were originally conceived in the 1960s. Yet in practice, lots in such initiatives are overwhelmingly allocated to middle- and upper-class buyers. The Kolkata Metropolitan Development Authority, the primary urban development agency in the region, has, for example, moved away from its stated "expertise and experience in slum improvement" towards housing, new area and commercial facility development with "significant cost recovery or surplus generation components" (KMDA 2006).

Social activists have also loudly decried other important manifestations of the developmental regime in Kolkata and West Bengal. In 1996, along with the municipal corporation, the state government launched the now infamous "Operation Sunshine" as a series of "urban renewal" and "regeneration" measures to remove and relocate hawkers and pavement dwellers from the city, to prevent them from "darkening the streets" (Dasgupta 2003). Almost a decade later, the campaign for the "beautification of Calcutta" (2003–5) included multiple evictions of "illegal slum and pavement dwellers": a total of 350,000 people were displaced, often with no notification and no provision for rehabilitation or resettlement. There was considerable mobilization within Kolkata and West Bengal against this (and within the Left Front coalition itself there was much debate over these measures), but in the end the evictions remain part of the price that – under a particular logic – must be paid in order to achieve the new Kolkata. Of course, the question of who makes these sacrifices is another issue. More recently, the city has proposed banning hand-pulled rickshaws as an unsightly and antiquated practice not in keeping with Kolkata's modern image (*Telegraph* 2007).

Such attempts are an important part of rehabilitating the city's image and standing. The view of Kolkata in the post-Independence period has not been a kind one. To many, the city has suffered for years from "a bad press" and "the poverty of representation", depicted overwhelmingly through the lenses of charity, poverty, pollution and tourism (Hutnyk 1996: vii). Chattopadhyay (2006: 2–3) similarly suggests that Kolkata has long served in urban studies as "a metaphor for urban disaster, as the worst possibility of urbanism, as modernity gone astray". To many of the "old Communists", Kolkata was indeed an ambivalent site. It remained the political and cultural capital of West Bengal, a grand old city of arts, education, commerce and connections to the outside world. But despite – and in some cases because of – such associations, Kolkata was viewed with suspicion by many within the leftist movements that swept West Bengal, especially from the 1960s onwards. To some it was the heart of a "horrible urban capitalism" (Roy 2003: 9), to others a poverty-stricken "black hole" whose colonial edifices and nationalist memories paled in contrast to an idealized Bengali countryside undergoing political mobilization and renewal. It was indeed towards the rural regions that the Left Front turned their attentions, with a program of extensive land reforms and political decentralization initiatives begun in 1977. Kolkata and many of its citizens were to complain in the years that followed that they had been forgotten and neglected; former Indian Prime Minister Rajiv Gandhi once famously dismissed it as a "dying city" (Bhattacharya 2007).

2 Calcutta becomes Kolkata: the city reborn

But the reports of its demise seem to have been premature. Over the past two decades, both the state of West Bengal and the city of Kolkata have been undergoing remarkable transformations. What the Left Front and the "business-friendly communists" who are its current leaders have attempted to fashion is a developmental model that mimics the Chinese experiments with capitalism and free market strategies. The aim is to develop urban and rural infrastructure simultaneously and to encourage further foreign investment in the region, with the aim of making Kolkata in particular a hub for new IT and service sectors as well as food processing, horticulture and agro-industries (Bhattacharjee 2007). Accordingly, Kolkata is experiencing a renaissance of sorts, with multinational corporations returning to the city and a real estate boom providing evidence of its supposed vitality and growth (Axiom Estates 2006; HDFC 2006).

The peri-urban fringes of the city in particular have seen remarkable changes in recent years. Many new industries have been located along the eastern edges of Kolkata, in or near publicly planned townships such as Rajarhat and Bidhannagar. There has also been a tremendous boom in new housing construction in these regions, primarily along the sides of or connected by the Eastern Metropolitan Bypass. The largest townships are designed to house up to half a million people, 78 percent of whom planners hope will be middle- and upper-income earners (KMDA 2006). Surrounding the housing complexes are office buildings, convention centers, gated green spaces, schools, private hospitals, multiplex theatres, luxury hotels, golf courses, restaurants and private roads. As one recent returnee to the region mentions, sitting in these new developments one can hardly believe that this is Kolkata (Bhattacharya 2007).

The majority of this new construction is along the eastern fringes of the city, though increasingly new buildings are cropping up within the city proper as well. The impact of the construction within the peri-urban fringes to the east can be seen in the following builder's map: we see hotels, IT parks, a convention center, new arterial roads, housing projects, flyovers and the highway, all pushing up against canals, water bodies and the East Kolkata Wetland itself.

Chakravorty (2005: 89) calls the developments along the city's eastern edges "the most significant new spatial component of the reforms in Calcutta", and sees Bidhannagar and Rajarhat as emblematic of the new development strategy that has gripped Kolkata's urban planners. These "international-style" luxury condominium complexes are designed and marketed towards an elite transnational diasporic subject – or for those who wish to live like them. Such assumed lifestyles and life-spaces are symbolic manifestations of the globalized future that has become a central aspiration for many urban planners, politicians, developers and even social movements. For many critics, these are spaces meant primarily for high-tech professionals and government workers, and are designed to facilitate Kolkata and West Bengal's active participation in the new information economy. As Samaddar (2011) notes:

Figure 6.1 Greater Kolkata.

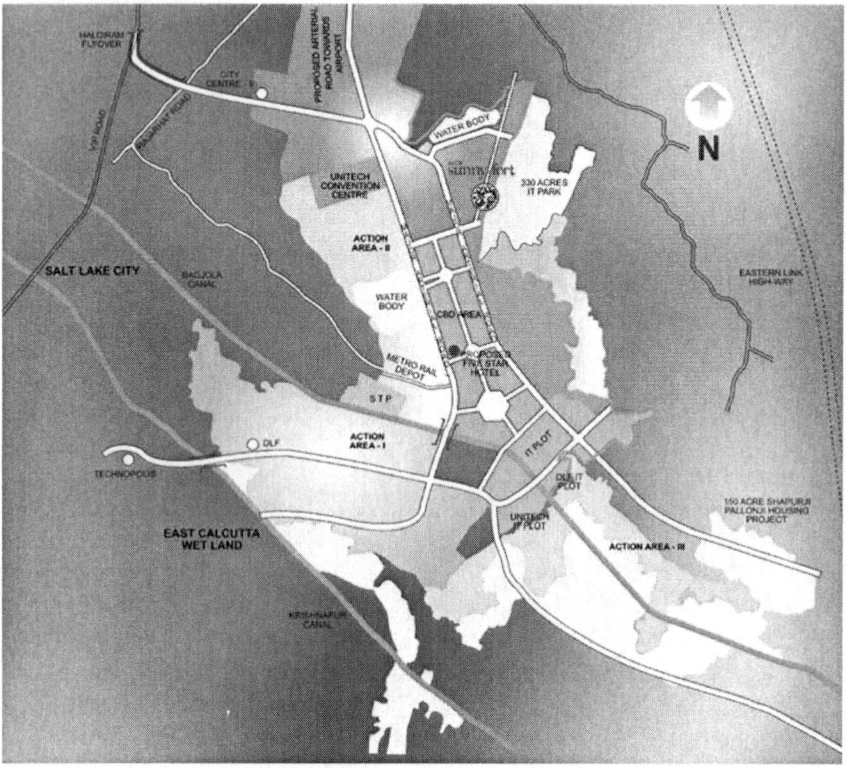

Figure 6.2 Builder's map showing proposed apartment complex.

> Kolkata has changed quite a lot in the last few decades. It wants to become Delhi. It must catch up with the flash and glitz found elsewhere. It too must have its high-tech township and must embody a new mode of circulation of money, information, human resources, and power. It does not think that its old organic character is worth retaining. If discarding the old organic character is the necessary sacrifice to make in order to develop, let that be. If road space increases while the space for human interaction decreases, that price Kolkata must pay.

If the new developments are indeed meant primarily for IT workers and global Indians, what room then remains for the many millions of Kolkatans who do not fit into this particular vision of the future? What impacts are these new industries and building projects having on the local economy, ecology and politics?

Many other Kolkata-based environmental and social justice groups have asked just such questions, raising serious concerns regarding the transformation of the city's peri-urban fringes. The eastern edge of the city not only contains a wealth of new housing complexes, townships and industries, it also borders (and often extends into) one of the last major green spaces within Kolkata's metropolitan area, the East Kolkata Wetlands. The Wetlands have long been seen as one of the only logical expansion routes for Kolkata's burgeoning population (CMDA 1977); the city is bounded to the west by the Hooghly River and has over its life crept north and south in what is seen by many planners as an unsustainable and haphazard fashion. There has therefore been an ongoing project throughout the postcolonial period to reclaim the wetlands in order to ease congestion and chaos in the city, with the building of the township of Bidhannagar in the 1960s hailed as a particular success (Chatterjea 1990). Wide paved roads, orderly blocks of townhouses, government offices, corporate headquarters and new industrial projects are all hallmarks of Bidhannagar. Many middle-class Kolkatans have indeed already moved away from the heart of the old city and resettled in Bidhannagar. Buoyed by the success of its earlier reclamation experiment, the government of West Bengal initiated the neighboring township of Rajarhat (also known as New Town Kolkata) in the 1980s. Designed to eventually become three times as large as its predecessor, Rajarhat is a prime new location for housing as well as a site for many of the newly arrived IT and BPO (business process outsourcing) companies' offices (KMDA 2006; Dey 2011).

Yet the East Kolkata Wetlands is not simply an empty space and resource awaiting the developmental designs of local authorities. It is also an important social and ecological site for Kolkata. A substantial green space bordering a metropolis with precious little otherwise, the Wetlands also acts as the primary waste management system for the city (Ghosh 1998). Its many wastewater-fed fisheries and farms provide foodstuffs for the markets of the urban poor. Between 60,000 and 100,000 people live in the Wetlands, in what are considered semi-rural villages and settlements (Bunting *et al.* 2001). The region is also home to a substantial number of informal communities and colonies, including refugees from as far back as the 1947 partition; indeed, the peri-urban fringes of Kolkata formed one of the first bases of political power for the Left Front within the vicinity of the city (Chakrabarty 1990). Farmers, fishers, migrant laborers, rag pickers, garbage workers and refugees both old and new live in the Wetlands (Bunting and Mukherjee 2002). In the rush to reclaim the swamps, scant provision has been made for most of their displacement or loss of livelihood.

For these reasons, several public-interest lawsuits have been brought by a variety of social and ecological justice groups to halt development in the Wetlands. Mass protests led to the shelving of a proposed plan to build a World Trade Center in the mid-1990s, while a West Bengal Supreme Court case ordered a stop to further urbanization in the Wetlands on the basis of potential environmental degradation (*All India Review* 1993). Bowing to public pressure, the government of West Bengal and the Kolkata Municipal Corporation in 2002 signed the Ramsar Treaty on the Conservation of Wetlands and declared the East Kolkata Wetlands a protected area (Ramsar Convention on Wetlands 2002).

Despite such protection, the development of the Wetlands continues unabated. For West Bengal's leftist leadership and its successors, the East Kolkata Wetlands has long represented an opportunity to aid in the transformation of the city. Yet unlike the famously noisy, polluted and congested core of the old town, with its long history and masses of inhabitants, the Wetlands holds the possibility and promise of the undiscovered country, a *tabula rasa* perfect for what Roy (2004: 153) calls the "territorial rewriting of the city's edges". It is here, therefore, that the most striking changes can be seen. Yet it is remarkable not only that the Wetlands has now been invaded by gleaming new office towers, theme parks, golf courses and shopping malls, but also for the manner in which this transformation is occurring, as well as its broader purpose. Kolkata's reshaping does not reprise the Left Front experiments with land and political reform in the countryside; instead, New Communism, argues Roy, manifests key elements of liberalization. In her view this has meant locating the state itself as a site of market activity, positioning real estate development as a centerpiece of the city's productive economy, capitalizing on state assets such as public land, taking on the mantle of entrepreneurialism and creating a distinct new geography, whose prime territory is the metropolitan region rather than the core city (Roy 2004: 153).

But perhaps the most visible sign of the transformed and intensified process through which Kolkata and its periphery are being developed is in the nature of the agents of change. Unlike the previous eras mentioned, the current context has seen a veritable explosion of partnerships doing the developing. No longer are state agencies and organizations taking the lead on their own. Instead, a complex matrix of institutional and informal linkages has taken their place. In some cases these are partnerships between government agencies and private corporations, especially in order to build new housing complexes and luxury apartments. The Left Front government had previously identified this strategy as a priority: its Department of Urban Development stated that in order "to mobilize more resources critically needed for a higher pace of development, participation of the Private Sector in the infrastructure projects are being encouraged by way of joint ventures" (GoWB 2007b). Another agency, the West Bengal Housing Board, similarly "realized the urgency and need to encourage public private partnership in the housing sector by forming Joint Sector Companies" (ibid.). The Kolkata Metropolitan Development Authority – a nominally non-partisan entity – has joined with private investors and corporations to build several apartment towers, townships, sports venues, IT logistics hubs (in Bidhannagar and Rajarhat) and entertainment complexes. The Kolkata Municipal Corporation launched a joint venture with United Credit Belani Group, an asset financing and real estate development company, in order to form the Calcutta Metropolitan Group (CMC). The CMC has been responsible for highway improvement projects, building hospitals, aquatic and adventure theme parks, high-tech and industrial parks, hydroelectric dams, the New Kolkata Township at Rajarhat, hotels and port facilities. These are but a few of many examples of such partnerships that are helping to transform Kolkata. And while the leftist governments at various levels that have led this charge towards an entrepreneurial and neoliberal development program may have fallen, there is no indication that their successors will dismantle this approach.

In some cases, foreign government agencies and development banks have partnered with municipalities and state departments to undertake specific projects. In yet other projects, diasporic capital – in the form of investments – are mobilized. Witness the description of one of the housing developments in Rajarhat:

> It would be a slice of Americana on the fringes of the city. We would call it Rosedale Garden, an exclusive NRI habitat in Action Area III of New Town, Rajarhat. It would offer NRIs the standard of living they are used to in the West. The complex would combine the best of both worlds, a blend of the conveniences of the West and the earthy charm of Bengal. It's not just about returning to one's roots, but also an urge, a commitment to make Bengal a better place.
>
> (Saha 2006)

A township on the western side of the city – developed jointly by the KMC and international investors (originally Indonesian, now replaced by investors from Singapore) echoes a similar sentiment:

> There are lots of rich Indians overseas. We intend also to sell to them. They will invest in this project and bring in the dollars. The NRIs [non-resident Indians] want to buy a second home at home.
>
> (Ciputra Group of Indonesia 2006)

As we see in such statements, while produced through a process of neoliberalization, the legitimation of the contemporary urban developmental vision in Kolkata is often tied to a particular notion of globalization, or at least of global identity. Kolkata's position and success as an entrepreneurial city is in this sense dependent on its ability to materialize its connections to the networked global economy through the presence of transnational capital and transnational subjects – both where they live and where/how they work. As Chakravorty (2005: 90) says of the transformation of Kolkata's eastern fringes, "clearly the planners want this development to contain new (rather than relocated) industry of the type that is high-tech or global in nature". Similarly, if one looks at the ideal inhabitant of the reclaimed Wetlands region, images of a transnational subject – or those who can live such an imagined lifespace – are invoked. South City Projects, an enormous complex just inside the city but on the edges of the Eastern Metropolitan Bypass, invites prospective buyers to "live the way the world does" and "promises a lifestyle of international standards" (South City Projects 2007). Bengal Silver Spring, another joint venture, tells you that it is "just the kind of place you would see in foreign glossies and regret you'd never get" and that living there will allow you to "make a world of difference" (Bengal Silver Spring 2007). Vedic Village, a more explicitly "Indian-themed" gated community located in the midst of the Wetlands, tells us that it "is what reflects the best that Kolkata has to offer to an international audience – a veritable paradise on Earth!" (Vedic Village 2007).

These new developments are thus a potent symbol of many things. For those who dream of recovering Kolkata's former glory, the building of skyscrapers and luxury dwellings in the city's peripheral marshland signifies an ability to "live like the world does". For those who talk of becoming a "world-class" city once again, the construction of dwellings for those who work in IT is only appropriate. For the CPI(M), the development of the Wetlands seemed for the better part of a decade proof positive that New Communism was working (Roy 2004). Invested with such signifiers, it has become that much more difficult to challenge and oppose the logic of development in the peri-urban fringes of Kolkata. The diffuse nature of those doing the developing itself makes the task all the more Herculean: where once one might have taken on the municipal government, regional developmental authority or state agency, now one must confront a plethora of parastatal agencies and public-private partnerships. Indeed, despite the storms of debate surrounding the controversial projects that have been at the root of the Left Front's woes – the well-publicized cases of Singur and Nandigram, attempts to set up special economic zones to build car and chemical factories respectively (Majumdar 2010) – these cases do not necessarily belie a disenchantment by the general public with the neoliberal developmental model adopted by the Left Front. Indeed, the struggles and debates seem often located outside of the city itself. It is the "golden countryside of West Bengal" that needs protecting, from the reach and avarice of neoliberalism and the excesses of the developmental state. It is the village – model of Left Front political reforms and economic growth in its first two decades in power – that is under threat. The focus, therefore, remains primarily on the Bengal out there, rather than the Kolkata close to home. This is not to suggest that opposition to Kolkata's redevelopment is absent within the city itself. But, for example, in the same weeks and months that the controversy regarding the violent expulsion and deaths of villagers in Nandigram – proposed site of an Indonesian-funded chemical plant – dominated local and national talk shows and news, and saw public intellectuals, artists and politicians join thousands of others in protest marches against the government of West Bengal, precious little attention was paid to the fact that evictions and violence *within* the city continued apace. The targets of the current evictions were primarily hawkers – 2,000 of whom were to be cleared from Howrah and Sealdh train stations and a further 10,000 of whom were to be removed from a 3 km radius around the Victoria Memorial, a prominent tourist attraction in Kolkata (*Times of India* 2007).

These efforts follow similar expulsions of these so-called "stains" on the face of the city; a KMC report described hawkers in 2006 as "a menace" and contributors to pollution and congestion – an oft-repeated theme – in Kolkata. Such evictions were not new: from Operation Sunshine in 1997, through the forced displacement of tens of thousands for environmental projects, the extension of the Metro subway system and new housing developments, the past decade has witnessed an ongoing cycle of dislocation of the urban poor in Kolkata, though the number of hawkers has actually multiplied in the years since (*Times of India*

2010). The attempted removal of such bodies indeed plays an important role in governmental visions of a world-class city. As one local activist puts it:

> The government wants big malls and giant retail markets to develop. It does not see any role for hawkers … it wants to virtually signal an end to hawking in urban areas.
>
> (HA4, 2007)

What is perhaps most interesting in the displacement from the railway stations and Victoria Memorial, however, is that the driving force behind the implementation of these policies is not restricted to the state government, the municipality or the police. Indeed, in some cases the initiatives originate with public-interest litigation, filed by a lawyer's association in the case of the railway stations, and by the Calcutta Chamber of Commerce and Industry to improve the tourism aspects of Victoria Memorial (*Times of India* 2007). The complaints are couched once again in the language of inconvenience and danger: hawkers impede pedestrian and vehicular traffic, they increase crime and violence, they represent a blight upon the changing city. The evictions, were, at any rate, carried out with their usual brutal force, though, as with many of the earlier attempts, most of the hawkers simply drifted back to their old haunts within months, weeks and even days (*Times of India* 2010).

3 Bourgeois environmentalism and the struggle over space

In the final section of this paper I would like to consider the terrain represented by the struggle over the right to the changing city of Kolkata, embodied by the contest between middle-class and poor inhabitants in these recent attempts to clear certain public spaces of hawkers. In a similar fashion, I would like to focus on several emerging debates regarding specific conceptions and uses of the local environment in Kolkata, and in particular the debate between middle-class aesthetics and the survival strategies of the urban poor in the context of the transforming landscape. I follow here, to some extent, Guha and Gadgil's (1995) reading of Western environmentalism and the Global South, Mawdsley's (2004) discussions of the Indian middle classes and the environment and especially Baviskar's (2004, 2006) important work on what she has termed "bourgeois environmentalism" in urban India. As she and several others have shown, a complex interaction between global environmental discourses – by state actors, NGOs, transnational activists and multinational corporations, amongst others – have led to the targeting and scapegoating of the urban poor in particular, for a range of environmental problems.

This has led in turn to the violation of the rights of workers through the closing down of industries, the displacement of slum dwellers for creating public nuisances – or in one famous case even the killing of a slum dweller by a middle-class mob for the "crime" of defecating in a "public" park – the displacement of people from protected areas and the banning of forest dwellers and tribal groups from using local resources, all in the name of conservation. Baviskar suggests that what we see emerging in India is "a complex interrelationship between two forms of environmentalism, with the green agenda of the rich leading to greater social and economic marginalization of the poor and their concerns over fair distribution of resources and safe working conditions" (Baviskar 2004: 89). She argues furthermore that "middle class assumptions of the working class as inferior shaped the articulation of the 'public interest' in Delhi at the expense of the interests of the poor" (ibid.). It is important at the same time to recognize, as Mawdsley reminds us, that there is a great deal of "diversity and dynamism within the middle classes in relation to the environment" (Mawdesley 2004: 94). She argues furthermore that we need to avoid the tendency to reduce actors to caricatures of "ecosystem people and omnivores", as some have labeled them, and instead to "develop situated understandings of what constitutes 'the environment' among different middle class groups; and underline the ways in which environmental issues reflect and are often emblematic of wider social and political debates" (ibid.: 88–9).

Returning to the case of Kolkata, then, what we see in several of the emerging environmental debates is a fundamental clash between visions of the city as it is, and the city as it is yet to come, to borrow from Simone's elegant language (2004). On the one hand is the attempt to renovate and recreate the gentleman's city: worldly, cosmopolitan, genteel and, above all, civilized (Roy 2010). On the other is an attempt to recognize and struggle for social as well as ecological justice in everyday Kolkata. I want to look at these struggles by focusing on one particular project that brings together multiple themes and discourses – on global environmentalism, middle-class sensibilities, transnational life-spaces, international development and local politics.

This project, entitled the Kolkata Environmental Improvement Project (or KEIP), is designed to do many things: the laying down of new sewer lines and the restoration of old sewer pipes and infrastructure; the desilting of canals and the relocation of squatters currently living along them to newly built housing projects; and the construction of new eco-tourism and environmental education facilities along the periphery of the city. The KEIP states its own mandate as "a multi-agency endeavour to arrest environmental degradation and improve the quality of life in Kolkata" (see www.keip.in) – a tall order indeed. But how does this project fit into the broader scheme of middle-class/urban-poor conflict that I have laid out previously? Let me begin by describing the project in greater detail. The research on which these arguments rest is culled from fieldwork, including a series of interviews with a variety of stakeholders such as project planners, managers, local activists and politicians in Kolkata, throughout the fall of 2007 and again in the spring of 2012.

The initial project arose out of longstanding efforts by regional planners – particularly at the Kolkata Metropolitan Development Authority (KMDA) – to secure international funding for an infrastructure project. Much of Kolkata's existing sewage and drainage infrastructure has a colonial pedigree, with precious little investment since. In the late 1990s the government of West Bengal, in conjunction with the KMC (Kolkata Municipal Corporation), was able to secure a US$250 million loan from the Asian Development Bank to undertake a massive redevelopment plan, in order, as they put it, to "save the city suffering from pollution, water-logging and malaria". By 2007, a series of cost overruns had led to the need for additional funding; the ADB provided an extra US$80 million, with the West Bengal government putting up US$19.5 million and the KMC adding US$14.10 million. The cost overruns are due in large part to interagency squabbling – particularly between the KMDA and KMC – over project priorities. The overall coordination of the project is undertaken by an outside consulting firm hired by the ADB, led by a Dutch civil engineer and a New Delhi-based team of experts, assisted by local planners and NGOs. This has resulted in tremendous duplication and an extremely inefficient approach (PP1 2007). KMDA and ADB planners complained, for example, about the fact that politicians would often call up the project through KMC intermediaries and insist that a small section of sewer pipe – sometimes as little as two feet in length – be laid so as to satisfy a particular constituency (PP2 2007). The KMDA planners seemed wholly unconvinced about the competence of both the KMC decision-makers (whom they saw as corrupt) and the ADB experts (whom they saw as outsiders). The KMC (PP3 2007), on the other hand, was particularly exercised about the local NGOs hired to carry out social impact assessments, monitoring and resettlement of those displaced by the canal desilting efforts. They worried that such NGOs were in fact fronts for political parties, though my interviews with the ADB staff who hired them seemed to indicate that "professionalism" – including the maintenance of a glossy website – was a higher criterion for hiring them than any local experience or background (PP4 2007).

In the months to come, through my interviews and observations of development sites I learned much more about what constituted environmental priorities in these tangled and often contradictory visions of the new Kolkata. I had been curious to uncover what discourses and practices, beyond rebuilding sewage and drainage systems, sought to shape what the KEIP called "one of the nation's most prestigious development projects" (KEIP 2012). In particular, did efforts to remake the gentlemanly city find any resonance in this well-funded and high-profile project? Do forms of bourgeois urban environmentalism intersect with this project, and if so, how? Is there evidence not only of competing interests and visions but also of internal contradictions and challenges within the various stakeholder groups?

There is indeed, I would contend, evidence of such dynamics within this effort to renew Kolkata's environment, especially apparent if we examine two particular facets of the KEIP:

1 the restoration of parks and water bodies for leisure purposes; and
2 the development of the East Kolkata Wetlands into an eco-tourism and environmental education attraction,

While neither of these project components has anywhere near the dollar-value investment of the larger parts – desilting the canals and laying new infrastructure

– their importance and connection to larger environmental debates within Kolkata remains significant. If we look at them closely, moreover, we see evidence of a clash between very specific understandings of who has a right to inhabit and use these particular environments.

In terms of the proposed restoration of parks and water bodies, for instance, we see a focus on swimming pools and ponds, primarily in middle-class residential neighborhoods.

The plan calls for landscaping, widening of pavements and construction of boundary walls, with railings, retaining walls, gates, electrical work, repainting and the installation of security and surveillance systems (KEIP 2012). Accompanying these efforts would be environmental education focusing on ecological restoration for local schoolchildren and other users: the water bodies in this vision needed to be understood and retained primarily as a park space, for users to "enjoy and experience" nature. Yet within Kolkata urban water bodies perform a host of important functions. One local organization has estimated that at least 3,000 such ponds play a pivotal role in the lives of the urban poor, serving their bathing, cleaning, pisciculture and religious needs (Vasundhara 2006). The pollution of such water bodies is undeniable and a pressing concern. However, simply to reinvent these ponds as sites of leisure and contemplation, effacing their various other uses, does not resolve the issue of where the (primarily urban poor) will turn in order to satisfy their survival needs. We have moreover seen conflicts arise between attempts to "preserve" these natural spaces and attempts to use them for daily needs.

In a case that echoes some of Baviskar's examples in Delhi, one of the organizations I was working with talked of conflicts with the WWF and several local

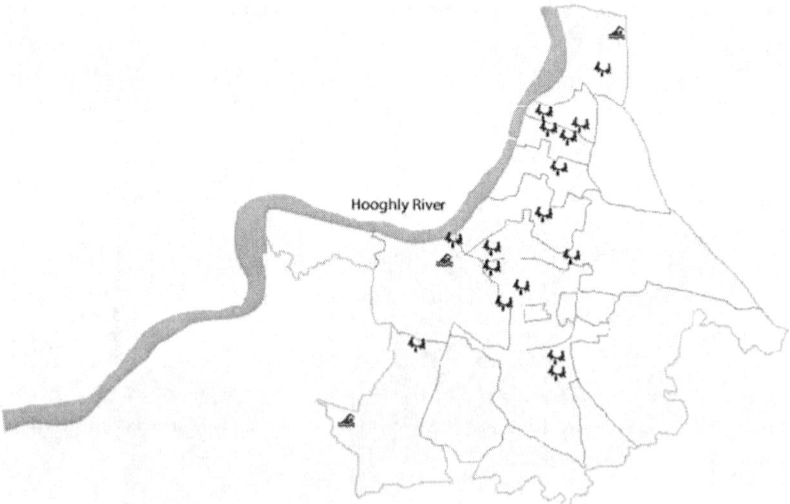

Figure 6.3 Proposed restoration of parks and water bodies, KEIP.

NGOs and community organizations that sought to protect several ponds by banning poorer users from them (EA 2007). They had gone so far as to ask the police to demolish slum dwellings that were within a particular radius of several ponds so as to make it more inconvenient for the urban poor to encroach upon the water bodies. Only the intervention of several social justice groups and a long set of negotiations resolved the issue to some extent. It is in this context that the KEIP's strategy must be read (EA 2007). Though limited at this point to only a handful of water bodies, the strategy is meant to provide a model of how ponds might be protected. But in the end, whose right to the city is being protected and whose is being abrogated? At the same time that the urban poor are being targeted for their misdeeds with regard to water bodies, large-scale developers are left unscathed despite their much greater destruction of ponds (Vasundhara 2005; Basu 2005).

In 2007 a court challenge contested South City's fourth luxury tower for encroaching on an important local pond. While this case was still being heard, the developer of "international-style" apartment complexes, which are part and lot of Kolkata's dream of regaining global status, broke ground on yet another tower, one that most certainly infringed on the same water body (Vasundhara Annual Survey Report 2006). And yet precious little controversy ensued, despite the flouting of environmental regulations. But then, South City and similar complexes play a very different role in the vision of the new city. The urban poor are part of Kolkata's past; South City, where one can "live the way the world does" is very much part of a differently imagined future. Interestingly, this case also emphasizes the note of caution that Mawdsley (2004) sounds for us when considering the contradictory and multifaceted nature of environmental actors: several of the denizens of South City (the apparent embodiment of bourgeois transnational and environmental identities) were later to approach local activists to help rehabilitate the same pond infringed upon by their new home (HA5 2012).

Finally, I want to end by talking about how the project fits more generally into the ongoing transformation of the East Kolkata Wetlands. The Wetlands is, as noted earlier, an important social and ecological space along the edge of the city.

As described previously, in recent years the Wetlands has emerged as a contested zone, where "international-style" developments and various other markers of global lifespaces – shopping malls, private schools, country clubs, golf courses, entertainment complexes – abound. Even though the region as a whole is technically protected by the Ramsar Convention on Wetlands, a treaty signed by the government in 2002, it has seen considerable encroachment, illegal construction and the shrinking of its borders, mainly due to the proposed development of new middle-class housing projects. Yet for many of these new developments, the very nature of the Wetlands is a prime selling point. The advertising copy for many projects emphasizes the ability to live "close to nature", "in more natural surroundings", and "with nature" – the lure of these new dwellings thus allows the inhabitant to simultaneously connect with and live

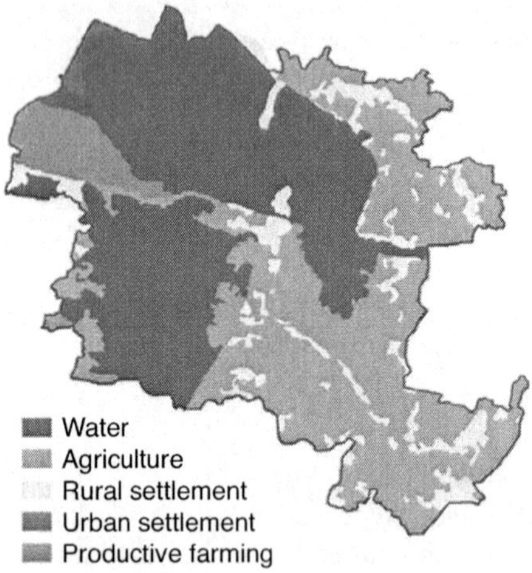

Land use	Area
Water body	5,852.14 ha of which fish farming roughly constitutes about 3,898.70 ha (45.93%)
Agriculture land	Approximately 4,959.86 ha (38.92%)
Garbage disposal site	602.78 ha (4.73%)
Urban settlement area	91.53 ha (0.73%)
Rural area	1,234.99 ha (9.69%)
Total area	12,500.00 ha*

*Additionally 241.30 ha are being added to make the system integral.

Figure 6.4 The diverse demands on the East Kolkata Wetlands, KEIP.

beyond the city (Vedic Village 2007). One Canadian owner whom I interviewed "couldn't believe how quiet it was – I could barely hear the city. I felt a breeze that I've never felt in Kolkata before, I never even knew it existed. I felt as though I wasn't in Kolkata at all" (DM 2009).

Yet the KEIP is not building luxury apartments in the East Kolkata Wetlands. How is this project therefore connected to the transformations taking place there? Ironically, for a project that is concerned with cleaning and saving Kolkata, one of its chief environmental impacts is with regard to the Wetlands, as outlined in its initial proposal and subsequent reports. The KEIP suggests that the "development of the area should not change the current pattern of land use, nor should it result in upsetting the ecological balance" (KEIP 2012). Yet the current plans call for considerable impacts, especially in terms of potential pollution from construction and outflow from the canal desilting efforts. The KEIP's solution is to set up a Nature and Wetland Interpretation Centre on 30 acres in the Wetlands, with an additional 2,000 acres set aside for the demonstration of wise use, primarily in the form of eco-tourism.

The center is designed as an ecological attraction, with exhibition halls, auditoriums, observation hides, office space, a bird sanctuary, boating rentals and walking trails, bicycle pathways and a non-polluting transport system, including cycle rickshaws. Potential further uses include eco-forestry, horticulture, medicinal plant cultivation, the training of local villagers in naturopathy and herbal treatment and eco-friendly, low-cost housing. The KEIP planners acknowledge that:

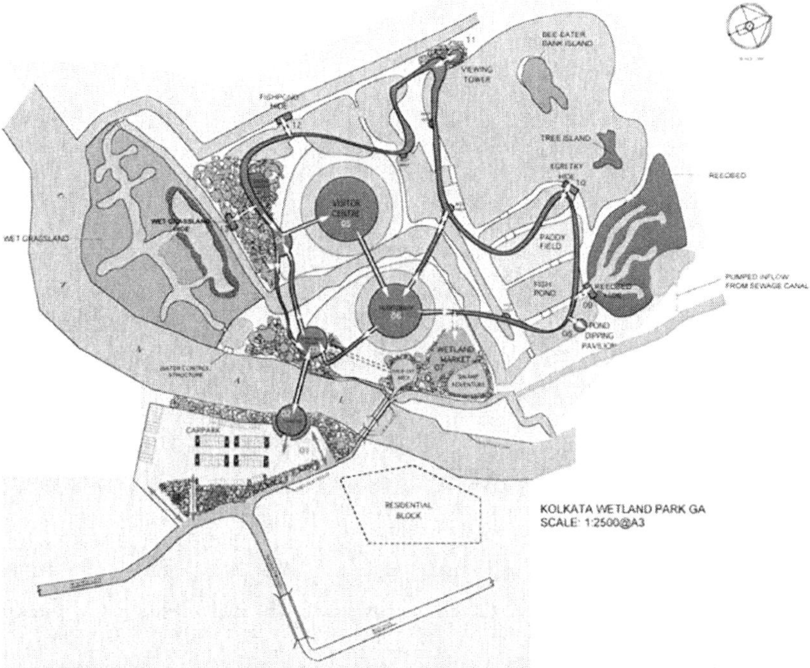

Figure 6.5 Proposed nature interpretation centre, KEIP.

> Such a Centre will provide a relatively new experience for Kolkatans which, on one hand, they may have difficulty understanding as a visitor attraction or, on the other, instantly adopt for its uniqueness and "edutainment" value.
> (KEIP 2012)

So for whom is this initiative designed? For the roughly 60,000–80,000 people that already live and work in the Wetlands? The working poor in the city who rely on the foodstuffs produced in the Wetlands fisheries and farms for part of their sustenance? What will this center achieve for them? What will it teach them about the environment in which they already live? Or will it serve instead as another "edutainment" facility, complementing the other attractions and theme parks designed to turn the Wetlands from a working space and lived ecosystem into a park and playground for those with a "developed" environmental consciousness?

So when we consider the Kolkata Environmental Improvement Project against the backdrop of these larger discourses and local politics, we see that the project as a whole is an important part of remaking Kolkata as a worldly city: the new global city in the twenty-first century must embrace a green sensibility or aesthetic. Sustainability, at least as a form of middle-class consumption, is an essential element of the newly urbane and civilized. The KEIP is in this sense an important part of coalescing and concretizing the dreams of the new urban middle class in Kolkata.

Conclusion

The year 2011 seemed to offer a new era in West Bengal postcolonial politics and life, with the defeat of leftist parties at municipal, state and federal levels. Over three decades of Marxist government collapsed in the face of scandals over the CPI(M)'s adoption of neoliberal developmentalism in the countryside that was once the hallmark of their reforms and success. And yet in the capital city of Kolkata similar displacements and dispossession of marginalized groups has not resulted in parallel political fallout; instead, the project of urban redevelopment continues apace. Intellectuals and activists alike thus embrace a neoliberal cityscape – full of privatized services, the diminishing of public spaces, increasing speculation in land and property and the growth of public-private partnerships – in the heart of one of India's remaining leftist strongholds. Such contradictions, however, are easier to reconcile if one draws on the framework offered by theorists such as Tickell and Peck, who point to both the adaptive nature of neoliberalism and its widespread adoption by a diverse array of ideological movements and institutional arrangements. Through such a lens we see that a neoliberal transformation in Kolkata makes perfect sense to elites and those who aspire to return the city to its former glory, an acceptable price to pay for reconnecting to global circuits of capital, mobility and culture.

This paper has argued that as part of that project of reclaiming some role within the world system of cities and regions, Kolkata has over the past two decades embarked on a sustained campaign to remake itself. Primarily under the guidance of Marxist leaders at the municipal and state levels and with active support and investment from various federal governments – both centrist and right of center (and all of them fully committed to liberalization) – Kolkata has been aggressively courting foreign investment and embarking on a growth strategy that bears considerable resemblance to the process Atkinson and Bridge (2005) have termed "global gentrification". The emergence of new skyscrapers, malls, flyovers, highways, movie theaters, conference complexes, gated communities, schools and restaurants are all testaments to the power of this new vision. So too is an ongoing cycle of displacement and dispossession, as both urban and rural poor are pushed further and further to the periphery (or into various parts of the informal economy) and can access the newly renovated spaces primarily through their labor in either building or servicing them.

The key question that one must ask in viewing this dynamic is, of course, what kind of Kolkata is being built here? Who can live in it, and who is kept out of it? How do we understand the changes that are taking place in terms of the desires and demands of the various residents of this megacity? As the final section of the paper illustrates in greater depth, we see all too often in the emerging struggles over the city and its spaces that the neoliberal vision of the city aligns well with the bourgeois desires of the middle class – as articulated by a clean, healthy and entertaining (rather than necessarily productive) environment, meant to be enjoyed by some but paid for and restricted from others. As the post-Left Front era in Kolkata and West Bengal unfolds, we will soon begin to see whether the neoliberal vision of a reborn Kolkata remains as generally appealing as it seems to have been so far or whether it will clash with another ideal, one perhaps articulated by the many who are left out of the plans of the global City of Joy.

References

All India Review (1993) "Calcutta High Court Ruling for Matter #2851 of 1992 D/-24–9–1992, People United for Better Living in Calcutta (PUBLIC), Umesh Chandra Bannerjee, Presiding Judge", *All India Review*: 215–32.

Atkinson, R. and Bridge, G. (2005) *Gentrification in a Global Context: the New Urban Colonialism*, New York: Routledge.

Axiom Estates (2006) "New Property Developments in Kolkata". Online: www.axiomestates.com/kolkata-property.htm?gclid=CK-pqIG1qIQCFRDOPgodvjb9jw (accessed 16 May 2006).

Banerjee, D. (1997) "The Political Economy of Imbalances across Indian States: Some Observations on 50 Years of Independence", Centre for Studies in Social Sciences, Calcutta Occasional Paper No. 166, November, Calcutta: CSSSC.

Banerjee, T. (2005) "Understanding Planning Cultures: the Kolkata Paradox", in B. Sanyal (ed.) *Comparative Planning Cultures*, New York and London: Routledge, pp. 145–64.

Bardhan, P. and Mookherjee, D. (2005) "Political Economy of Land Reforms in West Bengal, 1978–1998", working paper series, "Effects of Inequality on Economic Performance", Institute of International Studies, University of California, Berkeley. Online: http://globetrotter.berkeley.edu/macarthur/inequality/papers/MookhBengal.pdf (accessed 12 January 2007).

Basu, J. (2005) "Green Panel Wants Work Halt at Tallest Towers", *Telegraph*, 5 June.

Baviskar, A. (2006) "Demolishing Delhi: World-Class City in the Making", *Mute*, 2 (3): 88–95.

Baviskar, A. (2004) "Between Violence and Desire: Space, Power and Identity in the Making of Metropolitan Delhi", *International Social Science Journal*, 5 (175): 89–98.

Bengal Silver Spring (2007) homepage at www.thesilverspring.com (cited 10 September 2007).

Bhattacharjee, B. (2007) "From Agriculture to Industry: an Interview with Chief Minister Buddhadeb Bhattacharjee", in Government of West Bengal, *30 Years of Left Front Government in West Bengal – Issues and Reflections*, Kolkata: Department of Information and Cultural Affairs, Government of West Bengal.

Bhattacharjee, A. (2003) "Seminar on Accelerated Housing and Construction Industry in Bengal", keynote address by Minister of Urban Development, Government of West Bengal, to Federation of Indian Chambers of Commerce, 25 August, Kolkata.

Bhattacharya, D. (2007) "Return of the Native", *Telegraph* Online, 1 April. Online: www.telegraphindia.com/1070401/asp/7days/story_7592256.asp (accessed 2 September 2007).

Bose, P. K. (2000) *Refugees in West Bengal: Institutional Practices and Contested Identities*, Calcutta: Mahanirban Calcutta Research Group.

Bunting, S., Kundu, N., Punch, S. and Little, D. (2001) "East Kolkata Wetlands and Livelihoods: Workshop Proceedings", Land-Water Interface Production Systems in Peri-Urban Kolkata DFID-NRSP Project working paper, Stirling, UK: Institute of Aquaculture.

Bunting, S. and Mukherjee, M. (2002) "Peri-Urban Farming and Poor Livelihoods: Developing Plans for Action", Land-Water Interface Production Systems in Peri-Urban Kolkata DFID-NRSP Project working paper, Stirling, UK: Institute of Aquaculture.

Byres, T. (ed.) (2002) *The Indian Economy: Major Debates since Independence*, 4th edn, New Delhi: Oxford University Press.

Chadha, V. (2005) *Low Intensity Conflicts in India: an Analysis*, Thousand Oaks: Sage.

Chakrabarti, A. (2006) "Left Front: 235", *Telegraph* Online, Friday 12 May. Online: www.telegraphindia.com/1060512/asp/frontpagestory_6214770.asp (accessed 15 May 2006).

Chakrabarty, P. (1990) *The Marginal Men: the Refugees and the Left Political Syndrome in West Bengal*, Calcutta: Lumiere.

Chakravorty, S. (2005) "From Colonial City to Globalizing City? The Far-From-Complete Spatial Transformation of Calcutta", in N. Fyfe and J. Kenny (eds) *The Urban Geography Reader*, London and New York: Routledge, pp. 84–92.

Chatterjea, D. (1990) "Bidhan Nagar: from Marshland to Modern City", in S. Chaudhuri (ed.) *Calcutta the Living City: Volume II – the Present and the Future*, Delhi: Oxford University Press, pp. 176–80.

Chattopadhyay, S. (2006) *Representing Calcutta: Modernity, Nationalism and the Colonial Uncanny*, London and New York: Routledge.

Chattopadhyay, S. (2005a) "A Left Front Sweep", *Frontline*, 22 (12), 4–17 June. Online: www.hinduonnet.com/fline/fl2212/stories/20050617004002100.htm (accessed 14 July 2006].

Chattopadhyay, S. (2005b) "Capturing Kolkata", *Frontline*, 22 (4), 2–15 July. Online: www.hinduonnet.com/fline/fl2214/stories/20050715003003500.htm (accessed 14 July 2006).

Chief Electoral Officer (WB) (2011) Winning Candidate List, Assembly Elections. Online: http://ceowestbengal.nic.in/ (accessed 3 August 2011).

Ciputra Group of Indonesia (2006) "Kolkata West International City". Online: http://rc5a.ciputra.com/projects.php?p=kolkata accessed 3 July 2006).

CMDA (Calcutta Metropolitan Development Authority) (1981) *New Area Development at Extension of Baishnabghata-Patuli Township, Report 156, December 1981*, Calcutta: CMDA Directorate of Planning and Development.

CMDA (Calcutta Metropolitan Development Authority) (1978) *Report on Baishnabghata-Patuli Area Development Project (Revised), Report 58, April 1978*, Calcutta: CMDA Directorate of Planning and Development.

CMDA (Calcutta Metropolitan Development Authority) (1977) *East Calcutta Area Development (Summary), Report 46, April 1977*, Calcutta: CMDA Directorate of Planning and Development.

CMDA (Calcutta Metropolitan Development Authority) (1976) *Baishnabghata-Patuli Area Development Plan (Preliminary), Report 19, January 1976*, Calcutta: CMDA Directorate of Planning and Development.

CMDA (Calcutta Metropolitan Development Authority) (1975) *Salt Lake Office Complex (Preliminary), Report 7, March 1975*, Calcutta: CMDA Directorate of Planning and Development.

CMPO (Calcutta Metropolitan Planning Organization) (1966) *Basic Development Plan – Calcutta Metropolitan District: 1966–1986*, Calcutta: CMPO.

CPI(M) (Communist Party of India (Marxist)) (2007) "Response of the Left Parties to the Note on WTO Negotiations". Online: www.cpim.org (accessed 9 September 2007).

CPI(M) (Communist Party of India (Marxist)) (2000) "Globalization: Impact on Indian Economy". Online: www.cpim.org (accessed 9 September 2007).

Dam, M. (2005) "Left Front Sweeps Kolkata Civic Elections", *The Hindu* (online edition), 22 June. Online: www.hindu.com/2005/06/22/stories/2005062206200100.htm (accessed 4 September 2007).

Das, S. (1990) "The Politics of Agitation: Calcutta 1912–1947", in S. Chaudhuri (ed.) *Calcutta the Living City: Volume II – the Present and the Future*, Delhi: Oxford University Press, pp. 15–26.

Dasgupta, K. (2003) "Evictions in Calcutta: Creating the Spaces of 'modernity'", *City 4*, September: 31–43.

Dey, I. (2011) "New Town and Labour in Transit", *Labour in Transit*. Online: www.transitlabour.asia/blogs/newtown (accessed 9 August 2011).

DM (2009) interview with diaspora member, 7 July.

EA (2007) interview with environmental activist, 21 October.

Fort Group (2011) "NCCF Sunny Fort, New Town Kolkata", Monday 8 August. Online: www.fortgroup.com/sunny_fort.html (accessed 8 August 2011).

Ganguly, D. (2005) "City Planners in Save-Wetlands Cry", *Telegraph India* (online edition), Friday 11 March. Online: www.telegraphindia.com/1050311/asp/calcutta/story_4472351.asp (accessed 29 July 2006).

Ghosh, D. (1998) "Calcutta Wetlands: Turning Bad Water into Good", Ashoka's Changemakers. Online: www.changemakers.net/journal/98october/ghosh.cfm (accessed 15 August 2006).

GoWB (Government of West Bengal, Department of Housing) (2007a) Department details. Online: www.wbgov.com/e-gov/English/Departments/DeptDetails.asp?DPId=110&LinkId=1&Type=1 (accessed 10 September 2007).

GoWB (Government of West Bengal, Department of Urban Development) (2007b) "Major Initiatives". Online: www.banglarmukh.com/e-gov/English/Departments/DepartmentFrameNew.asp?DpId=255 (accessed 10 September 2007).

Guha, R. and Gadgil, M. (1995) *Ecology and Equity: the Use and Abuse of Nature in Contemporary India*, New York: Routledge.

HA1 (2007) interview with housing activist, 24 October.

HA2 (2007) interview with housing activist, 24 October.

HA3 (2007) interview with housing activist, 24 October.

HA4 (2007) interview with housing activist, 24 October.

HA5 (2012) interview with housing activist, 9 March.

HDFC Realty (2006) "Trends in Important Cities, Kolkata". Online: www.hdfcrealty.com/general/nri-trendincities.asp#5 (accessed 3 July 2006).

Hutnyk, J. (1996) *The Rumour of Calcutta: Tourism, Charity and the Poverty of Representation*, London and New York: Zed Books.

KMC (Kolkata Municipal Corporation) (2007) "Vision". Online: www.kolkatamycity.com/about_kmc_vision.asp (accessed 4 September 2007).

KMDA (Kolkata Metropolitan Development Authority) (2006) "KMDA New Area Development Projects". Online: www.cmdaonline.com/new_area.html (accessed 8 September 2007).

KMDA (Kolkata Metropolitan Development Authority) (2000) *Calcutta Development: Programmes and Projects*, Kolkata: KMDA Publications.

Kohli, A. (1987) *The State and Poverty in India: the Politics of Reform*, Cambridge: Cambridge University Press.

Majumdar, S. (2010) "The Nano Controversy: Peasant Identities, the Land Question and Neoliberal Industrialization in Marxist West Bengal, India", *Journal of Emerging Knowledge on Emerging Markets*, 2 (5): 40–66.

Mawdsley, E. (2004) "India's Middle Classes and the Environment", *Development and Change*, 35 (1): 79–103.

Mishra, S. (2007) "Agrarian Reforms, Panchayats and the Way Ahead", in Government of West Bengal, *30 Years of Left Front Government in West Bengal – Issues and Reflections*, Kolkata: Department of Information and Cultural Affairs, Government of West Bengal.

Mitra, A. (2007) *A Prattler's Tale: Bengal, Marxism and Governance*, transl. S. Bhattacharya, Kolkata: Samya.

Mukherjee, S. (2004) "Fast Forward on Flyovers", *The Telegraph* (online edition), Friday 27 August. Online: www.telegraphindia.com/1040827/asp/calcutta/story_3678779.asp (accessed 15 May 2006).

Nayyar, D. (1994) *Migration, Remittances, and Capital Flows: the Indian Experience*, New Delhi and New York: Oxford University Press.

Ong, A. (2006) *Neoliberalism as Exception: Mutations in Citizenship and Sovereignty*, Durham and London: Duke University Press.

Pal, A. (2005) "Metropolitan Decision-Making – a Case Study of Kolkata", Life in the Urban Landscape: International Conference for Integrating Urban Knowledge and Practice, Gothenburg, Sweden, 30 May–4 June.

PP1 (2007) interview with project planner, 12 November.

PP2 (2007) interview with project planner, 14 November.

PP3 (2007) interview with project planner, 17 November.

PP4 (2007) interview with project planner, 24 November.

Racine, J. (ed.) (1990) *Calcutta 1981: the City, its Crisis and the Debate on Urban Planning and Development*, New Delhi: Concept Publishing.

Ramsar Convention on Wetlands (2002) "India Designates Eleven New Wetlands of International Importance". Online: www.ramsar.org/wn/w.n.india_11new.htm (accessed 14 June 2005).

Ray, R. (1988) *The Naxalites and their Ideology*, Oxford: Oxford University Press.

Roy, A. (2010) "Re-Forming the Megacity: Calcutta and the Rural-Urban Interface", in A. Sorensen and J. Okata (eds) *Megacities: Urban Form, Governance and Sustainability*, New York: Springer, pp. 93–110.

Roy, A. (2004) "The Gentleman's City: Urban Informality in the Calcutta of New Communism", in A. Roy and N. AlSayyad (eds) *Urban Informality: Transnational Perspectives from the Middle East, Latin America and South Asia*, Lanham: Lexington Books, pp. 147–70.

Roy, A. (2003) *City Requiem Calcutta: Gender and the Politics of Poverty*, Minneapolis: University of Minnesota Press.

Roy, P. (2007) "There is No Alternative to Industrialisation Pledge in West Bengal", in Government of West Bengal, *30 Years of Left Front Government in West Bengal – Issues and Reflections*, Kolkata: Department of Information and Cultural Affairs, Government of West Bengal.

Roy, S. (2004) "Even the 'City of Joy' is Shining", Rediff.com. Online: http://in.rediff.com/money/2004/apr/21guest.htm (accessed 19 July 2006).

Saha, S. (2006) "A Slice of West: an Exclusive Residential Complex on the Fringes of

Calcutta Promises NRIs the Standard of Living they are Used to Abroad", *The Telegraph* (online edition), 28 April 2006. Online: www.telegraphindia.com/1060428/asp/propertt/story_6156981.asp (accessed 22 June 2006).

Samaddar, R. (2011) "Rajarhat, the Urban Dystopia", *Transit Labour*. Online: www.transitlabour.asia/blogs/dystopia (accessed 9 August 2011).

Sanyal, R. (1993) "Public Societies and Civic Institutions in Renaissance Calcutta", in H. Chakrabarti (ed.) *An Urban Historical Perspective for the Calcutta Tercentenary, Papers Presented at a Seminar at Victoria Memorial Calcutta, 15–16 December 1988*, Calcutta: General Printers and Publishers, pp. 71–6.

Sarkar, A. (1990) "The East Calcutta Wetlands", in S. Chaudhuri (ed.) *Calcutta the Living City: Volume II – the Present and the Future*, Delhi: Oxford University Press, pp. 173–5.

Sen, K. (2005) "Financial Policies and Investment in Post-Reform India: Macro and Micro Perspectives", Workshop on Indian Economy: Policy and Performance 1980–2000, Center for India and South Asia Research, University of British Columbia, 23–24 June.

Sen, S. (2011) "Bizarre Urbanism", *Transit Labour*. Online: www.transitlabour.asia/blogs/bizarre (accessed 9 August 2011).

Simone, A. (2004) *For the City Yet to Come: Changing African Life in Four Cities*, Durham: Duke University Press.

Sinha, P. (1991) "The Genesis of a Colonial City", in B. Dasgupta, M. Bhattacharya, D. Basu, M. Chatterjee and T. Bannerjee (eds) *Calcutta's Urban Future: Agonies from the Past and Prospects for the Future*, Calcutta: Sree Saraswaty Press, pp. 1–7.

South City Projects (2007) homepage, www.southcityprojects.com (accessed 10 September 2007).

The Telegraph (2007) "End of the Road for Rickshaws, *The Telegraph* (online edition) 2 August. Online: www.telegraphindia.com/1070802/asp/calcutta/story_8136738.asp (accessed 10 September 2007).

The Telegraph (2006) "Green Body Admits Gaffe", 6 June.

Tickell, A. and Peck, J. (2003) "Making Global Rules: Globalization or Neoliberalization?", in J. Peck and H. Yeung (eds) *Remaking the Global Economy: Economic-Geographical Perspectives*, London: Sage, pp. 163–81.

Times of India (2010) "Kolkata Sold Off, Piece by Piece", *Times of India*, 20 April. Online: http://articles.timesofindia.indiatimes.com/2010–04–20/kolkata/28135085_1_hawker-sangram-committee-new-hawkers-shaktiman-ghoshTimes of India (last accessed 3 October 2012).

Times of India (2007) "KMC Plans Mega Eviction Drive", *Times of India*, 30 November. Online: http://articles.timesofindia.indiatimes.com/2007–11–30/kolkata/27971302_1_hawker-eviction-drive-hatibagan-hawkers-sangram-committee (last accessed 3 October 2012).

Vasundhara (2006) *Annual Survey of Environment*, Kolkata: Vasundhara.

Vasundhara (2005) *Annual Survey of Environment*, Kolkata: Vasundhara.

Vedic Village (2007) "Homes with an International Resort". Online: www.thevedicvillage.com/introduction.html (accessed 10 September 2007).

WBIDC (West Bengal Industrial Development Corporation) (2006) "Gateway to the East". Online: www.wbidc.com/old/overview/gateway.htm (accessed 17 May 2006).

7 Public space versus tableau

The right-to-the-city paradox in neoliberal Bogotá, Colombia

Rachel Berney

Introduction

For the past twenty-five years, the city of Bogotá, Colombia, has labored to overcome its dystopian image as the world drug capital. Prior to this effort, the city environment was so hostile that residents were used to negotiating life only in their own self-interest. Throughout the 1990s and 2000s the city combatted this individualism by re-imagining public space as an educational ground for citizen interaction and learning. Civic administrations sought to integrate dispersed territories within the city, humanize public space and encourage socioeconomic integration. The mayor's office instituted public space projects to forge links between the city's prosperous north and the poorer south. The broad goal, according to officials, was to supersede the income disparities and hostilities among residents by forging a sense of shared citizenship and, as a result, generate attitudes of civility and common purpose.

Establishing orderly spaces was paramount to the success of these strategies to reinvent public space and citizenship. In support of this effort, citizens' right to public space was codified into the 1991 Colombian Constitution, and urban interventions in Bogotá's public spaces focused on improving equity of access. These policies assumed that by providing access to a key communal resource, public space, and to free programs there, new Bogotano citizens could be constructed from mutual social interaction and from pride of place. Paradoxically, while social programs and physical changes to public spaces themselves were effective in opening the city to all citizens, a strong tension developed between unfettered access and the imperative of orderly public space.

While officials focused diligently on expanding access to the city, they also created programs to scrutinize and direct behavior in public spaces (Berney 2011). The right to public space existed, but only if exercised with what officials deemed proper behavior, in line with a global neoliberal agenda that favors investment-worthy, stable cities. City leaders offered a sense of belonging and a right to the city, but under particular conditions. This paradox carried through to how citizenship was formed and inured by limits to behavior in the city's public space. While these programs of the 1990s and 2000s hoped that participation in public space would remake a person as a citizen, the actual reality was that, given how behavior was monitored and constrained by efforts to keep public space orderly, only certain people were allowed this promised citizenship.

This chapter explores how Bogotá's goal to produce citizens through access to public space played out against the need for orderly conduct within the neoliberal agenda. This is most visible in the activities of the two *public space mayors*: Antanas Mockus (1995–7 and 2001–3) and Enrique Peñalosa (1998–2000). While these two mayors reshaped governance in Bogotá and created expansive public space improvements, the new public citizens' behavior was policed and monitored, ultimately limiting the right to the city upon which the public space improvements were founded. This turned public spaces into tableaux – striking scenes of public order. Part and parcel of the state-building project, where public space was likened to the right to the city, were troubling trends in which neoliberal capital redirected and reshaped public space in undemocratic ways. This created conditions where some citizens' right to public space was worth less than others. In the following section I will expand on the connections between public space and the right to the city in the Bogotá case and how the (re)production of citizens was shaped by those ideals and by the need for foreign investment.

Public space, citizenship and the right to the city

Jürgen Habermas (1974) defined the public sphere as "a realm of our social life in which something approaching public opinion can be formed. Access is guaranteed to all citizens. A portion of the public sphere comes into being in every conversation in which private individuals assemble to form a public body" (Habermas 1974: 49). The potential for the "public body" of which Habermas speaks to occur in Bogotá depended on the quality, availability and types of public spaces, as well as how and whether they allowed people to come together.

Public space can be defined essentially as a space of unmitigated encounter where strangers may freely meet: its essential quality is *publicness*. The strength and possibility of reinvention via public space in Bogotá depended greatly on the concept of public space as the fundamental *punto de encuentro* (point of encounter) necessary for democracy (Borja and Muxí 2003: 25; Velásquez Carrillo 2004: 1; Carrión 2004: 60–1). Ideally, this point of encounter for Bogotá would be a "classless space". Castro Jaramillo writes, "public space has no *estratos* (socioeconomic divisions) and any investment that is made in it is for the benefit of the entire city"[1] (2003: 86). In this way, the public space mayors saw the creation of public space as a means of mitigating Bogotá's lawlessness and poverty, which played a large part in the city's global image.

Bogotá's public spaces became a crucible for the hopes and dreams of local leaders and were employed as a comprehensive fix for the city's problems (Berney 2011). Public space was seen as a territorial integrator and socioeconomic defragmenter, meant to empower and connect people. Local leaders viewed public space as an effective policy arena for delivering communal resources. Additionally, public space projects were viewed as easier, cheaper and more visible than other types of infrastructure and social projects (Salazar Ferro 2003: 72). As such, they were ideally suited to reflect the competency of the period's politically independent mayors, Mockus and Peñalosa (Berney 2008). These mayors viewed public space as the most effective platform for reaching and transforming Bogotá's citizens, while increasing equity regarding public resources in the city (ibid.). In this sense public space became a planning ideal, a symbol and a solution (Berney 2011). Making this space truly *public* also meant that all citizens would have a stake in maintaining, in the broadest sense, the newly constituted public space.

The creation of public space also represented a shift to understanding citizens' right to the city. The basic right of Bogotanos to use public spaces within the city is well protected by the 1991 Colombian Constitution. The constitution lays out a precise definition of public space and establishes the right of access to public space for all citizens (Martin and Ceballos 2004: 267). In Colombia, therefore, access to public space became a right to which each citizen was guaranteed. This constitutional change at the national level became integrated into a crucial period of policy and legal reform around public space in Bogotá during the public space mayors' time in office.

Latin American cities have engaged in a process of re-envisioning public space both as symbolic, symbiotic space and as an instrument for exchange and learning (Velásquez Carrillo 2004: 20). For example, in 2004, at the end of the public space mayors' time in office, the director of Bogotá's Instituto Distrital de Cultura y Turismo (IDCT, District Institute of Culture and Tourism), wrote:

> Parks are privileged sites for the creation and strengthening of citizenship and for the development and expression of citizen culture. The essence of citizen culture is based on tacit agreements, including self-regulation and comprehension of formal norms of co-existence.
>
> (Londoño Botero 2003: 11)

Another city institute, the Insituto Distrital de Recreación y Deporte (IDRD, District Institute of Recreation and Sports) exemplified this relationship by stating that "parks help determine how Bogotanos are citizens today" (Vargas 2003: 14). Citizenship and the right to the city were inextricably linked during the public space mayors' tenure: being a citizen meant a right to public space, while acting within public space taught Bogotanos how to be citizens.

The Bogotá case sheds light on key rights within the right to the city, in this case namely first- and second-generation rights as defined by Waldron, and individual or minority rights in relationship to the majority, as defined by Dworkin (Attoh 2011: 671–2). Waldron's work on generational rights specifies that first-generation rights "refer to the traditional liberties and privileges of citizenship", while second-generation rights "refer primarily to socio-economic entitlements" (ibid.). Through Bogotá's expansion of the right to the city through the right to public space, I contend the city intended to grant the right to the city in a literal and symbolic way as a right of citizenship via access to and mobility within the city's public spaces. Furthermore, the public space mayors sought to address socioeconomic rights, ones that Waldron writes "are indispensable in efforts to combat economic inequality and material poverty" (ibid.) through public space, a strategic, visible and cost-effective means of transforming the city.

This was a particularly successful project: when asked in a 2006 survey in Bogotá's public spaces who is responsible for those spaces, over one-third of respondents indicated that they were, and over one-half believed that they were the collective owners of public space. Furthermore, 82 percent of those surveyed believed that public space ought to belong to all Bogotanos (Berney 2006). The link between citizenship and public space recast public space as the ideal site for educating and creating citizens.

One way in which this citizenry was created was through a new mode of development that I call *pedagogical urbanism*, a cycle of planning interventions that conveyed and shaped notions of a healthy society through public space access and programming (Berney 2011). Pedagogical urbanism privileges education and reform; it is the merging of social and spatial planning to produce new norms. In the case of Bogotá's commitment to public space, pedagogical urbanism took shape in teaching the city's populace how to be citizens through public programs and monitored behavior. Pedagogical urbanism responded to some of the most serious challenges faced by Latin American cities. In contrast to the downward spirals of fear, misuse and lack of public space use that other Latin American cities saw during this period, the changes worked in Bogotá demonstrated a strong interest and investment in the city's public space and an enormous change from the recent past.

Bogotá's mayors, however, wanted to do more than educate citizens and create public space; their visions for improving their city were far grander. The city's planners and officials tied the goals of socio-spatial integration and renewed citizenship to their success at attracting foreign investment. The transformation of public space, in a very real sense, was meant to serve the larger goal of attracting international capital. In order to compete in the global market, however, Bogotá's public space needed to portray citizens who fit in with the neoliberal ideal. Thus, public space became less of a truly classless meeting point and more of a tableau, a "spectacle of the everyday".

In contrast to grand spectacles in the landscape, Bogotá provides an example of a new type of spectacle that highlights the importance of the everyday and the communal. This everyday tableau functions "like a theater in which a pacified public basks in the grandeur of a carefully orchestrated corporate spectacle" (Crilley 1993: 153). The "spectacle of the everyday" can be sold as part of the narrative of Bogotá as a safe, desirable place to do business, to live and to visit. The irony here lies in the fact that Bogotá's well-loved and used spaces may inevitably and increasingly become commodities to help sell the city to tourists and investors (Berney 2011). Or, if they fail due to lack of vision of their importance and/or maintenance, they will reveal the city as unsuccessful. I now turn to the background situation of Bogotá that set up the need for this public space project, before moving on to the specific ways in which the project was enacted.

Bogotá's reinvention

Dystopian Bogotá

Colombia's history illustrates the structural forces that contributed to dystopian conditions in its capital city of Bogotá. These forces include a power imbalance between the governors and the governed that was inherent to the political system – a system that was traditionally paternalistic, white, monied and classist; a history of collusion between the two dominant political parties that resulted in the repression of other sources of political power; an abdication of responsibility to the governed coupled with a lack of progressive political and social reforms; and a history of violent, bi-partisan civil wars that led to repeated breakdowns in the social order (Livingstone 2004; Sánchez 2001; Bergquist et al. 2001).

These breakdowns were mirrored by dire economic conditions related to rapid population growth. As Colombia's capital, Bogotá was and remains a major destination for internal migration and a powerful draw for those seeking work and trying to escape violence in the countryside. Of Colombia's population, 70 percent lives in urban centers (Velásquez Carrillo 2004: 7), resulting in a concentration of poverty and exacerbating other negative urban conditions. During the time period studied, for every ten Bogotanos employed, there were four who were underemployed and two who were unemployed (DAPD 2006: 55). These conditions led to "poverty, social-spatial exclusion, violence, and lack of participation in politics" (Velásquez Carrillo 2004: 9), exacerbated by negative trends facing other Colombian cities, including exclusion and fragmentation, increased perceptions of insecurity, a fear of public spaces, lack of access to healthcare, education and open space and un/underemployment (Velásquez Carrillo 2004: 9; Carrión 2004: 69–73).

While Bogotá experienced unprecedented population growth and subsequent urbanization, there was very little city control over the urban planning process and significant interference from local land developers and politicians (Livingstone 2004: 40).[2] There was an explosion of car ownership by a small minority of the population who used public space to park. This consumed open space, leaving no room for pedestrians (Violich 1987) and destroying sidewalks (Braun 1994). Public space became increasingly fragmented, exacerbated by illegal closures of public space (privatizations) enacted by individuals on publicly owned land (interview with Defensoría del Espacio Público [DADEP], 2006).[3] Throughout this period, the mayor enjoyed limited power, and the constant changes in administrations impeded planning or any coherent vision for development (Gilbert and Dávila 2002: 37).

Additionally, Bogotá maintained a difficult relationship with the National Congress (Gilbert and Dávila 2002: 41–2). The elites in Congress were from different regions of the country and wished to direct the country's funds to their own regions, rather than to the Capital District. The lack of symbolic ownership over the capital led to the lack of a shared sense of responsibility, and there was reluctance on the part of the city to organize itself. Tax collection became difficult because "the city was reluctant to raise its own taxes and the tax system was very disorganized and corrupt" (ibid.: 45). The entire system was outdated in terms of record-keeping and methods of collection, and the overgrown bureaucracy ate up what little money there was. As a result, by the mid-1980s the city had deteriorated due to lack of effective management: it was dangerous and in disarray. By 1992, Bogotá was bankrupt and owed US$2.6 billion to lenders. This was due in large part to the lack of leadership described above, as well as to widespread corruption and the refusal of the national government to fund Bogotá's administrative and upkeep needs adequately (ibid.: 41–5).

These dystopian conditions registered viscerally for Bogotanos in the city's public spaces. There was no management of public resources or enforcement of rules and norms, and whoever wanted to assert themselves over others could do so. Daily life in Bogotá became a struggle, characterized by a personal pursuit for survival and well being without regard for public welfare (anonymous interviews 2006). Braun describes the state of personal interactions between Bogotanos at this time as "the peculiar combination of easygoing camaraderie and cold aggression" that left one not knowing whom or what to trust (Braun 1994: 22).

Bogotá reinvented

Despite these stubborn challenges, the international press began touting Bogotá as an eminently livable city by the beginning of the twenty-first century. Pedagogical urbanism emerged from and, in some respects, triumphed over, the seemly intractable problems of previous decades. This success was due in large part to the administrations of Antanas Mockus (1995–7 and 2001–3) and Enrique Peñalosa (1998–2000), who, through social programs and changes to public space, took steps to improve safety, create a sense of community and improve equity regarding the use of collective public space resources, reworking the physical fabric of the city in the process. These programs and projects represented a shift from a privately focused urbanism to one that was public. Due to these strategies, in 2006 the prestigious Venice Biennale awarded Bogotá the Golden Lion award for cities. The Biennale is a longstanding international art and culture exhibition held in Venice, Italy, which aims to showcase the best thinking in several areas of the arts, including architecture. The 2006 Biennale was organized around the theme of "Cities, Architecture and Society". The Biennale's organizers said this about Bogotá:

> This city has in the last decades addressed the problems of social inclusion, education, housing and public space especially through innovation in transport. Bogotá has applied Mies van der Rohe's dictum "less is more" to the automobile: less cars means more civic space and civic resources for people. The city provides a model for streets which are pleasing to the eye as well as economically viable and socially inclusive. Bogotá is, in short, a beacon of hope for other cities, whether rich or poor.
>
> (Biennale di Venezia 2007: 1)

Bogotá's visibility continued to increase on an international scale during the following year. The American Planning Association's 2007 World Planning Keynote presentation was the "Miracle of Bogotá", presented by Carolina Barco Isakson, Colombia's Ambassador to the United States from 2006 to 2010 and former National Director of Planning for Colombia. Her presentation "described how that Colombian city was lifted from violence and dysfunction to the world capital status it enjoys today" (MacDonald 2007: 1). Barco further described the "Bogotá Project" as one that "restructured administration and finance, and engaged citizens to help the city rise from its low point in the 1980s to the titles of 'America's Cultural Capital' and the 'World Book Capital of 2007' ".

Today, Bogotá has grown increasingly popular as a place to visit for "takeaway" lessons. Urban planners, designers and city officials continue to descend on the city to learn from its public space and transportation changes with the intention of going home and creating a similar project there. Bogotá transformed itself, becoming an urban exemplar through its public space initiatives. This was done in an effort to recreate the mayoralty and adhere to a neoliberal vision, which, while creating public space, also limited what could be performed within it.

Recreating the mayoralty

While some local governments retain strong linkages with the national government, Bogotá and other Colombian municipalities enjoy a significant amount of autonomy (Enrique Peñalosa, personal interview 2008). Furthermore, a large majority of Bogotá's residents are consolidated within one administrative boundary, a condition that helps streamline governance (ibid.; Martin and Ceballos 2004: 256). Over the last few decades, a twofold focus on decentralization has occurred, which has strengthened Bogotá's local government. One focus was a more general Latin American trend towards liberalization and decentralization, and the other was specific to Colombia; leaders hoped that decentralizing power to local governments would reduce violence (Gilbert and Dávila 2002: 33).

These trends created a window of opportunity for the public space mayors. Previously Bogotá lacked both leadership and opportunities to lead, creating a backlog of intellectual elites fully capable of leading the city but with no access to local leadership roles. As the situation changed, Peñalosa and Mockus saw opportunities to enter the political sphere. Both ran independently, separate from party platforms, and both decided to focus on visible urban planning projects. Enrique Peñalosa (mayor from 1998 to 2000) commented, "I had studied urban issues for decades before I became mayor. It was my passion … I had written a lot about urban problems and solutions, and was obsessed with the historical importance of what was to be done" (Enrique Peñalosa, personal interview 2008). Mockus, who ran administrations from 1995 to 1997 and from 2001 to 2003, campaigned in an untraditional manner. Lucía Peña wrote of Antanas Mockus,

> Neither the candidate nor the campaign was viewed as an end in itself. Rather, the candidate presented himself as an instrument and the campaign

> as a process that should be placed at the service of a project for the city, committed to the defense of the collectivity and the development of a culture of citizenship.
>
> (Lucía Peña 1995: 30)

These running platforms strongly influenced the public space campaigns that both mayors undertook during their tenures.

Despite the profound impact of the specific activities of the two mayors, there were other factors at work as well. Taking the argument of Bogotá's uniqueness one step further, I argue that the city's singularity can be found more in the political economy of the city rather than in the statutory authority of the mayor. A key dynamic in Bogotá during the years discussed here is that Mayors Mockus and Peñalosa won elections on independent platforms. This contrasts with other major Latin American cities, where, at the time, initial elections (following the decentralization of power to the local level) went to the opposition parties. While the leaders in other countries were paying attention to their capitals so as to govern the country, in Bogotá, leaders were paying attention to the capital because no one could govern the country. There was also less ability, in Bogotá, to lead the country because the city is only one of four major metropolitan centers in Colombia. So, rather than leading in an outward direction, the city's leadership used inwardly focused strategies to (re)invent the city and, in doing so, had the freedom to focus on governing and creating change. Ironically, this tapped into a deep Colombian tradition of expert-led government.

From failed government to neoliberal vision

In Colombia generally, there has been a "failure of local government, even with reforms and greater autonomy to successfully meet city challenges" (Velásquez Carrillo 2004: 11). The challenge for the Colombian city was and remains to create "a collective space for enjoying a dignified life" and leave behind the exclusion and fragmentation of the "disarticulated city" (ibid.: 9). This created a challenge for Bogotá to rise to – a challenge that the city appeared to have answered by the end of the public space mayoral administrations. In 2006 city planning documents, Bogotá claimed to be the only Colombian city poised to meet the Millennium Goals benchmarks laid out by the United Nations Development Program for 2015 (DAPD 2006: 54).

Despite its well-regarded standing, Bogotá was not exempt from the complex of problems that other Latin American cities faced. Latin American cities are caught between two fundamental concerns that push investment in opposite directions. In one direction lies poverty, the amelioration of which requires resources, and in the other direction concerns about investing in productivity dominate. Bogotá's mayoral administrations sought to balance both concerns. According to the Plan Regional de Competitividad Bogotá – Cundinamarca 2004–14 (Regional Plan for Bogotá's Competitiveness), the city's vision for regional competitiveness by 2015 is to become:

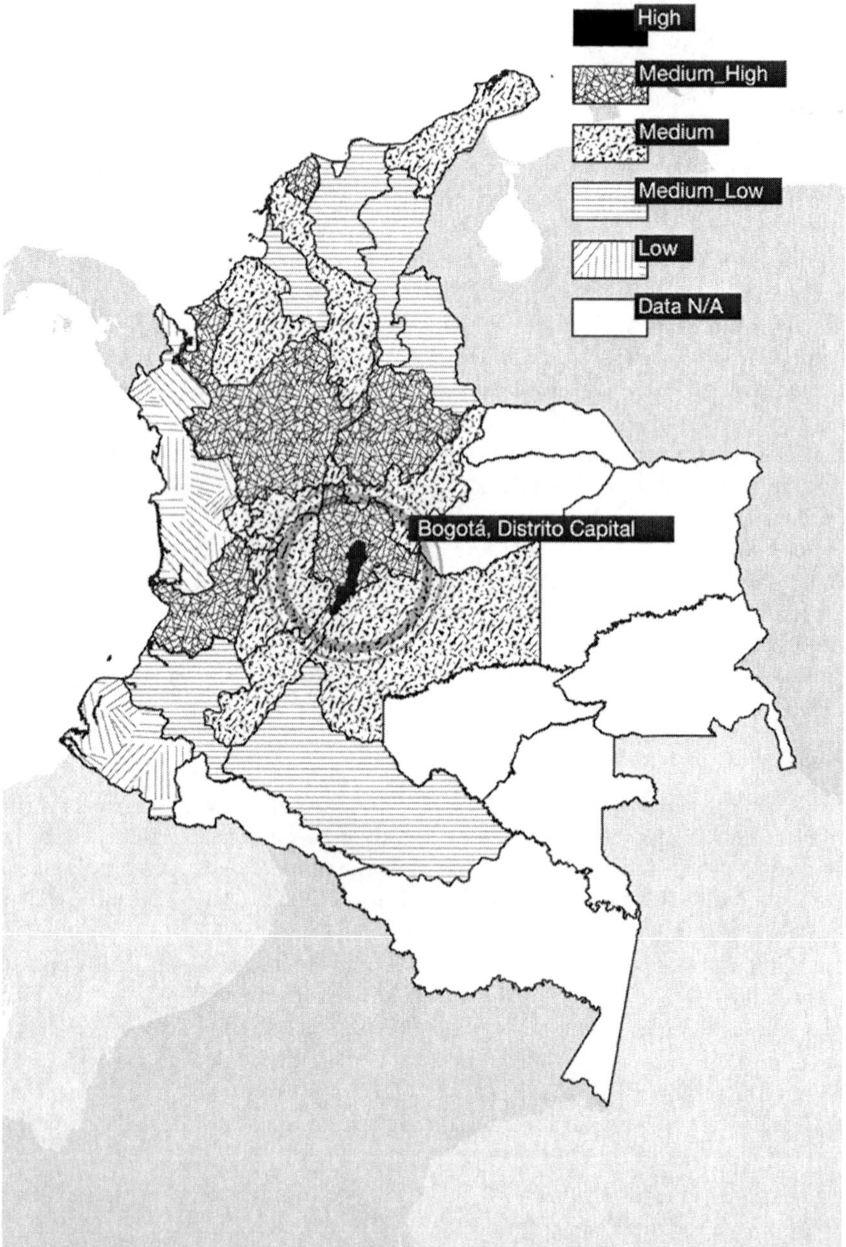

Figure 7.1 Bogotá depicted as the only municipality in Colombia in the high (*alto*) category for achieving the UNDP millennium goals.

> [T]he most institutionally, territorially, and economically integrated region of Colombia, with a productive and diversified base that emphasizes specialized services and agro-industry, articulated in the world market, in order to be one of the five primary regions of Latin America in terms of quality of life.
>
> (DAPD 2006: 61)

Bogotá's Mesa de Planificación Regional (MPR, Regional Planning Board) therefore linked productivity with several key concepts as a strategy for global competitiveness. Gilbert reflected, "Given that the neoliberal agenda now means that cities across the globe are in fierce competition with one another, how different Latin American cities manage to perform will determine their economic prospects" (Gilbert 1975: 46). The city's planning department cited the MPR in its own promotional plans: "The challenge is to incorporate in a successful manner the Bogotá city region in the dynamics of globalization, guaranteeing at the same time the inclusion and improvement of living conditions for the whole population" (DAPD 2006: 73). The MPR and the city planning department policies show that the contradiction of using public space for people and for profit was built into the planning goals themselves, creating a tension between the right to the city for all people and the need for space to represent capitalistic and neoliberal goals.

What is certain is that the further Bogotá travels down the path towards becoming fully articulated in the world market (DAPD 2006: 61), the city's public space must be used locally to improve quality of life and globally to present an image of a "world-class" city. Bogotá's continued success became visibly linked to the particular spatial logic created through public space changes (Berney 2008, 2011), especially from the perspective of the international gaze turned upon the city. In order to be a viable global competitor, the encounters within Bogotá's public space must portray attitudes and behaviors in line with neoliberal ideals. With this in mind, I will now turn to the specific ways in which the mayors enacted their public space plans and the tensions between the right to the city and neoliberal ideals.

State building through public space

The public space mayors

Bogotá has not been the subject of a singular development vision such as those that characterized some successful Latin American city remaking projects, such as Jaime Lerner's vision for Curitiba, Brazil (Irazábal 2005: 94–7). However, Antanas Mockus and Enrique Peñalosa had several aligned goals, which created a sense of continuity throughout their administrations. As the primary source of development vision for the city, the mayoral office distinctly identified itself as a political and social agent, and the mayors saw themselves as positioned in a heroic role of responsibility for making positive changes. They saw the city as the body on which to work, as well as a figure on which to make a mark. They took advantage of the greater legitimacy, flexibility and sense of agency in social and cultural issues (Borja and Castells 1997: 6) created by the changes in and empowerment of local governance described above. As previously stated, in this chapter I refer to Peñalosa and Mockus as the public space mayors due to their use of public space to enact their reforms.

The Bogotá public space mayors' vision was, in general, oriented around *lo social* (the social) and *lo público* (the public). Each mayor emphasized a different component of the construction of public space during his tenure in office: Mockus focused on changing behavior, creating opportunities for learning and creating community spirit; while Peñalosa sought to recreate the physical territory by focusing on creating more, and better, public space. Martin and Ceballos write, "despite differences in ideas, visions, and forms of management", Mockus and Peñalosa contributed to "a logic of constructive community" (2004: 146).

Mockus was primarily focused on "reconstructing citizenship" (Martin and Ceballos 2004: 146). He did this through various unconventional and often theatrical means, in line with pedagogical urbanism. He focused on everyday activities to bring residents out into public space, including the "Night for Women", in which men stayed in to care for their families and 700,000 women came out to enjoy the city's public spaces, bars and restaurants over three nights (Caballero 2004). In 1996 he distributed plastic cards depicting thumbs, which citizens were encouraged to aim up or down when they saw a fellow citizen partaking in civil or uncivil behavior. In a similar move, taxi drivers were encouraged to take etiquette classes that emphasized pedestrians' right of way; this was combined with an increase in sidewalks and mass transit. In 1996, rather than hire a larger police force to deal with traffic problems, Mockus hired mime artists to admonish unsafe drivers (Schapiro 2001: 28). While these moves were humorous and unconventional, they also sent a not-so-subtle message about behavior. These efforts were part of a move to create a "citizenship culture" that would bring moral and cultural behavior more in line with the law (Mockus 2012: 145). About his reforms, Mockus told an interviewer,

> if you want to change society's habits, law is only one of the means. Most people prefer internal mechanisms for determining for themselves what is right and what is wrong, but perceive other people as needing to be regulated by laws. The question I asked was how to reduce the difference between the laws and cultural and moral means of self-regulation.
>
> (Schapiro 2001: 28)

Mockus' reforms took place within the public sphere, with an emphasis on citizens' public behavior. Pedagogical urbanism was clearly at work in the mix between urban planning choices (creating new public spaces) and moral and social education within those spaces (programs and social projects). While Mockus' reforms had the positive benefit of increasing the sense of public safety and responsibility, they also constricted public behavior, as will be examined in more detail later in this essay.

In a similar vein, Peñalosa's tenure in office was concerned with "reconstructing the city" (Martin and Ceballos 2004: 146). He proposed that public space was one of the most important tools for creating equal opportunities for happiness, or satisfaction, in living there. Peñalosa also stated that creating better public spaces would allow people of different social classes to interact on an equal basis; he saw the right to the city and class struggles as inextricably linked. He built a bus system and increased the city's pedestrian-only streets. Many of his reforms centered around removing or limiting vehicle traffic to create space for human encounters. He created the longest pedestrian-only street in the world, which runs through some of Bogotá's poorest neighborhoods, declaring that "this pedestrian infrastructure shows respect for human dignity" (Peñalosa and Ives 2004: n.p.). He saw deep connections between access to public space and class disparities, writing of his public space projects:

> At first it may seem that in Third World cities with so many unmet needs, high quality pedestrian spaces would be a frivolity. On the contrary, where citizens lack so much in terms of amenities and consumption, it is quicker and more effective to distribute quality of life through public goods such as parks, plazas, [and] sidewalks than to increase the personal incomes of the poor. It is impossible to provide citizens certain individual consumer goods and services such as cars, computers, or trips to Paris. It is however possible to provide them excellent schools, libraries, sidewalks and parks.... Parks, plazas, pedestrian streets and sidewalks are essential for social justice. High quality sidewalks are the most basic element of a democratic city.
>
> (Enrique Peñalosa, personal communication 2007)

The neoliberal agenda, however, is apparent here: "the park would also favor competitiveness and economic growth, by making the city more attractive to highly qualified individuals and corporations interested in setting up shop in the region" (Enrique Peñalosa, personal communication 2007). Thus, public space is both for the poor as a right and for global capital. While his focus on public and pedestrian space was part of a move for human dignity, it also reflected the need to project a particular image of Bogotá to the rest of the world. In his administration, as in Mockus', the right to the city and public space as a free space of encounter clashed with global capital, as can now be examined.

Design and policy decisions as political tools

Public space was used to support the development and recreation of healthy civil society, but it was also used to sell the city as "world class", in line with the neoliberal agenda. This dual role of public space in Bogotá created an inherently conflicted situation. A successful process of supporting healthy civil society depends heavily on the ability to allow for conflict, and behavior on the fringes of, or even outside of, societal norms, while adhering to neoliberal ideas means behaving in an orderly, preconceived way. Staeheli and Mitchell write, "differences in the kind of rights and their deployment shape not only the nature of political conflict, but also the kinds of publics and even the kinds of cities that are created" (2008: 106). In shedding its previous negative image, Bogotá became a city in which citizens were required to behave in certain ways in public. The homeless and poor were removed from public areas, and civil behavior became monitored and enforced by other citizens. A true Bogotano, in a sense, was one who behaved well and fit within neoliberal class ideals (housed, gainfully employed, and with a strong sense of civic duty). Cosgrove writes that landscapes functioning as stage sets for human activity can play a small or large role in human "performances" (1993: 1). Bogotá's public spaces became overt and strong participants in the spectacle contained within them. Rather than being a free point of encounter, public space became a tableau, a space in which to perform the image of citizenship that the public space mayors needed to project to the world at large.

Behavior in public spaces was monitored at several different levels, from passive signage to civic guides, police and programs to manage behavior more actively. In many parks, city workers were employed to provide directions, lead

Figure 7.2 Bogotá envisioned in 2001 as "a city that constructs spaces of citizen encounter" – Plaza San Victorino in the *centro*.

recreational activities and direct people to pick up their trash and refrain from inappropriate behavior. This is also in line with Mockus' civic behavior programs, in which citizens were encouraged to police each other. This active behavioral control was further aided by Bogotá's application of a "new vocabulary for the citizens" as seen in public space signage systems, which provided more cues for proper behavior. Examples of such signage include "parks for learning how to live" and "the success of the park depends on good use" (Berney 2011), creating an obvious connection between the right to public space and the necessity for appropriate action within it. Learning how to live meant learning how to behave.

In addition, while access to public space was a right, the question of who was truly allowed access became a telling feature of citizenship in a city upping its world-class status. Bogotá's Camara de Comercio (Chamber of Commerce) reported in 2002 that parks in lower-income neighborhoods were far more likely to be in poor condition than those in richer neighborhoods. As well, my previous research (2008, 2011) has shown the inequitable treatment of public spaces across the city. While everyone might have a park to go to, not all parks are equal. Bogotá provides an example of Wastl-Walter and Staeheli's comment that "urban policy and urban design are increasingly implemented in ways that are undemocratic, that exclude the poor and that create cities that 'prioritize the needs of business and the wealthy' over the vast majority" (Wastl-Walter and Staeheli 2005: 2).

While the poor received lower-quality parks, they were also unofficially forbidden access to public space in other parts of the city. Many citizens took on unofficial roles as caretakers for public space, and some of these volunteer caretakers reported that they expelled *indigentes* (homeless people) from the spaces or prevented them from entering, especially in the northern part of the city (Berney 2006). In most of these instances, low-income vendors verbally and physically expelled the *indigentes* because the vendors worried they would hurt their sales.

This highlights another interesting tension between capital and the right to the city. Vending in public spaces is a traditional and, today, controversial, use of public space in Latin American cities, a means of earning income for those "outside" of the formal labor market. Mockus and Peñalosa both supported the curtailment of vending in public spaces; Peñalosa was particularly clear that vendors who were engaging in profit-making (private) activities should not be allowed to do so in public spaces. In the 2000s, many vendors were kept out of public spaces and were forced to sell their goods immediately outside parks and public spaces. While nearly half of all vendors reported police or other harassment, different types of vendors received different treatment. For example, licensed vendors who set up stalls in the open-air markets generally report that they are not harassed by the police, and neither are those who sell lottery tickets. At the other end of the spectrum, those who sell food from mobile carts in public spaces frequently report police harassment. It can be seen here that the police work in collusion with capital, protecting those "allowed" to engage in capitalism while expelling those who are more unruly and who serve a different demographic by feeding people rather than selling goods to those whose class privilege allows them discretionary income. In addition, while vendors are policed for not being "good" citizens, many respond by removing the homeless, stratifying themselves in the pecking order of who deserves access to public space and codifying appropriate behavior within it.

While the right to the city is part of the discourse of public space in Bogotá, it is clear that the right to the city is not evenly applied. The homeless are persecuted, as are vendors. It appears that "private" activities in public space engender opposition because they make people uncomfortable and do not provide a "correct" and positive image of the city. Watson describes private activities as those "framed with reference to bodies and corporeality" (Watson 2006: 161), such as sleeping, working, eating and living. These activities are increasingly seen as interfering with other peoples' use of Bogotá's public spaces, as they portray a messy and uncontrolled physical reality. The "democratic" management of public space in Bogotá, both by citizens and by officials, precludes individuals carrying out these functions in public. Thus, it can be seen that in Bogotá there is a contradiction in the right to the city between the expansion of the rights of citizenship and improvement of quality of life and place versus the contraction that occurs by limiting the rights of certain minorities who are in conflict

Figure 7.3 Vendors waiting for customers at one of the entrances to Parque Tercer Milenio, 2006.

with the real or perceived values of the "democratic" majority. What further complicates this is that the right of the majority typically mirrors what neoliberal capital and entrepreneurial government seek. This creates a built-in conflict between the majority view and the socioeconomic-related (second-generation) rights. In Bogotá, the tyranny (or right) of the majority results in the diminution of second-generation, socioeconomic rights for certain people, mostly those with lower incomes.

The city has largely failed to negotiate these conflicts, and as a result, the right to the city has contracted. As Attoh describes, "the right to the city, if understood as a second generation right (a right affecting socio-economic status), must also rely upon an explicit theory of justice" (Attoh 2011: 672). By excluding the poor and monitoring behavior, it seems that Bogotá has failed to develop a clear or consistent theory of justice, despite the public space mayors' rhetoric connecting justice with public space. Former mayor "Lucho" Garzón, who was elected in 2004, has been known to call Mockus and Peñalosa's work – especially Peñalosa's – *maquillaje* (make-up), charging that it changes the "façade" of the city without truly addressing underlying social injustices. There are cracks now in the city's fabric, and these cracks point to the question of how and if a theory of justice can be developed to mend them.

The public space project proposed significant cultural and social changes for a city that, well into the end of the twentieth century, was considered a dangerous place to be. I have previously questioned the stability of these changes (Berney 2011), given that they reflect the public space mayors' desires to support dual and conflicting goals. The dual role of public space in Bogotá as a social crucible and the face of an entrepreneurial city creates a paradox: in extending the right of public space and using it to address socioeconomic inequities while privileging the values of the majority and capital, a stalemate was created that precludes the possibility of real change. Public space cannot function as truly public if it is messaged, staffed and policed, and branded as tableau. As Mitchell writes,

> landscape is where one recreates – it is literally a resort – and where one basks in the leisure of a well-ordered scene. Public space is a space of conflict, of political tussle, of social relations stripped to their barest essentials. A place cannot be both a public space and a landscape, at least not at the same time.
>
> (Mitchell 2003: 136)

Losing *lo público*

This chapter presents one history of the many that are possible to recount of Bogotá. This history is focused on public space, as public space was a central policy focus and, in a sense, a state-building project for the Bogotá mayors who are examined here. Part and parcel of this state-building project, where public space was likened to the right to the city, is the troubling nature of neoliberal capital to redirect and reshape public space in undemocratic ways, causing some citizens' rights to this space to be worth less than others'. Public space is appropriated in spectacles of the everyday, creating a tableau that works against the true reality of the right to the city. Bogotanos' right to the city in the form of public space is directly impacted by the loss of "publicness" in the form of loss of rights by the minority – the homeless, the drunk and certain vendors – in public spaces in favor of the majority's values and preferences, as well as the "preferences" of neoliberal capital that direct and constrict behavior. The citizens allowed access to and created by public space are forced into certain types of citizenship at the expense of others, undermining the right to the city and remaking Bogotá's social and class fragmentation in less visible and more complex ways.

So, is the city better off after the public space mayors and their project of remaking the city? Or has the city "returned" to the same set of circumstances that existed prior to reform? Certainly, new mayor Gustavo Petra (2012–15), who ran on the platform "A Politics of Love", believes that socio-spatial fragmentation is as bad as ever. There appear to be new and growing inequalities, and a reduced capacity to integrate marginal populations. Has the city truly changed, or has public space just covered the issues with make-up? The city appears to be fracturing; it does seem as though the make-up has cracked.

Acknowledgment

I would like to thank the Office of the Provost and Advancing Scholarship in the Humanities and Social Sciences grant program at the University of Southern California for their support in preparing this chapter. I would also like to acknowledge *Sage/Society*, where select sections of this chapter were first published.

Notes

1 Bogotá's population can be broken down into relative levels of income with reference to its stratified taxation (*estrato)* system. Each household in Colombia is assigned a number from one to six according to its socioeconomic status (Martin and Ceballos 2004: 82). Ones and twos are classified as low income; threes and fours are classified as middle income; and fives and sixes are classified as upper income. This system in used, for example, to determine ability to pay for utilities. The utility payment structure is set up so that the threes and fours pay their own way, so to speak, while the fives and sixes pay more to help subsidize the reduced amount that the ones and twos pay.

2 Little of the land invasion or spontaneous popular settlement that has occurred in some other Latin American cities has taken place in Bogotá (Violich 1987: 224). Rather, a process of pirate (illegal) subdivisions has proliferated, in which private developers created subdivisions outside the legal approval process. These subdivisions were extremely popular, but were typically created without public space, services or utilities, and were often constructed on the least expensive and most undesirable land (ibid.; Observatorio de Cultura Urbana and Zambrano 2003: 51).

3 Illegal settlement continues in Bogotá. However, a majority of this, according to the Departamento Administrativo Defensoría del Espacio Público, is occurring on privately held land on the periphery of the city (email communication, DADEP, 15 June 2010).

References

Attoh, K. A. (2011) "What *Kind* of Right is the Right to the City?", *Progress in Human Geography*, 35: 669–85.

Bergquist, C., Peñaranda, R. and Sánchez, G. G. (eds) (2001) *Violence in Colombia: Waging War and Negotiating Peace*, Wilmington, DE: Scholarly Resources.

Berney, R. (2011) "Pedagogical Urbanism: Creating Citizen Space in Bogotá Colombia", *Planning Theory*, 10 (16): 16–34.

Berney, R. (2008) "The Pedagogoical City: How Bogotá, Colombia, Is Reshaping the Role of Public Space", PhD dissertation, University of California, Berkeley.

Berney, R. (2006) *Bogotá Public Space Survey*, Berkeley: University of California.

Biennale di Venezia (2007) "*La Biennale di Venezia, Biennale Architecture: Tenth International Architecture*". Online: www.labiennale.org/en/news/architecture/en/67078.1.html (accessed 1 October 2007).

Borja, J. and Castells, M. (1997) *Local and Global: Management of Cities in the Information Age*, London: Earthscan Publications Ltd.

Borja, J. and Muxí, Z. (2003) *El Espacio Público: Ciudad y Ciudadanía*, Barcelona: Electa.

Braun, H. (1994) *Our Guerrillas, Our Sidewalks: A Journey into the Violence of Colombia*, Niwot, CO: University Press of Colorado.

Caballero, M. C. (2004) "Academic Turns City into a Social Experiment", *Harvard Gazette*, 11 March. Online: http://cyber.law.harvard.edu/blogs/gems/culturalagency1/HarvardGazetteAcademicturnsc.pdf (accessed 20 May 2012).

Carrión, M. F. (2004) "Espacio Público: Punto de Partida Para la Alteridad", in F. Velásquez Carrillo (ed.) *Ciudad e Inclusión: por el Derecho a la Ciudad*, Bogotá: Fundación Foro Nacional por Colombia.

Castro Jaramillo, L. (2003) 'Espacio público: más que maquillaje urbano', in Universidad de los Andes (ed.), *Bogotá y lo Público*, Bogotá: Universidad de los Andes.

Cosgrove, D. (1993) *The Palladian Landscape: Geographical Change and Its Cultural Representations in Sixteenth-Century Italy*, University Park, PA: Pennsylvania State University Press.

Crilley, D. (1993) "Megastructures and Urban Change: Aesthetics, Ideology and Design," in P. Knox (ed) The Restless Urban Landscape. Englewood Cliffs, NJ: Prentice-Hall.

Departamento Administrativo de Planeación Distrital (DAPD) (2006) *Curso PNUD la ciudad y sus limites: hasta donde llega Bogotá?* Powerpoint presentation, versión larga, Bogotá: DAPD.

Departamento Administrativo de Planeación Distrital (DAPD) (2001) *Plan de Ordenamiento Territorial de Bogotá (POT)*, Bogotá: Alcaldía Mayor de Bogotá, DC.

Dietz, H. A. and Shidlo, G. (1998) *Urban Elections in Democratic Latin America*, Lanham: Rowman and Littlefield.

Gilbert, A. G. (1975) "Urban and Regional Development Programs in Colombia since 1951", in W. A. Cornelius and F. M. Trueblood (eds) *Latin American Urban Research*, Beverly Hills and London: Sage.

Gilbert, A. G. and Dávila, J. D. (2002) "Bogotá: Progress within a Hostile Environment", in D. J. Myers and H. A. Dietz (eds) *Capital City Politics in Latin America: Democratization and Empowerment*, Boulder, CO and London: Lynne Reiner Publishers.

Habermas, J. (1974) "The Public Sphere: an Encyclopedia Article", *New German Critique*, 1: 49–55.

Irazábal, C. (2005) *City Making and Urban Governance in the Americas: Curitiba and Portland*, Burlington, VT: Ashgate.

Livingstone, G. (2004) *Inside Colombia: Drugs, Democracy and War*, New Brunswick: Rutgers University Press.

Londoño Botero, R. (2003) "Presentación", in ODCT and F. Zambrano (eds) *Construcción del Espacio Público, Tres Parques de Bogotá: Nacional, Simón Bolívar, El Tunal*, Bogotá: Alcaldía Mayor.

Lucía Peña, S. (1995) "Rito y Símbolo en la Campaña Electoral para la Alcaldía de Bogotá", *Análisis Político*, 24 (January–April): 22–35.

MacDonald, J. (2007) "2007 World Planning Keynote Carolina Barco Isakson". Online: www.planning.org/conferencecoverage/2007/tuesday/worldkeynote.htm (accessed 20 May 2012).

Martin, G. and Ceballos, M. (2004) *Bogotá: Anatomía de una Transformación, Políticas de Seguridad Ciudadana 1995–2003*, Bogotá: Editorial Pontificia Universidad Javeriana.

Mitchell, D. (2003) *The Right to the City*, New York: Guilford Press.

Mockus, A. (2012) "Building 'Citizenship Culture' in Bogotá', *Journal of International Affairs*, 65 (2): 143–6.

Observatorio de Cultura Urbana and Zambrano, F. (eds) (2003) *Construcción del Espacio Público, Tres Parques de Bogotá: Nacional, Simón Bolívar, El Tunal*, Bogotá: Alcaldía Mayor.

Peñalosa, E. and Ives, S. (2004) "The Politics of Happiness", *Yes! Magazine*, 20 May. Online: www.yesmagazine.org/issues/finding-courage/the-politics-of-happiness (accessed 20 May 2012).

Salazar Ferro, J. (2003) "Que Hemos Aprendido del Espacio Público?", *Bogotá y lo Público* (Universidad de los Andes, Bogotá) 1: 69–76.

Sánchez, G. G. (2001) "Introduction: Problems of Violence, Prospects for Peace", in Bergquist, C., Peñaranda, R. and Sánchez, G. G. (eds) *Violence in Colombia 1990–2000: Waging War and Negotiating Peace*, Wilmington, DE: Scholarly Resources.

Schapiro, M. (2001) "All the City's a Stage", *Atlantic Monthly*, 288 (2): 28.

Staeheli, L. and Mitchell, D. (2008) *The People's Property: Power, Politics and the Public*, New York: Routledge.

Vargas, L. S. (2003) "Presentación", in ODCT and F. Zambrano (eds) *Construcción del Espacio Público, Tres Parques de Bogota: Nacional, Simón Bolívar, El Tunal*, Bogotá: Alcaldía Mayor.

Velásquez Carrillo, F. (ed.) (2004) *Ciudad e Inclusión: por el Derecho a la Ciudad*, Bogotá: Fundación Foro Nacional por Colombia.

Violich, F. (1987) *Urban Planning for Latin America: The Challenge of Metropolitan Growth*, Boston: Lincoln Institute of Land Policy.

Wastl-Walter, D. and Staeheli, L. (2005) "Introduction", in D. Wastl-Walter, L. Staeheli and L. Dowler (eds) *Rights to the City. International Geographical Union, Home of Geography Publication Series Volume III*, Rome: Societa Geografica Italiana.

Watson, S. (2006) *City Publics: The (Dis)Enchantments of Urban Encounters*, New York: Routledge.

8 Resisting the neoliberalization of space in Mexico City

David M. Walker

Mexico City is currently in the midst of a neoliberal socio-spatial transformation guided by municipal policies and transnational capital. This process of the neoliberalization of space is occurring in tandem with cultural and economic globalization. Mexico began implementing neoliberal policies and restructuring the country's economy in 1982. Thirty years of actively applying neoliberal polices has resulted in a reconfiguration of governance in Mexico City that has allowed for the partnership of city government and global capital interests to implement a municipal city policy called the Programa de Rescate, or the recovery program (PR). This program aims to regulate directly who has access to public urban space and how urban space in Mexico City is produced and consumed.

In order to meet the goals of the program the Government of the Federal District of Mexico City (GDF) is required to "normalize" economic activities in the historic center, which includes relocating street vendors, called *ambulantes*, from the streets and plazas of the center and into markets and warehouses in working-class neighborhoods that surround the historic district. Driven by the logic of roll-out neoliberalization, the GDF believes that the vendors must be relocated to "normalize" urban space in the center. "Roll-out-neoliberalization" consists of aggressive intervention by governments around issues such as crime, policing, urban policies and surveillance. Roll-out neoliberalization often involves state collaboration with the private sector to implement urban policies and to revitalize and police derelict areas that were thrown into disarray during the earlier period of roll-back neoliberalization (Barajas and Zamora 2002, Hackworth 2002, Portes and Hoffman 2003, Soederberg 2001). This new form of neoliberalization underwrites much of the logic behind the PR.

The normalization of space is being carried out with the goal of preparing the historic district for would-be investors to purchase and live in the refurbished palaces and apartments, and to attract direct foreign investment in hotels, restaurants and real estate schemes. The removal of the vendors is a key component to achieve the goals of the normalization of urban space. Despite the implementation of this urban revitalization policy by the GDF, local street vendors and Mexico City's citizens have risen up to confront the GDF. Vendors oppose the program as it prevents them from exercising their livelihood strategies. Mexico City residents in the Molina del Rey neighborhood in the upper-class Delgación

Miguel Hidalgo resist components of the program that would deteriorate their quality of life.

In this chapter I describe the revitalization policies that are being implemented in Mexico City and tell two stories of resistance to the program. The first discusses how and why the working-poor street vendors resist the normalization of space, a key component to the revitalization of the historic center. In the second I explain how and why elite residents resist the construction of what would have been Latin America's tallest skyscraper, a symbolic component of the program meant to represent Mexico City as a truly "globalized city".

I begin by describing "roll-back" and "roll-out" neoliberalism as relevant to socio-spatial transformation in Mexico City's historic center. I then provide an analysis of the actors who fund and manage the reconfiguration of Mexico City's urban space. I continue by critically describing the changes that are occurring in the urban built environment, and I analyze the ongoing material effects this has on street vendors and residents' lives. Penultimately, I map out how street vendors resist displacement and how elite Mexico City residents prevent the construction of an environmentally hazardous skyscraper. These two distinct reactions to the revitalization program provide a nuanced understanding of how vendors and the elite imagine the openness of the city and neoliberalism. The divergent reactions to the program show how elites and the working poor shape urban space within the context of Mexico City, and how street vendors and elite residents imagine distinctly what Mexico City stands for. I conclude by recognizing that street vendors and elite resistance strategists must continue their efforts to confront neoliberal policies, as actors putting such policies into place show no sign of resting.

Neoliberalization à la Mexicana

Mexico City's population hovers at around 21.2 million people with approximately 120 million living in the entire country. From colonialism to the present, Mexico City has rested at the epicenter of economic and cultural power. It is the administrative and cultural capital of the country, housing many of its most symbolic sites (Brunn 2012; Walker 2011). Mexico's extremely centralized post-Revolution system of governance meant that the bulk of the country's budget went towards the capital city (Gilbert 1996). The economic crises that ushered in the neoliberal era in Mexico, which began in 1982, negatively impacted the city. The hitherto majority budget that Mexico City enjoyed shrank, while industrialization and certain administrative duties moved to the US–Mexico border region and second-tier cities such as Guadalajara and Monterrey.

The year 1982 marked the beginning of the neoliberal experiment in Mexico. Under then President Lopéz Portillo, Mexico defaulted on its external debt, leading to an economic crisis that spread throughout Latin America and prompted what many scholars have called the lost decade of the 1980s (Gwynne and Kay 2004). The introduction of roll-back neoliberal policies in Mexico in the 1980s resulted in the deterioration of Mexico City's built environment in the form of cutting budgets for the maintenance of roads, parks, plazas and street lighting, limiting garbage collection (resulting in the informalization of garbage collection during the neoliberal crisis of the 1980s and 1990s that is only now being "regularized") and the elimination of budgets for urban beautification, public festivals and concerts and cultural events (PUEC 2008).

The crisis and subsequent implementation of neoliberalism enhanced informal economic activities such as clandestine manufacturing and street vending. While street vending has been practiced in Mexico for centuries, it increased during the 1980s. The economically active population looked more towards the informal sector of street vending as a form of livelihood in the face of massive unemployment and little government assistance. This increase, which persists today, also resulted in the change of urban land use in the center, where urban space became a site of uncontrolled and untaxed capital production. GDF officials consider this informal economic activity a chaotic form of production in urban space that requires normalizing under the current phase of roll-back neoliberalism. The PR plan employs the language of "normalization", which in this case involves the removal and relocation of street vendors from the historic center.

Socio-spatial changes in Mexico City's historic center since the 1982 debt crisis

Since the inception of the debt crisis and the adoption of neoliberal policies, Mexico City's economy shifted from relying heavily on manufacturing to depending more on the tertiary and informal sectors of the economy. The onset of neoliberalization across Mexico has also spurred the privatization of public space in Mexico City and the decentralization of economic and political power (Canclini 2000; Borja 2001; Davis 1994). As decentralization progressed and industry moved to offshore manufacturing centers in the US–Mexico borderlands, Mexico City experienced an increase in the construction of buildings for financial and information services and the development of tourism (Canclini 2000). Canclini states that these industries have

> [C]hanged the urban landscape in a number of areas (for example, the transformation of the Santa Fe dump into a high-tech business center or along the Paseo de La Reforma and the Historic Downtown of the city). The festival of Mexico City and the festival of the Historic Downtown Area, events whose goals include increasing the city's allure for tourists and turning it into an international metropolis, belong to a set of large development projects by means of which Mexico City's current administration is redefining the city's image.
>
> (Canclini 2000: 59)

One of the consequences of the introduction of neoliberal policies in Mexico has been the decentralization of the Mexican state apparatus and the emergence of transparent political affairs. The 1998–2004 administration of Mexico City was a product of decentralization and neoliberalism. Due to the federal government's obligations to finance the foreign debt, funding to the GDF was cut, with the result that the city's budget is not sufficient to allow the municipal government to oversee management of the built environment. Because of this situation the relatively weak city government has chosen to work with private capital to decide how residents will have access to the city and participate in the reproduction of urban space.

The Programa de Rescate, Mexico City's recovery program, is led by the municipality and a private corporation, Grupo Carso. The PR arose out of a partnership between federal and city authorities, together with private investors, in August 2001. It aims to renovate an area three times the size of the historic districts of Lima, Barcelona or Madrid (Valdes 2004). Its mandate covers an area of nine square kilometers, covering 668 blocks containing 1,500 buildings, including all of the sixteenth- and seventeenth-century colonial palaces still remaining from the viceregal era: 196 civic monuments; sixty-seven religious monuments; fifty-three museums; sixty-eight plazas; nineteen cloisters; twenty-eight fountains; and twelve public murals (PGDDF 2001). Already the PR has renovated thirteen streets and 615 buildings, initiating the transformation of the buildings and public spaces of Mexico City's historic center from a working-class stronghold into an aesthetically and socially more appealing upper-middle-class residential area and a promising site for local and global investors (Cardoso and Ramirez 2003; Gonzalez 2003). The urban transformation of the historic center under the PR will cost US$450 million and be implemented in three stages over a six-year period (PGDDF 2001).

The first stage consists of replacing water and sewer infrastructure and razing buildings damaged in the 1985 earthquake, and the development of a commercial corridor to connect the Zócalo to the Paseo de La Reforma business district (see Figure 8.1). The PR plans to improve tourist facilities during the second phase, including the installation of modern hotels and a visitor's center, and constructing Latin America's tallest skyscraper. The third stage involves improving and renovating buildings and housing into multi-use complexes consisting of retail shops, restaurants and upscale residences. In order to meet the goals of third stage of the PR, the GDF announced plans for the *uniformización*, or "normalizing" of the historic center. These call for increasing the number of green areas, park benches, public garbage bins and public phone booths, as well as running an anti-graffiti campaign and augmenting security.

The PR forms part of a larger set of strategies concerned with the image of Mexico City. In the last decade, key local institutions have developed strategies to promote the city to both local and global audiences. Central to this promotion is the slogan, *La Ciudad de la Esperanza*. The promotion of the capital as the "City of Hope" aims to overcome bad publicity and to newly represent the city as a good, safe place to visit, live and invest (Sanchez 2007). The backers of the PR are attempting to re-invent urban space in Mexico City by developing a nostalgic, imagined "regal" colonial history, which will be materialized through the renovation of colonial buildings, streets and plazas in the historic center.

Table 8.1 Objectives of the Programa de Rescate

Objectives of the Programa de Rescate	*implementation of objectives*
improve the built environment	Renovation of urban infrastructure (replacement of sewage lines, water lines); improvement of roads and pedestrian walkways; renovation of historical buildings; construction of parking structures.
Increase security	The sanitation and increased safety of streets; CCTV systems; special Historic Center Police Force funded by Fideicomiso; new urban lighting systems (overlaps with "built environment" category – but is implemented in the name of safety).
Encourage economic reorganization	Consolidation and maintenance of economic activities (retail, restaurants, bars, clubs, discos); normalizing of urban space through the removal of street vendors.
Increase tourism in the historic center	Bilingual Tourist Police; increase in tourism sites (tourist buses, etc.).
Increase the renovation of upscale residential dwellings	Encouraging targeted upper-class residents to move to the center.

Source: Fideicomiso del Centro Histórico de la Ciudad de México. Compiled by author.

To meet the goals of "normalizing" outlined by the PR, the Federal District's government (GDF) plans to remove 30,000 *ambulantes* who sell their wares on the plazas and in the streets of the historic center. This removal process is aimed at encouraging tourists, middle-class professionals and students to repopulate the city center. The appeasement, inclusion, negotiation and/or relocation of the *ambulantes* represent one of the most trenchant political issues associated with the PR. They sell everything from clothing and cookware, produced in small factories located in the Tepito neighborhood in the historic center, to pirated music cassettes and DVDs made in China (Castells and Portes 1989; Cross 1998). The proliferation of the *ambulantes* has occurred in tandem with the neoliberalization of politics, economics and space in Mexico City over the last twenty-three years (Cross 1998; Davis 1994; Low 2000; Ward 1998). The *ambulantes* are central to the economic needs of the city's working poor, many of whom travel from all over the city to take advantage of the cheap merchandise they provide. Meanwhile, a great number of the *ambulantes* are residents of the historic center, especially in the Plaza de Santo Domingo and Tepito neighborhoods. The *ambulantes* are organized into distinctive groups with strong leadership. During the period of roll-back neoliberization in the 1980s and 1990s, the *ambulantes* literally picked themselves up by their own bootstraps through the creation of clandestine employment opportunities. The *ambulante* groups, such as Grupo Capitalista, participate in the fabrication of consumer goods in clandestine factories found in the Tepito neighborhood of the historic center, in addition to selling consumer goods on the street and plazas. The *ambulantes* are embedded in the

Figure 8.1 Important sites in Mexico City's historic center.

urban spaces of the historic district; it will prove difficult to remove the clandestine factories and vendors from the center permanently.

Implementation of the recovery program

René Coulomb emerged as the main actor to begin the energetic promotion and restoration of the historic center at the start of the roll-back neoliberal period in 1982. Coulomb argued that a site so charged with national symbolism, and replete with grandiose architecture representing the birth of the Mexican people, required stewardship and guidance in order to "rescue" the built environment from further deterioration and neglect. At the height of the debt crisis in 1982, in the Iglesia de San Francisco, a church not far from the famous Torre Latino Americano on Calle 16 de Septiembre, just behind the Palacio de Los Azulejos, Coulomb first presented his plan of how to salvage the historic center from further decay to a delegation of politicians, businessman, artists and journalists. Coulomb organized a dinner in the main cloister of the church to promote a program jointly funded by the federal and local governments and private investors. This event signaled the birth of an association whose main purpose was the preservation and restoration of the historic center, something that had not occurred before Coulomb's efforts. Coulomb's initiative led to city policies that eventually coalesced in the formation of the PR in August of 2001. In 2001 the political and economic climate converged after nearly twenty years of the implementation of neoliberal policies in Mexico. It was in this context that the PR was formed, to pilot urban revitalization through a public-private partnership.

In 2001, then President of Mexico Vicente Fox and the populist mayor of Mexico City, Andrés Manuel López Obrador, created a Consulting Council for the recovery of the historic center (*Consejo Consultivo*). The council consisted of Carlos Slim, CEO of Grupo Carso; the Cardinal of Mexico City, Norberto Rivera; the Archbishop of the Orthodox Church, Antonio Chedrahui; the historian Guillermo Tovar y de Teresa; the director of influential newspaper *El Universal*,[1] Juan Francisco Ealy Ortiz; and the renowned journalists Ricardo Rocha and Jacobo Zabludowsky. The composition of the Consulting Council's leadership reflects a larger set of strategies informed by concerns over the image of Mexico City, and the group began developing tactics to promote the city to both local and global audiences. The Consulting Council functions as a think tank that influences and frames the strategies implemented by the PR. The Consulting Council also provides a distinctive example of collaboration among politicians, right-wing capitalists, populists and religious and cultural representatives.

The majority of the funds that are used to "rescue" the historic center stems from Carlos Slim's company, Grupo Carso. The Fideicomiso – the trust fund for the recovery of the historic center – oversees the funds. The Federal-Municipal Council manages on-the-ground projects.

According to the director of the Fideicomiso, Adrian Pandal, who is also a personal assistant to Carlos Slim,

> many of the funds come from TELMEX [TELMEX is Slim's corporation that later became Grupo Carso]; all of the funding originally came from TELMEX, now the Fideicomiso is an independent organization with the goal of recuperating and recovering the historic center.
>
> (Adrian Pandal, personal interview, spring 2006)

Table 8.2 Organizations that supervise the implementation of the Programa de Rescate

Entity	*Function*	*Funding*
Fideicomiso del Centro Histórico de La Ciudad de México	Private trust fund of the historic center	Primarily Grupo Carso
Fundación del Centro Histórico de la Ciudad de México	Private real estate agency that is a branch of the Fideicomiso	Grupo Carso, direct investors
Gobierno del Districto Federal (GDF)	implements and oversees the municipal policy – the PR: Programa de Rescate del Centro Histórico	The GDF is funded at multiple scales, but the money it funnels to oversee the PR comes from Grupo Carso, the EU, Unesco and private investors

Source: organizations that supervise the implementation of the Programa de Rescate. Compiled by author.

In the early stages of the program, Slim, the EU and UNESCO provided all of the funding for the renovation of the center. Five years after the implementation of the program, individual investors began to emerge with small-scale projects in the center, such as bars, restaurants and galleries. Pandal's description of the structure of the fiscal organization of the PR demonstrates how convoluted the process is:

> In the project of recuperating the historic center there is an executive committee, consisting of 30 percent of the local government, 30 percent from federal government and 40 percent of social society. All the planning is done in these executive committees, and Fundación is a private foundation that works in the same direction in the health and social services that are needed. So it is a parallel work that we are working in the same direction, even the federal government has given money, grants for projects in the historic center – the money goes through the Fideicomiso del Centro Histórico, not the Fundación.
>
> (Ibid.)

The recovery program is managed by three entities (see Table 8.2) consisting of the Fideicomiso, which is the trust fund; the Fundación, a private real estate company; and the Government of the Federal District (GDF), which has all the usual responsibilities that a city government would have but also oversees how the municipal policy is implemented. The inner workings of the recovery program are convoluted and opaque in a way that is consistent with the Mexican political tradition (Barry 1995).

Pandal states,

> There are two things which oversee the rescue project. There is the private company called the Fundación Centro Histórico that has acquired over 75 buildings and those buildings are being renewed, and we have created more than 5,000 jobs in the area and have created more than 1,000 apartments in the area, and brought in restaurants, bars, etc. That is the private side. Then there is the Fideicomiso: it was created to work with the people in the area. So we have created hospitals, we have equipped hospitals with some things they do not have, we make many cultural events. There are two instruments, the Fideicomiso, which is to improve the social economic level of the people, and the Fundación, which is also private, but functions to recuperate the private building and make them more alive, with better use.
>
> (Adrian Pandal, personal interview)

Both of the entities described by Pandal were created and funded under the auspices of Grupo Carso. To see the continuation of these programs from one *sexenio*[2] to the next, they have been linked – indeed, monitored or overseen – by the GDF (Government of the Federal District). This is a very important component of the PR. Continuity through administrations at the level of Mexico City and the nation has been one of the key challenges to funding and implementing the PR. The Fundación, as a private company operating under the guise of a public institution, allows for continuity of implementation for the PR across administrations.

Transformation of the socio-spatial environment

> A well-used city street is apt to be a safe street. A deserted city street is apt to be unsafe.
>
> (Jacobs 1961)

The directors of the PR aim to transform Mexico City's image into a suitable site for national and international tourism and direct national and foreign investment, and to encourage targeted middle-to-upper-class residents to purchase renovated properties in the historic district. The reinvention of space under the guise of the PR, in the form of encouraging transnational corporations to invest or even relocate to Mexico City, often leads to direct competition and confrontation over the utilization of space. In the case study below, the construction of the Sheraton Centro Historico directly impacted low-income housing in the center. To create space for targeted investments from transnational corporations required the reconfiguration of space in the form of the removal of tenement housing surrounding the Alameda Park, one of the few ample green spaces found in the historic district.

The logic behind the implementation of the PR is to provide an image that Mexico City is a safe place to visit, live and invest. The construction of a mega-palace-cum-playground for the wealthy adjacent to the Alameda Park in the historic center provides an example of how the promoters of the PR have successfully attracted venture capitalists to invest in the built environment of the historic center. Many of the buildings that flanked the south side of the Alameda were low-income residential buildings, badly damaged by the 1985 earthquake and still in disrepair by 2001. Although the buildings were structurally unsafe, many of those adjacent to the south side of the park still functioned as homes for the working-poor residents. The successful continuity of the restoration of the Alameda Park under the PR signified the loss of homes for the working poor in this area. In order to meet the goals of the PR, the Council incorporated past efforts to revitalize the historic center and oversaw the Proyecto Alameda, a formal part of the PR consisting of the refurbishment of the Alameda Park and conversion of the surrounding residential areas into upscale restaurants and hotels.

The Sheraton Corporation, with its main offices in Canada, worked directly with the Government of the Federal District, the Fideicomiso and a local Mexico City architectural firm, Pascal Arquitectos. The Sheraton Centro Histórico Hotel was the first new edifice constructed in the area for forty years, and the first renovation in the Alameda to occur since the 1985 earthquake. The impressive hotel holds multiple restaurants, bars and an event and business center that can accommodate more that 5,000 people. It occupies nearly half the southern flank of the Alameda Park and consists of two large-scale city blocks.

The residential buildings adjoining the Alameda Park were razed to facilitate the construction of the mega-hotel. The Government of the Federal District expropriated the buildings, and since no property taxes had been paid on them for years, the owners did not come forward and no redemption payment had to be paid by the GDF. The working-poor residents that were still living in the buildings when demolition began had not paid rents in decades, if ever. They did not have leases and could not appeal for ownership under the vacant building laws of Mexico City as they had been pirating both electricity and water. According to Alfonso Hernandez, the director of Centro de Estudios Tepiteños, the residents who lost their homes along the Alameda were not paid any money for their inconvenience nor were they relocated by the authorities (the Fideicomiso or the GDF) to new housing.

The historic center provides housing and employment to Mexico City's working poor. Figure 8.2 shows an image of a street filled with street vendors and their customers juxtaposed with an image of the Cinco de Mayo Street that has been renovated and restored to its colonial regalia, albeit with McDonald's sponsorship. The first image displays the appropriation of urban space by street vendors and shoppers, the streets, sidewalks and plaza functioning as an outdoor

Figure 8.2 Calle Brasil filled with informal economic activities.

bazaar. The street vendors are at work selling their wares while their customers talk, eat, congregate around the stalls and shop. The streetscape is a site of economic production and consumption where vendors earn their livelihood and the working poor are able to make purchases of necessary goods that they would be unable to afford in department stores. There exist multiple street-vending groups that work in the center and, although they are not united, their collective act of street vending and appropriation of urban space frames this locale. The activities along the streets have multiple functions: notice that there are formal businesses (as opposed to informal vendors) to be found along the street, thus representing the functional dualism that exists in Mexico City in regard to street vendors and shop owners.

In Figure 8.3, the recovery program has "rescued" the area by normalizing this street through the removal of vendors. The façades of the church and palaces

Figure 8.3 McDonald's-sponsored trash bin amidst colonial-style Cinco de Mayo Street.

have been refurbished and the street now enjoys a trash bin. US tourists find the second streetscape to be more inviting for observing the colonial architecture (personal interview, 2005). The street filled with vendors and their working-poor patrons startled the US tourists, as they perceived the street scene to be dangerous due to so many people clustered in one spot (personal interview, 2005). Even US and European college students living and studying in Mexico City expressed similar reservations about streets filled with *ambulantaje* and preferred the "recovered" and normalized streets and plazas of the center.

In contrast to the tourists' perspective, Jane Jacobs finds in general that demolished working-class housing units – removed to make way for tourist infrastructure and streets void of city life: vendors, patrons, food stalls, general carousing and congregating – hardly produce the types of vibrant urban environments conducive to creating communities and safe sidewalks for city residents (Jacobs 1961). Indeed the reaction of US and European tourists and students who approved of the "normalized" spaces of the historic center can only indicate how the recovery program is attempting to reproduce a culturally smoothed-over elite international space, the kinds of spaces that have ruined communities and sidewalks in the Global North (Florida 2008). Below I turn to how *Chilangos* (residents of Mexico City) resist the top-down production of smoothed-over elite urban space for their right to the city.

Strategies of resistance: street vendors stand up to oppressive neoliberal policies

I will now turn to look at how street vendors employ political and spatial tactics – including acts of violence – to avoid being evicted from the alternative livelihood spaces they have appropriated and reproduced. They defy the poverty foisted upon them through the neoliberalization of Mexican economics, politics and state, in an industry that provides jobs for 30,000 people across the city. Street vending is in and of itself a reaction to the implementation of neoliberal policies in Mexico and a mechanism employed by the working poor to cope with the poverty that has worsened under them.

Street-vendor groups have forged multiple informal economic activities. They have created clandestine clothing and appliance factories and built studios that pirate electronic goods such as CDs, DVDs, books and computer software, and they sell their wares in the streets and plazas of the city. According to Alfonso Hernandez, the director of the CET (Center for Tepito Studies), and documents produced by this think tank, the act of street vending is not just a passive reaction to the neoliberalization of the economy and the state, it is also an active affront to the political and economic changes associated with the neoliberal paradigm shift in Mexico. Street vending functions as a popular economic movement that negotiates poverty and resists the reconfiguration and normalization of urban space under the auspices of the neoliberal recovery plan. The vendors embrace informal economic activities in order to continue living and working in the sites they choose.

Much of the scholarship analyzing resistance tactics in the Global South aims to contrast commonly conceived forms of rebellion, such as violent unrest, against forms of less-recognized protest incorporated in daily activities (Scott 1985). Through studying Malay peasantry, Scott realized that "the rare heroic, and foredoomed gestures of a Nat Turner or a John Brown are simply not the places to look for the struggle ... one must look rather at the constant, grinding conflict over work, food, autonomy, ritual – at the daily forms of resistance" (ibid.: xvi). The vendors provide an example of these daily acts of resistance in how they perform in popular informal spaces of economic production. Through place-based, daily politics involving surreptitious economic activities, the vendors have reconfigured the urban spaces of the city by re-inventing their livelihood strategies and effectively thwarting the goals of the PR.

On 17 February 2007, riot police conducted a pre-dawn raid on a building at number forty Tenochtitlan Street, in the Tepito neighborhood. Prior to the raid, the building had been nicknamed the "Fortaleza", or fortress, as the residents put up enormous opposition to prevent the authorities from carrying out the eviction process. The police invaded this eighteenth-century palace to evict 144 people living there, consisting of multiple families, who had turned the colonial building into a living space and warehouse to store the goods they sold in the streets. This is a site where vendors store stolen and/or pirated goods, while drug trafficking and prostitution also go on in and around the building. Many residents had been occupying the building for decades and not all of them were involved in disreputable activities. Police filled the alleyways – normally rugged paths filled with *ambulantes* – surrounding the old colonial palace, and occupied the communal spaces consisting of the inner patios of the colonial, Mozarab-influenced architectural jewel.

In the past, the residents had successfully prevented the intrusion of the authorities. Neighborhood associations with names such as Frente Francisco Villa and Frente Zapata barricaded themselves in their homes and surrounded their buildings with pregnant mothers and children to deter the police from violently evicting residents. However, in this instance, the pre-dawn forced entry caught the residents by surprise. Moreover, the recently elected mayor of Mexico City, Marcel Ebrard, approached the "rescate" of the center under the rubric of the PR with a much more aggressive stance than his predecessor, Andrés Manuel López Obrador. The residents were taken off guard.

The people who lived in the fortress immediately called the media and began to make placards denouncing the raids as a violation of their rights, citing the Constitution, and insisting that they were within the realm of the law to occupy the building. Residents complained, "the police know perfectly well how things run around here and which are the residences which are 'chuecas' [a slang word that signifies illegitimate or unlawful]" (Flores 2007). This reaction – to protest the eviction – is one of the many forms of resistance to the PR in the historic center. The vendors use political and spatial strategies and public demonstrations to protest the neoliberalization of urban space.

Another example: in February 2005, an *ambulante* organization called Grupo Capitalista took advantage of a large demonstration in the Zócalo (Mexico City's large main plaza) put on by a party opposing the Mayor's office (they were not affiliated with the vendors). The Grupo's leader knew that the police would be too much occupied in containing the demonstrators in the Zócalo to prevent the vendors from setting up their stalls. The vendors seized the opportunity to take back several streets from which they had been removed as part of the PR. One of their spatial strategies is to locate stalls near intersections, where they can quickly move from a "recovered" street (in which they are not allowed to sell their wares) to a street that has yet to fall under the PR.

The well-organized vendors carry cell phones and have lookouts surrounding the area. When the police patrols approach, lookouts warn vendors in advance so they can quickly move their carts to the safe street – only to return to their original location after the patrols have gone through. Figure 8.4 shows women selling therapeutic herbs on a "recovered" street. The women have set up their wares on a mere bedsheet rather than a formal stall, as this allows them to pick up their goods quickly, should the police arrive, and move to a street where they will not be hassled by the authorities. This style of vending has been labeled "*ambulantes-toreros*" – or bullfighter street vending – as the rapid lifting up of the bedsheet mirrors the movement of the matador as he raises his cap over the *toro*. Many vendors now prefer to risk peddling their wares in this manner on "recovered" streets and plazas in the center, as these newly renovated spaces attract more tourists who will pay more for the goods.

When authorities physically attempt to remove vendors, there are sometimes responding violent public demonstrations. For example, the young male protester in Figure 8.5 holds a sign that reads "Tepito until death". He is protesting the removal of vendors from the streets and plazas of Tepito and the relocation of residents from renovated buildings. Street vending is the lifeblood of this community. The residents believe that the end of vending will lead to the ruination of their livelihoods. Their right to the city is their very survival.

Violent protests in Tepito

In January of 2006, a combined force of 480 Mexico City police officers and Mexican soldiers conducted a raid in Tepito to remove street vendors and secure the area for recovery (Flores 2007). The special operation confiscated ten tons of pirated materials, mostly CDs, computer software and DVDs. This was the largest effort by the authorities to remove street vendors in preparing the zone for recovery (ibid.). The neighborhood erupted in mass protests. Residents stated that they would not allow authorities to remove any more street vendors, dislocate residents from their homes or permit continued confiscation of goods (ibid.). Protests lasted for three solid days. The vendors jeopardized their own street commerce in order literally to commandeer the areas, resisting removal by the authorities. Many incidents of violence ensued, against both police and residents.

Figure 8.4 Police officer questions *ambulante* women about their activities (source: photo by author).

Figure 8.5 People protesting the Programa de Rescate in Tepito (source: *La Jornada*, 2005).

The vendors refused to allow a continued presence of the authorities in Tepito, and called for Tepito autonomy to prevent further encroachment by the state. Figure 8.6 provides a poignant example of the strength of vendors to hold their ground and guarantee access to the streets and plazas of the city where they sell their wares.

Vendor groups also involve women in their protest against the PR. In contrast with the violent images above, group leaders will manipulate popular perceptions of vendors as violent male youths through the involvement of maternal-looking women.

Figure 8.7 portrays one such grandmotherly figure bearing a sign that reads, "the Centro just isn't the Centro without *ambulantes*". This sign is about the livelihood of vendors, suggesting that the historic center is the natural place for vendors: it is where they belong. Street vendors' presence in the center also provides cheap wares for the working poor of Mexico. How will mothers, aunts and grandmothers provide for their families without the cheap, pirated goods that vendors provide to these working-poor families?

The confrontation over the production and reproduction of space in the historic center, between the *ambulantes* and the promoters of the PR, provides an understanding of how distinctive groups imagine the utilization of urban space in Mexico City. In the following section I turn to an example of elite resistance to a component of the PR: the effort to construct Latin America's tallest skyscraper.

Resistance to the construction of the Torre Bicentenario: 2006–11

Neoliberal urban-governance projects in the Global South do not necessarily have the overarching support of elite interests. When elites are confronted with projects, intended to benefit society at large, that impede their interpretation of quality of life in the city, they will also resist urban revitalization programs. One of the Consulting Council's overarching goals is to convey to the wealthy of Mexico and beyond that the city has finished its modernization period: to let the

Figure 8.6 A protestor points the barrel of a pistol towards the camera while holding a placard that reads, "I am a street peddler".

Figure 8.7 A female protestor.

Figure 8.8 Digitized image of would-be Torre Bicentenario.

world know that the city is no longer an economic backwater, that it is a fun place to live and play and a safe place in which to invest. The Council aims to attract global finance and capital, increase foreign direct investment and promote Mexico City as a truly "global city". The construction of the Torre Bicentenario – intended to commemorate the bicentenary of Mexico's independence from Spain – was supposed to have been the grandiose neoliberal urban project that would catapult Mexico City into the ranking of the world's truly great cities. The successful movement by elites to prevent the construction of the Torre Bicentenario skyscraper provides an example of elites' non-complicity to a component of the PR.

The leftist mayor and Council believe this project would hurl Mexico City to the forefront of modernization as "no other city in Latin America will have a tower of this size now", which signifies to the mayor that, "we're [Mexico City] ahead of everyone else" (Mayor Ebrard, quoted in Malkin 2008). A communiqué promoting the construction of the tower furthers the developmental discourse voiced by the mayor:

> [T]he decision to build the Bicentennial Tower shows that there exists political and economic stability in Mexico ... the symbol of stability that the Tower invokes will increase direct national and international investment in development projects, which will increase Mexico's competitiveness.
>
> (www.torrebicentenario.com, last accessed 6 June 2012)

The promoters of the tower believe that

> with the construction of the Bicentennial Tower, Mexico City will rise to the same level as the great capital cities of the world. The construction of this project will bring benefits in terms of competitiveness to the city, which will allow for the creation of more wealth and the potential to confront the social and infrastructure barriers [in Mexico City].
>
> (Ibid.)

Much developmental cachet is being placed in the Torre Bicentenario. The digitized image created by the architecture firm portrays an ominous specter indeed.

The seventy-story skyscraper would have been built in the Polanco neighborhood, hovering above the Chapultepec Park. The park is home to the Chapultepec Castle, an important historical site that represents Mexican nationalism and resistance against "Yankee imperialism" as well as housing the world-renowned National Anthropology Museum. The Castle and museum are the dominant structures in the park. Residents from the neighborhood association, Colonos de Molino de Rey, claim that the seventy-story monstrosity would have been an eyesore for residents and culture-goers alike. On the proposed skyscraper site exists a building that provides an example of mid-twentieth-century functionalist architecture. The National Institute of

Anthropology and History, prompted by the neighborhood association Colonos de Molino de Rey, rushed through an upgrade of the building's protected status as a historical monument with the goal of slowing construction of the skyscraper. The Federal Agency for Culture and the Arts has also declared (since 13 August 2007) that the building standing on Torre Bicentenario's projected site, a petrol station known as Super Servicio Lomas built by architect Vladimir Kaspe in the 1940s, is catalogued as a building with artistic value.

The Fideicomiso purchased the site for US$18 million in 2006, and gathered private capital to cover the $600 million in construction costs required to build the Torre Bicentenario. The Council selected a foreign architect over the well-qualified Mexican architects who have gained worldwide recognitions, reflecting the Council preference for employing foreigners to help "recover" Mexico City. The utilization of foreign architects and foreign security firms to construct building in Mexico and regulate urban space increased protests against the PR and the skyscraper project. Those protesting the construction of the skyscrapers, however, are not the working-poor residents: rather they are professionalized elites. The members of the neighborhood association Colonos de Molino de Rey have the economic means to negotiate and resist the construction of the skyscraper through legal channels. For example, the community's members combined funds to hire a high-profile environmental lawyer to write an environmental report on the impacts the construction and placement the skyscraper would have on the Molino de Rey neighborhood and the Chapultepec Park. The report found that the project would absorb the resources currently allotted to the neighborhoods and overstrain the electrical grid and roadways, while the construction of a parking lot in the Chapultepec forests would eliminate much needed green spaces in this city of twenty-five million residents. Due to antiquated water systems, Mexico City loses up to 40 percent of the potable water that it provides to its residents (Pezzoli 2000). The 300-meter Torre Bicentenario would not only require large amounts of water but also enormous pressure to force it up through seventy stories, robbing neighboring residents of precious water. The electrical grid is also already strained, and according to the report, the presence of a building this size would cause blackouts in surrounding neighborhoods. It also claimed that the skyscraper would cast a shadow in the winter and summer months spanning 376 meters in distance, robbing the neighborhoods of the limited sun cast in the basin of Mexico City. Additionally, the sheer size of the prospective skyscraper proves a hazard in such a seismically active zone as Mexico City, where residents are still painfully carrying with them the memories of the set of 1985 earthquakes that devastated many parts of the city, crossing class lines and reaching across old and new neighborhoods.

Much of the resistance to the tower has been mobilized by the elected delegate who heads the Delegación Hidalgo, the would-be site of the Torre Bicentenario (a *delegación* is similar to a borough in New York – the Federal district is divided into sixteen *delegacións*). Until 2000, the heads of each *delegación* were appointed by the Mayor of Mexico City. With the advent of decentralization, the delegates from each *delegación* are now popularly elected. The autonomy of the *delegación* causes even greater messiness in local politics. The Mayor of Mexico City is from the PRD, a left-of-center party, while the delegate from Delegación Hidalgo is a PANISTA, from the right-of-center PAN party. The GDF is controlled by the PRD.

Colonos de Molino de Rey took the environmental impact plan to the federal government offices on land use and to the GDF. Although GDF regulations existed against the construction of buildings over three stories tall in Delegación Hidalgo, the GDF had originally waived these restrictions, arguing that the tower would achieve great status for the city. With the insistence of the neighborhood association and the opposition to the construction of the tower by the National Institute of Anthropology and History and the Federal Agency for Culture and the Arts, coupled with the environmental impact report, the GDF courts finally cancelled permits for the construction of the tower.

The political geographies embedded in the resistance to the construction of the Torre Bicentenario speak to the complexities of the creation, implementation, resistance and negotiation vis-à-vis the PR. Through the mobilization of environmental laws and local politics through the GDF court system, the residents accomplished their goals.

Future spaces of the globalization of neoliberal capitalism

The public-private materialization of neoliberal governance does not only prevail among political regimes headed by the doctrines of free trade and markets. The adoption of neoliberal accountability and social liberalism has seeped into new and unexpected places. From NGO directors, to former Marxist-Socialist politicians, to indigenous feminist leaders, the practices of neoliberal governance tactics can be witnessed in the form of political and economic transparency, free trade over the environment and individual rights over the commons, to name but a few. Indeed, in the case study offered above, the left-of-center Mayor of Mexico City, Andrés Manuel López Obrador, became bedfellows with neoliberal billionaire tycoon Carlos Slim in the creation and implementation of the PR. Actors from the supposed right and left who are putting such policies into practice show no sign of resting. The reason why access to public space remains more important than ever is precisely because of the prevalence of neoliberal governance and the smoothing over of public space that is occurring at an accelerated global rate. It is in the city streets where these confrontations will occur – even in this era of social networking and globalization.

Mexico is one of the few examples of countries in the region where popular politics has not rejected the Washington Consensus for a neostructuralist regime, perhaps due to its relationship and proximity to the United States. Nonetheless, while there sits in Los Pinos, the presidential palace in Mexico City, a president with degrees from Harvard in economics and public administration who is well versed in neoliberal governance, there are many examples of people rejecting the type of top-down, socio-urban control as that purported by the PR, which aims to normalize the historic center for tourists, international investors and the like.

In this chapter, I described the revitalization policies that are being implemented in Mexico City and I relayed two stories of resistance to the program. The first analyzed how and why the working-poor street vendors resist the normalization of space, a key component to the revitalization of the historic center. In the second, I explained how and why elite residents resist the construction of what would have been Latin America's tallest skyscraper. These two examples demonstrate how distinctive groups interpret neoliberal governance polices, in this case the PR. On the one hand, the street vendors aim to secure their access to the streets and plazas to produce and sell their wares, thus effectively resisting the neoliberalization of space in the center. The *ambulantes* imagine the historic center as a site for their own economic production, and as a site for consumption by working-poor residents who rely on the pirated products the vendors peddle. To date, $1 billion has been spent on the rescue and recovery of the historic center and the PR, but street vendors remain. Their enduring presence is a result of their fight to utilize public space in the way they see fit.

On the other hand, the elite residents of Molina de Rey do not necessarily support the *ambulantaje*. Indeed, a side argument against the construction of the Torre Bicentenario centered on how the tower may have attracted *ambulantes* who would sell food outside the skyscraper to employees and residents. Nevertheless, the residents of Molino de Rey do not blindly support each aspect of the PR. The residents believed the construction of the tower would negatively impact their quality of life, and they successfully argued that the tower's negative environmental impacts would degrade their neighborhood. The courts agreed, and the Council and GDF halted construction of the tower. These two examples of how distinctive groups interpret neoliberal urban policies and imagine the utilization of space speaks to the openness of the city. In Mexico City, urban space and active political participation directly impact the utilization of the city. It is in the public spaces of Mexico City that these struggles play out, and that the city is imagined and produced.

Notes

1 *El Universal* is a right-of-center newspaper published daily in Mexico City, and is part owned by Slim.

2 A *sexenio* is the six-year term of office that federally elected and municipally elected officials serve in Mexico.

References

Barajas, L. F. and Zamora, E. (2002) "Nuevas Formas y Viejos Valores: Urbanizaciones Cerradas de Lujo en Guadalajara", in L. F. Barajas (ed.) *Latinoamerica: Paises Abiertos Ciudades Cerradas*, Guadalajara: Universidad de Guadalajara, UNESCO.

Borja, J. (2001) "Centros y Espacios Publicos como Oportunidades", *Perfiles Latinoamericanos*, 19: 115–30.

Brunn, S. (2012) *Cities of the World: World Regional Urban Development*, New York: Rowman & Littlefield Education.

Camp, R. (1996) "Some Interpretations of the Mexican System", in R. Camp (ed.) *Politics in Mexico*, New York and Oxford: Oxford University Press.

Canclini, N. G. (2000) *Consumers and Citizens: Globalization and Multicultural Conflicts*, Minneapolis, MN: University of Minnesota Press.

Canclini, N. G. (1989) *Hybrid Cultures: Strategies for Entering and Leaving Modernity*, Minneapolis, MN: University of Minnesota Press.

Castells, M. and Portes, A. (1989) "World Underneath: the Origins, Dynamics and Effects of the Informal Economy", in A. Portes, M. Castells and L. A. Benton (eds) *The Informal Economy: Studies in Advanced and Less Developed Countries*, Baltimore: Johns Hopkins University Press.

Cross, J. C. (1998) *Informal Politics: Street Vendors and the State in Mexico City*, Stanford, CA: Stanford University Press.

Davis, D. E. (1994) *Urban Leviathan: Mexico City in the Twentieth Century*, Philadelphia, PA: University of Philadelphia Press.

Dezalay, Y. and Garth, B. (2002) *The Internationalization of the Palace Wars: Lawyers, Economists, and the Contest to Transform Latin American States*, Chicago and London: University of Chicago Press.

Esteva, G. (1991) "Tepito: No Thanks, First World This Inner-City Community in Mexico City Prefers to Manage its Own Affairs, Without Interference from the Forces of "Development", *Reclaiming Politics*, Fall/Winter: 38–46.

Flores, L. G. (2007) "Por La Noche de Primavera, Fiesta en Plazas y Calles del Centro", *La Jornada*, 14 March. Online: www.jornada.unam.mx/2007/03/14/index.php?section=capital&article=039n1cap.

Gilbert, A. (2004) "The Urban Revolution", in R. Gwynne and C. Kay (eds) *Latin America Transformed: Globalization and Modernity*, Vol. 2, Oxford: University of Oxford Press, pp. 93–116.

Gilbert, A. (1996) *The Mega-City in Latin America*, Tokyo, New York and Paris: United Nations University Press.

Gwynne, R. and C. Kay (eds) *Latin America Transformed: Globalization and Modernity*, Vol. 2, Oxford: University of Oxford Press.

Hackworth, J. (2006) *The Neoliberal City: Governance, Ideology, And Development in American Urbanism*, Cornell, NY: Cornell University Press.

Hackworth, J. (2002) "Postrecession Gentrification in New York City", *Urban Affairs Review*, 37 (6): 815–43.

Hackworth, J. (2000) *The Third Wave*, New Brunswick, NJ: Rutgers University.

Hamnett, C. (1973) "Improvement Grants as an Indicator of Gentrification in Inner London", *Area*, 5: 252–61.

Johns, M. (1997) *Mexico City in the Age of Diaz*, Austin, TX: University of Texas Press.

Jones, G. A. and Varley, A. (1999) "The Reconquest of the Historic Centre: Urban Conservation and Gentrification in Puebla, Mexico", *Environment and Planning A*, 32: 1547–66.

Jacobs, J. (1961) *The Death and Life of Great American Cities*, London: Random House.

Labotz, D. (1995) *Democracy in Mexico*, Council Bluffs, IA: Southside Press.

Low, S. M. (2000) *On the Plaza: the Politics of Public Space and Culture. Power, Domination, and Resistance in Public Space*, Austin, TX: University of Texas.

Malkin, E. (2008) "Latin America's Tallest Skyscraper", *New York Times*, 20 February.
Pezzoli, K. (2000) *Human Settlements and Planning for Ecological Sustainability. The Case of Mexico City*, Cambridge, MA: MIT Press.
PGDDF (2001) Programa General de Desarrollo del Distrito Federal.
Piccato, P. (2001) *Crime in Mexico City, 1900–1931*, Durham, NC: Duke University Press.
Pile, S. (1997) *Geographies of Resistance*, Oxford: Oxford University Press.
Poniatowska, E. (1995) *Nothing, Nobody: the Voices of the Mexico City Earthquake*, transl. A. C. de Schmisdt and A. Schmidt, Mexico DF: Mexico Temple University Press.
Portes, A. and Hoffman, K. (2003) "Latin American Class Structures", *Latin American Research Review*, 38 (1): 41–82.
PUEC (Programa Universitario sobre Estudios de la Ciudad) (2008). "El Ambientaje en la Ciudad de Mexico: Investigaciones Recientes", Mexico City: Programa Universitario sobre Estudios de la Ciudad UNAM.
Potuoglu-Cook, O. (2006) "Beyond the Glitter: Belly Dance and Neoliberal Gentrification in Istanbul", *Cultural Anthropology*, 21 (4): 633–60.
Profitt, T. D. (1996) *Tijuana: the History of a Megalopolis*, San Diego, CA: San Diego State University Press.
Programa de Recuperación (2001) *Consejo Consultivo: Entrevista con Carlos Slim Helu*, Ciudad de Mexico: Centro Historico.
Rama, A. (1996) *The Lettered City*, Durham, NC: Duke University Press.
Roberets, S. M., Secor, A. and Sparke, M. (2003) "Neoliberal Geopolitics", *Environment and Planning A*, 35 (12): 2094–6.
Ros, J. (2000) "Employment, Structural Adjustments and Sustainable Growth in Mexico" *Journal of Development Studies*, 36 (4 April): 100–19.
Saade, C. L. (2003) "De $450,824 Millones, la Participacion de Empresarios, *La Jornada, Mexico DF* (22 January).
Sanchez, A. B. (2007) "Preocupa Concentracion de Espacios por Grupo Carso en el Centro Historico", La Jornada: 28 December. Online: www.jornada.unam.mx/2007/12/28/index.php?section=capital&article=028n1cap.
Scott, J. C. (1985) *Weapons of the Weak: Everyday Forms of Peasant Resistance*, Newhaven, CT: Yale University Press.
Smith, R. G. (2003) "World City Actor-Networks", *Progress in Human Geography*, 27 (1): 25–44.
Smith, R. G. (2000) "World-City Network: a New Metageography?", *Annals of the Association of American Geographers*, 90 (1): 123–34.
Soederberg, S. (2001) "From Neoliberalism to Social Liberalism: Situating the National Solidarity Program within Mexico's Passive Revolutions", *Latin American Perspectives*, 118 (3 May): 104.
Teichman, J. (1992) "The Mexican State and the Political Implications of Economic Restructuring", *Latin American Perspectives*, 19 (2), Spring: 88–104.
Tuathail, G. Ó. and Roberts, S. M. (1998) *An Unruly World? Globalization. Governance and Geography*, London: Routledge.
Valaskakis, K. (1999) *Globalization as Theater*, Oxford, UK: Blackwell Publishers.
Venegas, L. (1999) *Los Centros Historicos en Nuestro Tiempo*, Mexico City: Conaculta.
Walker, D. M. (2011) "Urban Geography Past, Present, Future", in J. Stoltman (ed.) *Twenty-First Century Geography: a Reference Handbook*, Los Angeles, CA: Sage.

Walker, D. M. (2000) "Ejido Privatization and Urbanization: Neoliberalism and Landscape Changes in Tijuana Mexico", unpublished MA thesis, San Diego State Univeristy.

Ward, P. (1998) *Mexico City: the Production and Reproduction of an Urban Environment*, Boston, MA: G. K. Hall & Co.

Warnock, J. (1995) "The Economic Challenge", in Warnock, *The Other Mexico*, Montreal: Black Rose Books.

9 City ghosts

The haunted struggles for downtown Durban and Berlin Neukölln

Christine Hentschel

Urban futures emerge not only from visionary yearnings for the city yet to come, but also from dystopian depictions of the city to be avoided. This is especially true for cities that strive to free themselves of their troubled histories and leave behind the ghosts of the past. For activists of a democratic and just urban renewal in the South African coastal city of Durban, *segregation* is a ghost from the past that haunts the city's present, one that shapes debates around the possibilities of a truly open and democratic urban sphere. In the eyes of city activists in the German capital, the youngest ghost inhabiting Berlin is *gentrification*. The process of upgrading and selling off the former GDR district of Prenzlauer Berg threatens to repeat itself in the low-income district of Neukölln, which finds itself increasingly conquered by young artsy hipsters. The concerns with segregation in South Africa and gentrification in Germany share a focal point: the refusal to accept a city that curtails the poorest population's right to the city. In each case, the dominant critical narrative (segregation/gentrification) is embedded in a particular "spatial history" (Abbas 2008: 12) and in the city's own dreams. This chapter traces these dreams (and nightmares) and reflects on the possibilities of overcoming the particular narrowness inherent in the two cases by reading one case against the other.

Durban and Berlin seem to have little in common. The two cities assume completely different roles, both in the global system and in their respective countries. What they do share, however – apart from their population size of 3.4 million inhabitants – is their experience of a decades-long history of urban division, which in both cases began to crumble around 1990. The sudden political transformation in both cities was part of a radical political shift on a broader national scale and had far-reaching implications for the spatiality of the cities. These urban transformations incited hopes and visions for a newly open and integrated city – imaginations that only occur with such ardor during historical ruptures (Latour 2005: 59; Foucault 1991).

I will first sketch a methodological argument on the validity of comparing urban spatial transformations in the Global South and the Global North. I will then discuss the South African urban ghost of segregation, arguing that the political debates on (the dangers of) a new segregation are too narrowly focused and miss softer tendencies of "social sorting" (Lyon 2003) that are at the heart of the everyday organization of the right to the city. I will then turn to Berlin and the widespread concerns about gentrification in Neukölln, a district that bears the stamp of "failed integration" and a variety of social difficulties, but that has recently begun to attract students, artists and young international hipsters. I argue that the dominant critiques of gentrification in Berlin are driven by white arrogance and self-hatred and do not engage in any meaningful alliances with the district's longer-standing (i.e., Turkish, Arab) population. In the last section, I reflect on the limitations of such paradigms for the city's future. I argue that underestimation and overestimation hinder both debates. In the case of Durban, many activists who have raised their voices against new forces of segregation in South African cities seem not to have fully embraced new articulations of segregation. Segregation no longer materializes itself solely in walls and control posts, but has become sophisticated: soft, small-scale, aestheticized, driven by consumption and sensation. The other extreme is apparent in Berlin; gentrification has become a standard diagnosis for any change in the urban environment and a commonplace insult against a wide range of lifestyles. Both contestations have too narrow a horizon to address sufficiently the critical questions of their respective cities. Widening this horizon can succeed when the past does not serve as the lone and crucial reference point, but when, instead, the organization of the right to the city is always examined under contemporary conditions. Here the double perspective on the debates of Durban and Berlin will be useful for highlighting the respective blind spots in the fights against their city ghosts. I will show that Berliners could benefit from fearing segregation, while the Durbaners could do with a stronger suspicion towards gentrification and beautification.[1]

Postcolonizing urban research

My concern in this article is partially meta-methodological: towards a multisite (Marcus 1995) urban inquiry that neither assembles all urban developments under the narrative of neoliberal globalization nor assumes the incomparability of each case. I share the unease that scholars like Jenny Robinson, Ananya Roy, Edgar Pieterse, Achille Mbembe and Sarah Nuttall have expressed in recent years: urban theory, Robinson argues, has been divided between "Western and other cities", between "celebrations of 'modernity'" on the one hand and "promotion of urban development" on the other (Robinson 2006: 2). This division of the urban world has resulted in an impoverishment of inquiry and theory-building on cities in the South *and* the North. African cities were, and still are, viewed as "the intractable, the mute, the abject, or the other-worldly", the bad example of something else (Mbembe and Nuttall 2004: 348). It is time, Mbembe and Nuttall argue, to "write African cities into theory" (Mbembe and Nuttall 2008: 4), notably by moving beyond narratives of crisis, underdevelopment and ethnographic othering and reaching out to urban theory at large. This also means challenging Western-centric accounts of the innovation and worldliness of cities and diversifying geographical references for urban theory-building (Robinson 2006: 11, 168ff.; Roy 2009; Robinson 2011).

The call to postcolonize urban research needs to be brought to the North, especially when a place is branded as poor, violent, failing – as in Neukölln. The conceptual angle on the right to the city in the Global South pursued in this volume can help to destabilize narratives of what cities of the Global North and the Global South are, and with which vocabulary we can think about them (Pieterse 2011; Pile 2006). If, for example, "the world at large is looking ever more 'postcolonial' ", as John and Jean Comaroff (2006a: 6) argue, the usual a priori distinctions between the Global North and the Global South are up for debate. A new set of tools is needed in which uncertainty becomes a technique of inquiry, a condition for discerning where the points of connection and disruption lie, both within and between cities. As literary and film theorist Ackbar Abbas holds, one should

> see the city not through privileged moments of insight or revelation, i.e., through epiphanies (suddenly I understand); but rather through working with our uncertainty, puzzlement and confusion, i.e., through negative epiphanies (suddenly I understand that I don't understand). I understand that I have been seeing everything through a frame, and that what is necessary is not to frame what I see, but to deframe it and allow anomalous details to have their say.
>
> (Abbas 2008: 14)

Deframing Durban and Berlin also implies not remaining attached to "the meaning of Durban" and the "meaning of Berlin" to the world. South African cities have for decades been a "negative role model for urban planners in other parts of the world" (Simone 2005: 321); they became iconic for their brutal history of segregation, tremendous social inequalities and, more recently, staggering crime rates, but also for their unprecedented transition to democracy. Berlin, on the other hand, had been the capital of Nazi rule, the line dividing the world into socialist and capitalist regimes – made literal by the Wall – and became the iconic place for the sudden dissolution of the East–West divide (Ladd 1997).

Rather than just reflecting on what these cities stand for – how they are special – one can see Durban and Berlin simply as "ordinary cities" in a "world of cities" (Robinson 2006, 2011). In Durban and Berlin global trends of neoliberal governance take place and unfold along particular spatial histories, with their own ruptures and contradictions. Durban's legacy of urban segregation, pervasive urban fear and a very tangible involvement in the global flows of amusement and consumption through offering itself as a host city for the 2010 World Cup make it a particularly strong site at which to investigate neoliberal urban transformation. Similarly, Berlin, with its history of division, growing creative industry and fascination with being "poor but sexy" (Mayor Klaus Wowereit[2]) has a particular story of global (ordinary) neoliberal transformation to tell. Two ordinary cities, two locales in which the *general* and the *particular*, the *global* and the *local* can be conceptualized side by side and, at points, interwoven.

Segregation, Durban's ghost

"A powerful ghostly hand of segregation intertwined with inequality" is at work in South African cities (Pieterse 2006: 286). In this chapter, I trace the contours of this ghost of segregation in its different shapes, beginning with the "old" segregation. I then examine the material and psychological ruins left in its wake, reflect on a growing unease about a "new" segregation in South Africa and, finally, extend the notion of new segregation into mechanisms of urban ordering that are usually taken for granted. I argue that style, emotion and the agency of space need to be taken seriously as mechanisms of social sorting and cannot be subsumed under the workings of race ("old" segregation) or class ("new" segregation).

"Old" segregation

The South African system of apartheid made the black majority not only non-citizens (denying them the right to democratic participation in the state), but also non-city-zens (denying them the right to the city – Pieterse 2009). Governing (through) space in apartheid can be described, in Fanon's words, as compartmentalizing spaces in all spheres of urban life (Fanon 1991: 82). Apartheid divided "homelands" from the "real" South Africa, townships from the "real" towns and cities, thus reducing them to "scattered partial forms hanging on to the state's body" (Mbembe 2004: 389). It partitioned beaches into white, non-white and international. It isolated servants' quarters from the houses of the homeowners and separated public facilities. An apparatus of architectural devices, such as partitions and walls, control posts and gates, were the means through which space was ordered, coded and divided.

The compartmentalization of urban space did not mean total separation. Indeed, the racialized frontiers in many cases can be more succinctly imagined as membranes that allowed for highly regulated movement between the different demarcated territories under the regime's influx control. Influx control channeled and distributed African labor and restricted and meticulously administered the movement and journey of black people into white areas (Posel 1991: 75ff.).

Segregation's ruins

Post-apartheid cities set out to overcome segregation and become integrated, democratic and inclusive. Yet segregation persists. Apartheid urbanism has left a spatial morphology that cannot be undone overnight, despite the efforts of city councils to build inclusive cities (see Pieterse 2006: 286; Myers and Murray 2007; Terreblanche 2002). A Durban city official puts it succinctly:

> Spatial restructuring is extremely difficult. And it shows you [that] if you really want to wreck a city, design it wrong spatially and they'll hardly ever be able to get it right afterwards [...].
>
> (Interview, traffic planner 2007)

Apartheid's ruins are not only spatial. A particular (mainly white) proclivity to mourn a lost urban sphere has laid itself on an otherwise enthusiastic mood among city makers to build an integrated democratic city. Referring to "his" beach, a hotel safety manager, responsible for security issues on the wider Durban beachfront, expresses such a sense of loss:

D: Because, I don't know why, before 1994, it is not that I am a racist or from the old regime or whatever, I could walk on the beachfront. [...] It was safe. [...] I think, let's go back in national stats and look at mugging tourists in the country before 1994 and after the year 2000. Then, obviously, you see the difference. [...]

CH: Who was at the beachfront before [...] the 1990s? Who had the right to the beach actually?

D: You know, at one stage, you're right, I don't know if it was before 1994, only whites were allowed to the beachfront. I think that changed in 1994. Before '94, no, you're correct there; only whites had the right to go to the beach before 1994. [...] Which is maybe wrong. I don't approve of that. Surely that was wrong. But if we go to the beach, we go to the beach to enjoy ourselves and to have a good life, or do we go there to get killed? That is what's changed. We've changed this thing about racism and everything, but now people get killed.

(Interview, beach security manager 2007)

The manager expresses nostalgia for an urban space that no longer exists. In that space, "this thing about racism" appears as a less-than-concrete and slightly dubious narrative outweighed by more tangible "facts", such as that he "could walk on the beach", whereas "now people get killed". Racism, he implies, belongs to the old regime and has been replaced by killings in the new regime – a much more urgent concern. To be sure, his grief for a safe beachfront is a form of nostalgia for an urban sphere in which most of the population was denied the right to the city and "killing" was not debated, counted or visualized to the same degree, rendering it impossible to "go back in national stats" and compare the numbers (see Comaroff and Comaroff 2006b: 220; MacDonald 2006: 13; Shearing and Berg 2006: 197; Shaw 2002: 15). Through such an "apartheid nostalgia" (Kynoch 2003), as through the material ruins of apartheid, segregation reaches into the urban spheres of the "New South Africa".

Segregation's new Gestalt

Segregation persists in the material ruins of apartheid spatiality as well as in the grieving for a lost urban past. It also re-emerges in novel forms, driven by different actors. New fragmentations have arisen from the ruins of apartheid segregation, this time not directed by a state force but created in a polyphonic assemblage of private developers, house owners as well as a variety of informal initiatives and everyday deals. Driven by a deep fear of violent crime, the middle classes withdraw from the city into fortified cocoons of imagined safety on the outskirts of the city. They have given up on the city – an attitude that has further undermined the dreams of a truly public sphere in the New South Africa (Nuttall 2008: 198; Beall *et al.* 2002; Robins 2002; Bremner 2002).

Post-apartheid city-making, many deplore, is merely neoliberal spatial ordering (Miraftab 2007: 603; Robins 2002), a privatized urbanism (Murray 2008: 25) that creates exclusionary forms of enclave-style urban development. New urban forms in South African cities, from City Improvement Districts to priority development precincts, these critics argue, are driven by big companies, or municipalities that reconfigure themselves as companies ("municipality inc." – Miraftab 2007), and flourish at the cost of possible social and spatial integration (Murray 2008; Hart 2008; Bond 2004; Samara 2011; Pieterse 2009). At the heart of the analysis lies the emergence of a new segregation, as South African political economist and activist Patrick Bond writes:

> The reality is that South Africa has witnessed the replacement of racial apartheid with what is increasingly referred to as class apartheid – systemic underdevelopment and segregation of the oppressed majority through structured economic, political, and cultural practices.
>
> (Bond 2004)

This new class segregation reduces the achievements of universal citizenship in the New South Africa to a "contingent practice" (Murray 2008: 26), where the urban poor cannot benefit from their abstract right to the city (Miraftab 2004, 2007).

Most analyses of a new segregation focus on the social and political effects of boundaries, restriction and forceful exclusion, and scrutinize the agency of big business, the municipality and their various partnerships. They discover, as it were, apartheid's little brother in contemporary techniques of urban politics. I argue that this perspective is too narrow: this framing of neoliberal segregation prevents some of the equally important dynamics in city making from being addressed. It not only fails to do justice to genuine municipal attempts to democratic und socially fair urban re-ordering, but – even more importantly for this chapter – it is blind to the manifold (sometimes small-scale and informal) techniques of governance that do not speak the brutal language of exclusion. In the present phase of late capitalism, seduction and desire are sometimes more powerful mechanisms of urban ordering than restriction and prohibition. City makers around the world are increasingly attuned to the powers of affect and aesthetic, and this is not only true for brand-conscious city governors as a way of competing on a world scale, but also for the urban everyday of a barkeeper or a street trader. A critique of segregation needs to delve into the nitty-gritty of small-scale ordering and address the variety of social sorting in the city. I elaborate on this in the next section, using the example of a downtown bar owner who is trying to make his bar safer.

Between old and new segregation: flirty surfaces

Egagasini is a popular downtown bar in the former Point Road area (now renamed Mahatma Gandhi Road) – an area that has struggled with social problems and crime for a long time. Recalling the opening of his bar four years earlier, the owner, a young Zulu man, says that he "knew what kind of people [he] want[ed]". His plan was to "be expensive, because I need that type of people. Because, if you're cheap, you're inviting the wrong kind of people" (interview, bar owner 2008). In Egagasini, attracting the "right" people and keeping the "wrong" people out is done using visual and sensual cues. The display of liquor is crucial. In his newly opened and more luxurious downstairs lounge, the owner has opted for a particular arrangement of the liquor he sells in order to send a specific message:

> For the drinks that I've got, like Johnnie Walker, you've got five different types [...]. So, I cut the cheap one and then I start with the [more] expensive. [...] So, now, automatically a person communicates with the product and then he'd tell himself, "No, I am not suitable for this place, because I can't afford." You don't have to tell him.
>
> (Interview, bar owner 2008)

The barkeeper's enthusiasm for "not having to tell people" is striking. He instead outsources relevant communication to objects in his bar: the purchasable liquor, when aesthetically arranged, sends out messages that are inviting to some and repellant to others. He sees the liquor arrangement as *interacting* with customers, who, by viewing the display of liquor in the lounge, can decide whether they "fit in". No bouncer or barkeeper needs to tell somebody that he or she is unwelcome; they will quickly feel it for themselves.

The liquor sold is only one part of a wider ambiance project, in which music, color and decoration are expected to affect security on the premises. His DJ plays jazz, not "bubblegum music", as he calls the sounds in other popular clubs and bars in the area. In addition, colors (especially red, orange and yellow) "are most important, because they communicate with your heart" (interview, bar owner 2008). Asked about a dead lion hanging on his wall, the owner tells me that he shot it himself in the Zululand countryside. In response to my question as to why he chose a lion to decorate the wall, he replies with fervor: "I mean, you feel great" (ibid.). The right music, colors and wall decoration, according to his rationale, can affect human feelings and behavior *directly*, without having to turn to more costly (human) mediators such as security personnel.

The bar owner integrates music, color, wall decoration and liquor presentation into the safety plan for his bar. In addition to marking who is and who is not invited, these spatial clues are also implicated in shaping people's well being on the premises and contributing to the respect they show the place and other patrons. His bar's flirty surfaces attract and seduce, creating an atmosphere in which crime is unlikely. While dominant "criminologies of place"[3] (Herbert and

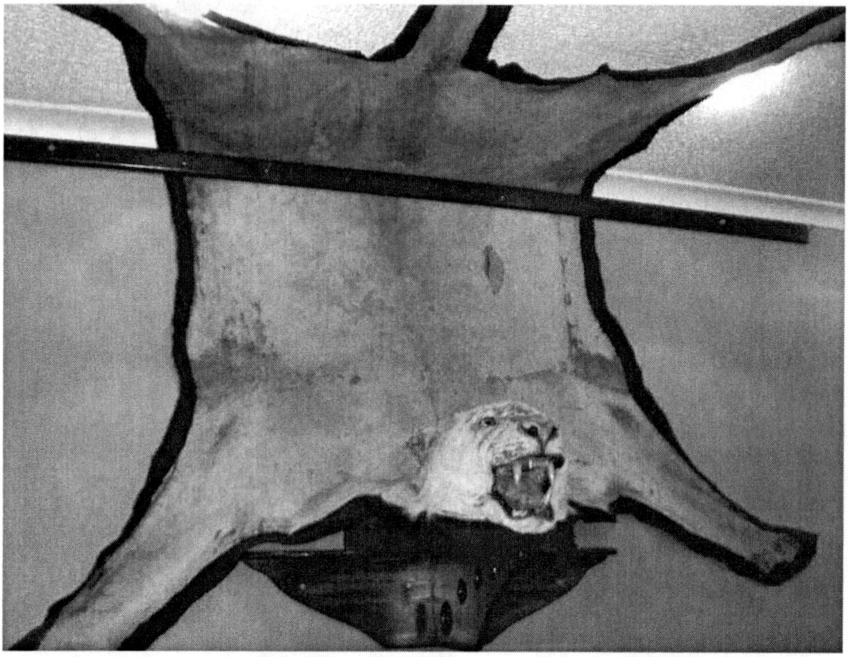

Figure 9.1 Lion decorating the wall of Egagasini (source: Christine Hentschel, 2008).

Brown 2006) have created techniques for "outdesigning crime" (Clarke and Newman 2005), I have proposed the term *outcharming crime* (Hentschel 2011) in order to highlight the emotional language through which potential dangers are addressed. Good habits are imagined to spread and to *outcharm* all potentially contagious bad behaviors. Outcharming crime, then, according to the bar owner's ideal, operates through a double mechanism: it entices the right kind of people to the place and makes them behave in the right way.

Seeking to charm and to seduce through the spatial language of music, decoration and liquor, the bar owner has created a spatialized "lovemark". The "lovemark" is branding specialists Saatchi & Saatchi's discovery that businesses have to make their consumers "love" their products. People should not just need, respect and value such products; they must love them, discover the mystery and the passion attached to them and plunge into the sensuality and intimacy they embody (Roberts 2007). This principle is increasingly applied to the urban sphere by a variety of urban planners, image specialists and businesses (Thrift 2004; Löw 2008; Landry 2007). The creator of Egagasini wants people to feel good in his place, not just to respect it: he competes for paying and peaceful clients by appealing to their senses.

Yet Egagasini is located in a rough and crimeridden downtown area. The entrance to the lounge leads through an underground parking area often used by informal hawkers to store their merchandise and to sleep. The day before the 2008 opening of the new lounge, graffiti on the wall gave a clear message as to what was not desired in the underground parking lot: "This is not a toilet. No sleeping on permisses [sic]." A few feet away, one could read: "Keep clean". The bar owner's flirty and non-verbal messages (to his customers) are limited to the bar's interior, while negative and verbal messages (towards those who could threaten the ambiance or the safety of the club) are written on the walls outside. At the entrance to his bar, a sign says, "No drugs, no weapons".

Interestingly, these "straight" messages do not directly exclude particular categories of people (the homeless, drug users, weapon holders), but particular items (weapons, drugs) and behaviors (defecating, sleeping, dealing drugs). Via signs, the owner decides to ban specific *actions* and *objects*, not *people* (at least not explicitly). In contrast to older logics of social sorting in South Africa, which were built on the violent restriction of particular spaces for racialized categories of people, these emerging spatial modes of urban ordering are more open, sophisticated, subtle and limited to a particular time-space. In such new surfaces for the exercise of power, mechanisms for urban ordering are inscribed in visual and atmospheric cues. This renders social sorting fairer in principle, but requires

Figure 9.2 Graffiti in the underground parking lot beside Egagasini (source: Christine Hentschel, 2008).

Figure 9.3 Graffiti in the underground parking lot beside Egagasini (source: Christine Hentschel, 2008).

a certain level of skill and knowledge in reading and understanding the codes. The bar owner, just like other guardians and managers of urban places, draws from a wide choice of messages – flirty or repellant, coded or straightforward – depending on whom he wants to reach. What is appealing to some is not so "nice or cuddly" (Thrift 2004: 58) to others – especially those who are too shabby and poor to fit into these handsome spaces.

The negotiation around the right to the city as witnessed in the case of the bar is subtle, indirect and built around people's "culture" and "style". Segregation today also needs to be understood through the seductive but prickly articulations of global "lovemark capitalism" – that is, consumption, taste, emotion. The endeavor, then, consists in disentangling how affective and seductive modes of ordering are intertwined with more rigid and exclusionary maneuverings of urban life.

Gentrification, Berlin's ghost

Berlin, too, is a haunted city (Ladd 1997: 1). It is "a city whose buildings, ruins and voids groan under the burden of painful memories" (ibid.: 3). The Nazi regime, the destructiveness of World War II and the division into East and West compose the troubled history of twentieth-century Berlin. In the shadow of this historical legacy, officials were eager to reinvent the reunified Berlin as a post-totalitarian, democratic and cosmopolitan city that displays both confidence and humility (Ladd 1997; Till 2005). Post-Wall Berlin became a large construction site: new governmental and corporate buildings catered to its regained role as the capital of reunified Germany; new memory sites were created to make visible (some of) the wounds of the past, while obscuring others. Decades-old ghosts mingled with a much younger specter not yet common in the German discourse: gentrification. While its older siblings claim their influence in the remaking of the inner city, gentrification is the ghost born of a massive urban upgrading in some of Berlin's eastern residential districts and now frames the dominant public discourse about neighborhood transformation. I will briefly trace the arrival of gentrification in Berlin via Prenzlauer Berg, then reflect on the current gentrification in Neukölln and its somewhat awkward "worlding" strategies. Finally, I will comment on the Berlin gentrification discourse as an overly self-conscious and self-reflexive ghost.

Berlin's youngest ghost

In the South African urban context, the sensitivity to and the struggle against segregation arises from the decades of apartheid; in Berlin, the ghost of gentrification is relatively young and stems from the aftermath of German reunification and the upgrading of particular areas of the former East, for example, Prenzlauer Berg. The paradigm of Berlin housing politics shifted in the 1990s. Until that time, both German states had a strong (even though very different) policy of social housing for the middle classes and the poor (Bernt and Holm 2005: 108; Blasius and Dangschat 1990). This changed after reunification. Facing an enormous budget deficit in the 1990s, the Berlin government sold off real estate and public services (housing, electricity, transport, water) in an attempt to meet the shortfall. This was the beginning of a "total politics of buyout", as a leading official in the department for city development describes it (interview, Berlin Senate Administration for City Development 2010).

A radical transformation simultaneously began in some districts of East Berlin. After the fall of the Wall, formerly GDR-owned real estate was returned to its original owners or their heirs, who often promptly resold them to professional real estate companies (Holm 2006: 117). With the housing stock in a very poor condition, Prenzlauer Berg was designated a redevelopment zone and the Berlin government invested millions in upgrading its housing and infrastructure (Bernt and Holm 2005; Holm 2006; Dörfler 2010). Within a couple of years, its demographic structure dramatically changed. A better-educated, higher-earning population with different cultural tastes in-migrated; the old East Berlin working class out-migrated. Prenzlauer Berg is a place where organic wineries, gourmet shops and baby clothes stores flourish today, and rents have risen sharply. The story of transition in the inner city of former East Berlin in the 1990s has created the dominant framework through which present-day developments all over the city are viewed and judged. The ghost of gentrification goes around the city, and this time, many think, it will hit Neukölln.

Neukölln is a low-income neighborhood located in former West Berlin. Around 33 percent of its population has an immigration background; most came from Turkey as so-called guest workers to Germany in the 1960s, while others arrived as political refugees from former Yugoslavia or Lebanon (Amt für Statistik Berlin-Brandenburg 2011; Bezirksamt Neukölln 2006). The district has always been poor, but in the 1990s, after the Wall fell, Neukölln was hit especially hard (Hüge 2010). The remaining factories moved to the cheaper outskirts in the former East, and municipal support for Neukölln was reduced and transferred to urban renewal initiatives in former East Berlin districts. The drain of public support and factories further deepened the district's social problems. Neukölln became the epitome of a depressed, angry neighborhood.

In the early 2000s, enthusiasm seemed to overlay Neukölln's agony. The district's roughness was suddenly charming, its empty buildings were remodeled as places of creative potential and its cultural frictions emerged as the necessary constituents of any exciting neighborhood. The city magazine *Tip* recently compared present-day Neukölln to New York's Lower East Side thirty years ago (Slaski 2010). Crucial to the district's transformations has been the state-funded agency Coopolis, which fights shop vacancies in Neukölln by helping artists, artisans and creative businesses to negotiate low rents and by persuading property owners to offer their premises for free on a temporary basis in exchange for renovation work. Within a couple of years, Neukölln had turned into a paradise for poor creatives.

Neukölln's "worlding"

While word had already begun to spread that "Neukölln rocks" (*Zitty* 2008), manifold image campaigns were initiated (both by state-funded and privately sponsored branding projects) to polish up the area's reputation. It is fascinating to what extent such campaigns make reference to spatial and cultural iconography from world cities like New York. A newly released free booklet advertising the low-income, economically struggling shopping street Karl Marx Straße is entitled "Broadway Neukölln". In a similar vein, Neukölln designers created a T-shirt that says "The Big Buschkowsky" with a photo of the district's mayor Heinz Buschkowsky, associating him with the protagonist of the American film *The Big Lebowski*. Buschkowsky, like Karl Marx Straße, is neither fancy nor appealing: he is a rough guy who speaks his mind, and whom many people respect for his honesty – in stating, for example, that the politics of "integration" in Germany have failed and that his district is the best evidence for such failure.

Such appropriation of cultural brands from distant world cities is a well-known phenomenon, but anthropologists have documented it mainly with respect to poorer cities in the Global South. In North Tanzanian Arusha, the anthropologist Brad Weiss writes that barbers, for example, name their shops "Brooklyn Barber's Shop", "California Hair Cutting" or "Paris Cuts" – associating their businesses with symbols of Western modernity and implying that life is better elsewhere (Weiss 2009: 72ff.). By writing itself into a Broadway or Hollywood world, Neukölln does not propel itself into the league of such iconic places but rather shows that itself to be on par with Arusha, Tanzania, and other "ordinary cities" (Robinson 2006) where people dream of "urban elsewheres" (Robinson 2011).

In Neukölln, it seems, public image campaigners, artists and shop and barkeepers enact their own versions of "worlding". With "worlding", Simone describes the efforts of urban Africans to "engage the city as a resource for reaching and operating at the level of 'the world' " (Simone 2001: 22). For Asian cities, Roy and Ong (2011) conceptualize worlding as the "art of being global": it comprises the "ambitious practices that creatively imagine and shape alternative social visions and configurations – that is: worlds beyond those that already exist (Ong 2011: 12). Worlding cities often involves challenging dominant global or national hierarchies of cities. The manager of a Neukölln TV station, for example, calls his area "the avant-garde. It has been through all that which the rest of Germany still has to go through" (interview, Neukölln TV station 2010). He deems Neukölln's misery a comparative advantage, because the district has for decades confronted the problems that German society has yet to solve: immigration and intercultural dialogue, the crises of unemployment and education. Indeed, John and Jean Comaroff's argument, that postcolonial polities are the "harbinger [...] of a global future, of the rising neoliberal age at its most assertive" (Comaroff and Comaroff 2006a: 42), could be extended to Neukölln, where mingle disparate narratives of ghetto apocalypse and gentrified dystopia, poor and sexy, troubled integration and celebrated internationalization.

Meanwhile, for many newcomers Neukölln's appeal lies not in its future but in its "now". The advertisement for a newly created flea market, "Nowkoelln Flowmarkt", nonchalantly renames the neighborhood in an illustration of this fascination with Neukölln's present. *Now* is the moment in which Neukölln is energetic, exciting and charming: it must be celebrated before gentrification devours it, before gourmet cappuccino and handmade chocolate spoil the district and before rents rise to the levels reached in other parts of Berlin. House owners already report that after decades of difficulty in renting out their flats, they can finally ask for the rent that they have long wished for (Gannott 2010; Gentrification Blog 2010; Berliner Mieterverein 2010; interview, house owner 2010).

The Nowkoelln advertisement itself gives a taste of the anticipated end of the *now*. The flea market is called *Flowmarkt* (a play on the German for flea market: *Flohmarkt*). On the back of the leaflet is the slogan: "The flow must go on!" After Broadway Neukölln, The Big Buschkowsky and the New York Lower East Side, here is another uniquely Neukölln interpretation of a world hit: Queen's

Figure 9.4a Advertisement for Nowkoelln Flowmarkt. Poster and web campaign.

Figure 9.4b Advertisement for Nowkoelln Flowmarkt.

"The Show Must Go On". Just as the show must continue, even after the singer's foreseeable death, Neukölln's flow must continue even against threats of gentrification. And here the replacement of "show" with "flow" is telling: what the hipsters love so much is the "authenticity" and honesty of Neukölln, the absence of glamor and mainstream entertainment. As a networking activist amongst in-migrating creatives in Neukölln states:

> Neukölln's charm is that it is raw and rough. Like a raw diamond. One can still jump around, can realize one's dreams. [. . .] There is the space and you can still occupy it.
>
> (Interview, Neukölln networking activist 2010)

While many newcomers to Neukölln have stories of the cheap apartments and vacant spaces that lured them, for many long-term residents this transformation remains obscure. One young man who grew up in Neukölln and now runs a fast food restaurant explained how his Turkish and Arab friends perceive the transformation. He describes that, for many, it was simply "incredible", "what actually happened" (meaning how empty shops became a paradise for young creatives):

> people [now] move into those corners and do something with them, where [not even] a foreigner wanted to do anything with them anymore. This, after all, is why they were empty, because there is always this hierarchy. Alright, so there comes a German and does something with it, then comes a Turk and tries to do something with it, then maybe comes an Indian who tries to do something with it and if nothing at all works anymore, then it remains vacant. So then it is really over. And then suddenly the people from Swabia [in Southern Germany] come and open a bar or a pub where for ten years nobody wanted to do anything anymore and it had remained vacant. And now they are putting shops in and, of course, it works fantastically. And now many [of the locals] have the feeling that they have missed something.
>
> (Interview, Neukölln *Imbiss* owner 2011)

Suddenly, something functions in the district that did not function before, a shift that eludes most locals' understanding. The hipsters have a narrative on how they found Neukölln, how they populated its "voids", while most of the long-term residents do not. This speechlessness vis-à vis the changes also means that the hipsters' Nowkölln is not the old-timers' Neukölln, even if some long-term residents ride on the momentum and sell beer and soy milk to their new neighbors.

Mourning the future: the gentrifiers' self-hatred

Meanwhile, the "Nowkoelln" hipsters cannot entirely enjoy "their now". Graffiti insults them as yuppies and tells them to go home: in anti-gentrification blogs, Neukölln is the most problematized district in Berlin's discourse. In Neukölln, they say, the "caravan of the young and beautiful" (Gentrification Blog 2010) moves through and "displaces the poor by constructing the habitus, latte by latte, of the new urban middle class", as Sharon Zukin phrased it in the context of New York (Zukin 2010: 18). They are not themselves rich, but with their "cultural power" (Zukin 2010: 4), these artists and students serve as the "foot soldiers of gentrification" (Smith 2007). At the end of this battle, an anonymous poster campaign warns that the neighborhood is dead – murdered by the kind of hipster depicted in the photo.

I spotted this poster (Figure 9.5) in a window of an antique shop with the name *arm und sexy* (Poor and Sexy). Targeting the hipster's consciousness, it asks: *"Bist du auch ein Kiezkiller?"* (Are you also a killer of the neighborhood?). The poster

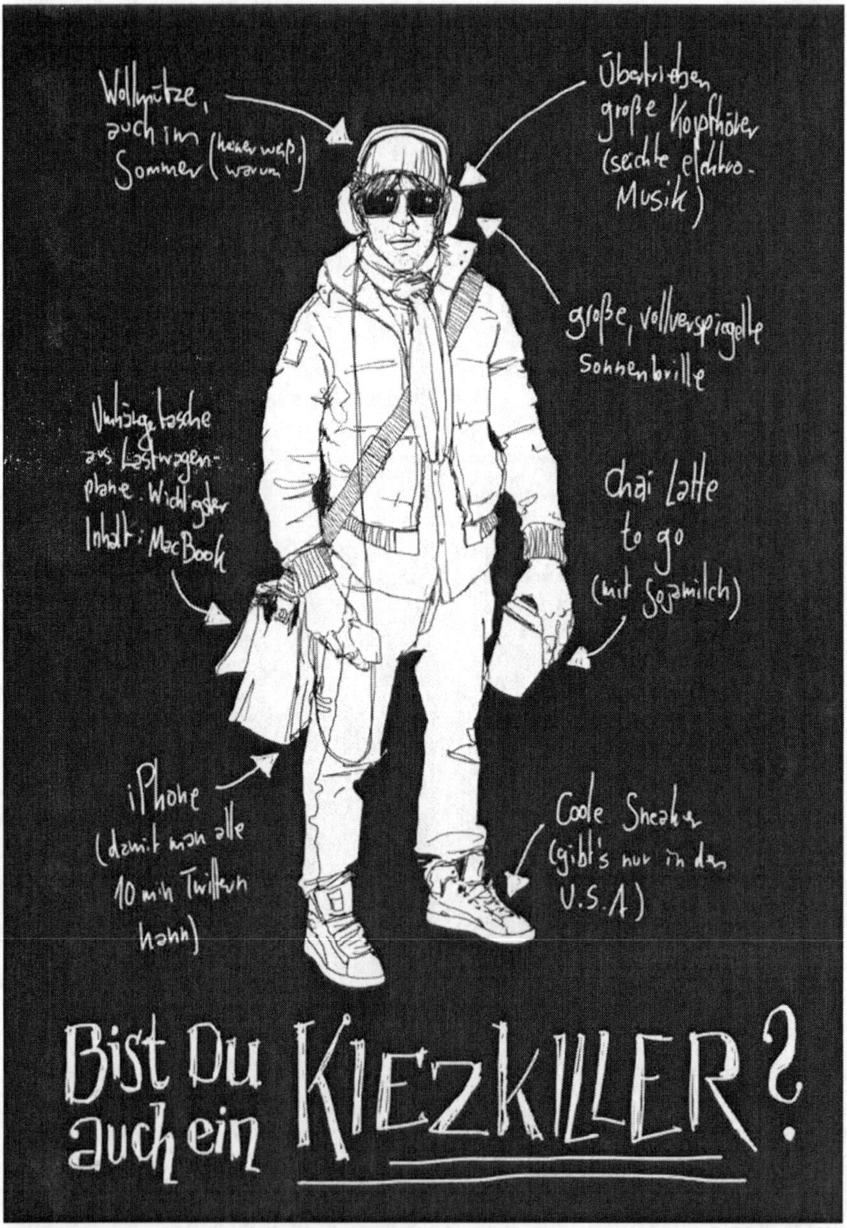

Figure 9.5 Are you also a killer of the neighborhood? Anonymous poster campaign in Neukölln, December 2010.

portrays the stereotypical Nowkoelln hipster, with iPod, trendy jeans, MacBook, cool American sneakers and a chai latte to go. With his humongous headphones, sunglasses and woolen hat, the *Kiezkiller* looks like an alien, as though he belongs to another planet. While walking through the streets of Neukölln, he stays encapsulated in his own world; his surroundings will not get through to him. At the same time he seems familiar. The poster's punch lies in the passerby's familiarity with the hipster codes while reinforcing such familiarity by involving him or her in the picture ("you", "also"). The poster bears no signature or label; the shop owner tells me that the designer wants to stay anonymous. The accusation is harsh: a person with a particular lifestyle and outfit is stereotyped and held responsible for the death of a neighborhood. What kind of death (death of authenticity? aesthetic death? displacement of the poor? death by boredom?), or what kind of neighborhood, the poster does not clarify; other, more structural drivers of gentrification are not mentioned.

The in-migrating students and artists have long internalized the accusation. They are self-reflective about their own and their peers' role in the neighborhood. A curious fight has arisen around who was here before the pull to Neukölln began, who is more of a neighborhood native, and hence, who is here legitimately. It is striking how far the debate is broken down to individuals, whose tastes and choices are depicted as key drivers of the dynamics that have fashioned Neukölln as Nowkoelln and that might turn Nowkoelln into a version of Prenzlauer Berg. Not the flows of capital (Smith 1979) or "the literal and figurative effacing of the proletariat in the city" (Wacqant 2008: 199), not urban politics that employ gentrification as a systematic and comprehensive method (Smith 2008; Bernt and Holm 2009) or international developments that force cities into competition with one another are held responsible here, but those "ready" to pay higher rents are blamed. The individualizing gentrification critique of the poster stands for a new genre of public discourse in which celebration and critique of gentrification are blurring.

An effect of the hipsters' self-hatred is that the debate itself takes place mainly between the young, white and educated. The voices of immigrants and lower-class old timers are rarely heard in gentrification slanders, and their absence in activists' alliances seems to cause little outcry amongst anti-gentrification activists. Seen against a South African context and a decades-long struggle in which blackness has been central, Berlin's lack of integration of the non-white population into the critical reflections on neighborhood transformations is particularly striking. In Berlin, it seems, the attention Neukölln's transition receives is not so much a concern with social diversity in the neighborhood as a minority worry about losing the excitement of a unique place. In this sense, the kind of gentrification critique encountered on Neukölln's walls and blogs remains vague: conscious of the battle over the district and the dangers of displacement (see the widespread slogan, "All of us stay!"), yet strikingly indifferent to and ignorant of the working-class population around it (see Slater 2006; Allen 2008; Wacqant 2008).

Conclusion: city ghosts in pasts and futures

Struggles against city ghosts are fights for a better city that take the troubled experience from a city's past as their key reference point to debate present and future city politics. I have explored segregation and gentrification as two such ghosts that frame the struggles for inner-city Durban and Berlin Neukölln. I have pointed to the narrowness of these frameworks in each case. Activist and scholarly debates in Durban focus on the (new and leftover) brutal architectural and technological forms of segregation (restrictions, surveillance, exclusion) and thereby miss softer techniques of governing whose impact on people's right to the city is equally severe. Hence, only segregation in its most rigid sense – as a socio-spatial exclusion of a particular group – the black, the poor – is recognized as a ghost to mistrust and combat. But contemporary techniques of social sorting have become small scale, fluid and lifestyle-based. They revolve around aesthetics and feelings for a place, and often avoid explicitly excluding groups of people but aim instead at shaping their behavior in a particular place in time.

While the case of Durban lacks a critical consideration of the workings of lifestyle and aesthetics in the spatio-political landscape of the city, in Berlin the critique of gentrification revolves exactly around these notions of lifestyle and aesthetics. Here, the fight against the ghosts of gentrification is often interpreted as a moral judgment of a particular lifestyle as misplaced in a district such as Neukölln. With such an individualized approach, the anti-gentrification campaigns examined in this chapter have depoliticized their own endeavor.

I do not mean to caricature the important struggles against segregation and gentrification in South African and German cities. Rather, I want to understand the respective frameworks through which urban activists look at critical issues of their city and ask what these frameworks (do and do not) allow them to see. By taking the example of the bar owner, I have explored a broad concept of social sorting – of behavior rather than identity – in which taste and style, emotion and sensation play an important role and are closely entangled with, yet not reducible to, class and race alone. Hence, anti-segregation activists need to include these affective and small-scale mechanisms of spatial ordering into their analysis and politics and ask, for example, how the everyday judgments of a security guard, the place-making practices of an informal trader, the aesthetic strategies of a big corporation or the decisions of a local planner are intertwined in their logics of ordering and how they are shaped by global rationales of governing cities. With the example of "Nowkoelln" I have pointed towards a standardized self-hatred among the footsoldiers of gentrification, who, in contrast, may take the affective and the small scale too seriously. Thus, the rising fight for Neukölln bears the stamp of a hipster-bourgeoisie that fears the loss of the very place where their Berlin feeling can best unfold.

The double perspective on the debates in Durban and Berlin sheds light on some blindspots in both discussions. This is most apparent in the Berlin case. Here, the social and political worlds of the different populations in the neighborhood of Neukölln rarely touch. Separated, even segregated social worlds do not seem to cause any embarrassment among much of the urban left, and the realities of working-class Arab and Turkish families or German welfare recipients often elude their apprehension and their interest. Berlin could use some of the long-grown consciousness vis-à-vis segregation that has developed in South Africa due to its painful history of apartheid and of persisting inequalities. On the other hand, South African urban activism might integrate into its anti-segregation critique a stronger suspicion of upgrading and inner-city beautification. After decades of experiencing space as a brutal and ugly mechanism of exclusion, there is a widely shared feeling in South Africa (also among the left) that handsome and charming spaces are the most powerful pieces of evidence for a successful urban transformation. In other words, by borrowing one another's ghosts from time to time, the respective struggles could embrace a wider vision for a right to the city.

I certainly do not mean to suggest that critics and activists should turn their backs on history, yet the specters of the past need not be their only or most powerful inspiration. Critical urban thought begins, in Doreen Massey's words, with an imaginary of space as "genuine plurality", and can unfold in a "radical commitment to the openness of the future" (Massey 2005: 281, 279).

Notes

1 This paper is based on field research conducted between 2005 and 2009 as part of my PhD thesis, "The Spatial Life of Security: Durban, South Africa" (Hentschel 2010), as well as ongoing research in Berlin Neukölln conducted in 2010 and 2011 in the context of my post-doctoral project, "KippCity", supported by the ICI Berlin. I would like to thank Katrin Pahl, Kit Heintzman, Ulrike Bialas and the editors for their inspiring comments on earlier versions of the text.

2 Klaus Wowereit used this expression first in an interview with the German magazine *Focus Money* in 2003; since then it has become a widely used Berlin slogan (see Frey 2003).

3 The term "criminologies of place" critically labels American and British approaches that address crime as a problem of spatial management – especially the "broken windows" theory (Wilson and Kelling 1982) and "situational crime prevention" (Clarke 1992). In these internationally highly popular crime prevention strategies, the relationship between urban space and crime control is depicted as negative and restrictive: crime can be "designed out" from particular places – for example, through "target hardening", surveillance systems and a built environment that makes the "rational offender" decide that this particular place is not worth the effort. Critical criminologists and urban sociologists have pointed to the exclusionary (side-)effects of these measures and their aggressiveness towards the poor (Harcourt 2001; Garland 2001).

References

Abbas, A. (2008) "Poor Theory and New Chinese Cinema: Jia Zhangke's 'Still Life'", lecture held at the Critical Theory Institute, University of California, Irvine, 3 December.

Allen, C. (2008) "Gentrification, Research and the Academic Nobility: a Different Class?", *International Journal of Urban and Regional Research*, 32 (1): 180–5.

Amt für Statistik Berlin-Brandenburg (2011) website: www.statistik-berlin-brandenburg.de/Regionales/r_sozh_g.asp (accessed 20 September 2011).

Beall, J., Crankshaw, O. and Parnell, S. (2002) *Uniting a Divided City: Governance and Social Exclusion in Johannesburg*, London and Sterling, VA: Earthscan Publications.

Bernt, M. and Holm, A. (2009) "Is It, or Is Not? The Conceptualization of Gentrification and Displacement and its Political Implications in the Case of Berlin-Prenzlauer Berg", *City*, 13 (2–3): 312–25.

Bernt, M. and Holm, A. (2005) "Exploring the Substance and Style of Gentrification: Berlin's 'Prenzlberg' ", in R. Atkinson and G. Bridge (eds) *Gentrification in a Global Context. The New Urban Colonialism*, London and New York: Routledge: 106–20.

Bezirksamt Neukölln von Berlin (2006) "Nationalitäten in Neukölln – differenziert nach Altersgruppen und Statistischen Gebieten", Deptartment of Youth, District of Neukölln.

Blasius, J. and Dangschat, J. (eds) (1990) *Gentrification. Die Aufwertung innenstadtnaher Wohnviertel*, Frankfurt and New York: Campus.

Bond, P. (2004) "South Africa's Frustrating Decade of Freedom: from Racial to Class Apartheid", *Monthly Review*, 55 (10): 45–55.

Bremner, L. (2002) "Closure, Simulation and Making-Do in the Contemporary Johannesburg Landscape", in O. Enwezor, C. Basualdo, U. Meta Bauer, S. Ghez, S. Maharaj, M. Nash and O. Zaya (eds) *Under Siege: Four African Cities. Freetown, Johannesburg, Kinshasa, Lagos. Documenta 11_Platform4*, Ostfildern-Ruit: Hatje Cantz, pp. 153–72.

Clarke, R. V. (1992) "Introduction", in R. V. Clarke (ed.) *Situational Crime Prevention: Successful Case Studies*, New York: Harrow and Heston, pp. 1–45.

Clarke, R. V. and Newman, G. R. (eds) (2005) *Designing Out Crime from Products and Systems*, London: Criminal Justice Press.

Comaroff, J. L. and Comaroff, J. (2006a) "Law and Disorder in the Postcolony: an Introduction", in J. Comaroff and J. L. Comaroff (eds) *Law and Disorder in the Postcolony*, Chicago and London: University of Chicago Press, pp. 1–56.

Comaroff, J. and Comaroff, J. L. (2006b) "Figuring Crime: Quantifacts and the Production of the Un/Real", *Public Culture*, 18 (1): 209–46.

Dörfler, T. (2010) *Gentrification in Prenzlauer Berg? Milieuwandel eines Berliner Sozialraums seit 1989*, Bielefeld: Transcript.

Fanon, F. (1991) *Les damnés de la Terre*, Paris: Gallimard.

Foucault, M. (1991) "Politics and the Study of Discourse", in G. Burchell, C. Gordon and P. Miller (eds) *The Foucault Effect: Studies in Governmentality, with Two Lectures by and an Interview with Michel Foucault*, Chicago: University of Chicago Press, pp. 53–72.

Frey, G. (2003) "Lassen Sie uns über Geld reden", *Focus Money*, 46. Online: www.focus.de/finanzen/boerse/aktien/money-talk-lassen-sie-uns-ueber-geld-reden-_aid_249988.html.

Gannott, S. (2010) "Vertreibungsgefahr ist nicht so groß", *Die Tageszeitung*, 1 January.

Garland, D. (2001) *The Culture of Control: Crime and Social Order in Contemporary Society*, Chicago: University of Chicago Press.

Gentrification Blog (2010) "Berlin Nordneukölln: es Gibt Sogar Schon Diplomaten Hier", 5 January. Online: http://gentrificationblog.wordpress.com/2010/01/05/berlin-nordneukolln-es-gibt-sogar-schon-diplomaten-hier/.

German Ministry of Transport, Building and Urban Development, Statusbericht (2008) "Zum Programm Soziale Stadt, Berlin", July.

Harcourt, B. E. (2001) *Illusion of Order: the False Promise of Broken Windows Policing*, Cambridge, MA: Harvard University Press.

Hart, G. (2008) "The Provocations of Neoliberalism: Contesting the Nation and Liberation after Apartheid", *Antipode*, 40 (4): 678–705.

Hentschel, C. (2011) "Outcharming Crime in (D)urban Space", *Social Dynamics*, 37 (1): 148–64.

Hentschel, C. (2010) "The Spatial Life of Security: Durban, South Africa", unpublished PhD thesis, School of Social Sciences, University of Leipzig.

Herbert, S. and Brown, E. (2006) "Conceptions of Space and Crime in the Punitive Neoliberal City", *Antipode*, 38 (4): 755–77.

Holm, A. (2006) "Urban Renewal and the End of Social Housing: the Roll Out of Neoliberalism in East Berlin's Prenzlauer Berg", *Social Justice*, 33 (3): 114–28.

Hüge, C. (2010) *Die Karl-Marx-Straße. Facetten eines Lebens- und Arbeitsraums*, Berlin: Karin Kramer Verlag.

Interview, Aktion Karl Marx Straße (2010) Berlin, 28 October.

Interview, bar owner (2008) Durban, 29 April.

Interview, beach security manager (2008) Durban, 23 April.

Interview, beach security manager (2007) Durban, 2 June.

Interview, Berlin Senate for City Development (2010) Berlin, 27 December.

Interview, house owner (2010) Berlin, 16 December.

Interview, Neukölln *Imbiss* owner (2011) Berlin, 11 May.

Interview, Neukölln networking activist (2010) Berlin, 8 October.

Interview, Neukölln TV station (2010) Berlin, 10 November.

Interview, traffic planner, Ethekwini Transport Authority (2007) Durban, 29 September.

Kynoch, G. (2003) "Apartheid Nostalgia. Personal Security Concerns in South African Townships, South Africa", *SA Crime Quarterly*, 5: 7–10.

Ladd, B. (1997) *The Ghosts of Berlin. Confronting German History in the Urban Landscape*, Chicago and London: University of Chicago Press.

Landry, C. (2007) *The Art of City-Making*, London and Sterling, VA: Earthscan.

Latour, B. (2005) *Reassembling the Social: an Introduction to Actor-Network-Theory*, Oxford: Oxford University Press.

Laumann, W. (2010) "Mietenentwicklung im Reuterkiez", presentation for the Berliner Mieterverein.

Löw, M. (2008) *Soziologie der Städte*, Frankfurt am Main: Suhrkamp.

Lyon, D. (2003) "Introduction", in D. Lyon (ed.) *Surveillance as Social Sorting: Privacy, Risk, and Digital Discrimination*, London and New York: Routledge: 1–9.

MacDonald, M. (2006) *Why Race Matters in South Africa*, Cambridge, MA: Harvard University Press.

Marcus, G. (1995) "Ethnography in/of the World System: the Emergence of Multi-Sited Ethnography", *Annual Review of Anthropology*, 24: 95–117.

Massey, D. (2005) "Spaces of Politics", in D. Massey, J. Allen and P. Sarre (eds) *Human Geography Today*, 2nd edn, Cambridge and Malden, MA: Polity: pp. 279–95.

Mbembe, A. (2004) "Aesthetics of Superfluity", *Public Culture*, 16 (3): 373–405.

Mbembe, A. and Nuttall, S. (2008) "Introduction: Afropolis", in S. Nuttall and A. Mbembe (eds) *Johannesburg: the Elusive Metropolis*, Durham, NC: Duke University Press, pp. 1–33.

Mbembe, A. and Nuttall, S. (2004) "Writing the World from an African Metropolis", *Public Culture*, 16 (3): 347–72.

Miraftab, F. (2007) "Governing Post Apartheid Spatiality: Implementing City Improvement Districts in Cape Town", *Antipode*, 39 (4): 602–26.

Miraftab, F. (2004) "Making Neo-liberal Governance: the Disempowering Work of Empowerment", *International Planning Studies*, 9 (4): 239–60.

Murray, M. (2008) *Taming the Disorderly City: the Spatial Landscape of Johannesburg after Apartheid*, Ithaca, NY: Cornell University Press.

Myers, G. A. and Murray, M. J. (2007) "Introduction: Situating Contemporary Cities in Africa", in M. J. Murray and G. A. Myers (eds) *Cities in Contemporary Africa*, New York: Palgrave Macmillan, pp. 1–25.

Nuttall, S. (2008) "Literary City", in S. Nuttall and A. Mbembe (eds) *Johannesburg: the Elusive Metropolis*, Durham, NC: Duke University Press, pp. 195–218.

Ong, A. (2011) "Introduction: Worlding Cities, or the Art of Being Global", in A. Roy and A. Ong (eds) *Worlding Cities: Asian Experiments and the Art of being Global*, Malden, Oxford: Wiley-Blackwell.

Pieterse, E. (2011) "Grasping the Unknowable: Coming to Grips with African Urbanisms", *Social Dynamics*, 37 (1): 5–23.

Pieterse, E. (2009) "Post-apartheid Geographies in South Africa: Why are Urban Divides so Persistent?", paper presented at Workshop Interdisciplinary Debates on Development and Cultures: Cities in Development – Spaces, Conflicts and Agency, Leuven University, 13 December.

Pieterse, E. (2006) "Building with Ruins and Dreams: Some Thoughts on Realising Integrated Urban Development in South Africa through Crisis", *Urban Studies*, 43 (2): 285–304.

Pile, S. (2006) "The Strange Case of Western Cities: Occult Globalisations and the Making of Urban Modernity", *Urban Studies*, 43 (2): 305–18.

Posel, D. (1991) *The Making of Apartheid, 1948–1961: Conflict and Compromise*, Oxford and New York: Clarendon Press.

Roberts, K. (2007) "Introduction – the New Consumer: Lovemarks and the Consumer Revolution", in K. Roberts (ed.) *The Lovemarks Effect: Winning in the Consumer Revolution*, Brooklyn, NY: Powerhouse Books, pp. 9–12.

Robins, S. (2002) "At the Limits of Spatial Governmentality: a Message from the Tip of Africa", *Third World Quarterly*, 23 (4): 665–89.

Robinson, J. (2011) "Cities in a World of Cities: the Comparative Gesture", *International Journal of Urban and Regional Research*, 35 (1): 1–23.

Robinson, J. (2006) *Ordinary Cities: Between Modernity and Development*, London and New York: Routledge.

Roy, A. (2009) "The 21st Century Metropolis: New Geographies of Theory", *Regional Studies*, 43 (6): 818–30.

Roy, A. and Ong, A. (2011) *Worlding Cities: Asian Experiments and the Art of being Global*, Malden, Oxford: Wiley-Blackwell.

Samara, T. R. (2011) *Cape Town After Apartheid: Crime and Governance in the Divided City*, Minneapolis and London: University of Minnesota Press.

Shaw, M. (2002) *Crime and Policing in Post-apartheid South Africa: Transforming under Fire*, Bloomington, IN: Indiana University Press.

Shearing, C. and Berg, J. (2006) "South Africa", in T. Jones and T. Newburn (eds) *Plural Policing: a Comparative Perspective*, London and New York: Routledge, pp. 190–221.

Simone, A.-M. (2005) "The Right to the City", *Interventions*, 7 (3): 321–5.

Simone, A.-M. (2004) "People as Infrastructure: Intersecting Fragments in Johannesburg", *Public Culture*, 16 (3): 407–29.

Simone, A.-M. (2001) "On the Worlding of African Cities", *African Studies Review*, 44 (2): 15–41.

Slaski, J. (2010) "Spielplatz der Avantgarde", *TIP Berlin*, 39 (6) , 4 March: 29–35.
Slater, T. (2006) "The Eviction of Critical Perspectives from Gentrification Research", *International Journal of Urban and Regional Research*, 30 (4): 737–57.
Smith, N. (2008) "On 'The Eviction of Critical Perspectives' ", *International Journal of Urban and Regional Research*, 32 (1): 195–7: www.bmgev.de/mieterecho/324/09-gentrifizierung-neil- smith.html.
Smith, N., interviewed by A. Holm (2007) "Der Zusammenschluss zwischen Anti-Gentrifizierungskämpfen und den Aktivitäten der weltweiten Bewegung für soziale Gerechtigkeit kann extrem bedrohlich werden", *Mieterecho*, 324, Oktober.
Smith, N. (1979) "Toward a Theory of Gentrification: a Back to the City Movement by Capital, Not People", *Journal of the American Planning Association*, 45 (4): 538–47.
Terreblanche, S. (2002) *A History of Inequality in South Africa, 1652–2002*, Pietermaritzburg and Sandton: University of Natal Press.
Till, K. E. (2005) *The New Berlin. Memory, Politics, Place*, Minneapolis and London: University of Minnesota Press.
Thrift, N. (2004) "Intensities of Feeling: Towards a Spatial Politics of Affect", *Geografiska Annaler*, 86 (1): 57–78.
Wacquant, L. (2008) "Relocating Gentrification: the Working Class, Science and the State in Recent Urban Research", *International Journal of Urban and Regional Research*, 32 (1): 198–205.
Weiss, B. (2009) *Street Dreams and Hip Hop Barbershops: Gobal Fantasy in Urban Tanzania*, Bloomington, IN: Indiana University Press.
Wilson, J. Q. and Kelling, G. L. (1982) "Broken Windows", *Atlantic*, March: 1–16.
Zitty (2008) "Neukölln rockt", 20 March. Online: http://service.zitty.de/magazin/4135/.
Zukin, S. (2010) *Naked City. The Death and Life of Authentic Urban Places*, Oxford: Oxford University Press.

Part III

Governance and counter-governance

The shape of urban conflict and the urban future

10 Insurgency and institutionalized social participation in local-level urban planning

The case of PAC *comuna*, Santiago de Chile, 2003–5[1]

Ernesto López-Morales

Introduction

From the 1990s onwards, most large Latin American inner-city areas have experienced different forms of urban restructuring. Santiago is no exception to this rule. Since 1991, the property-led urban renewal market of Santiago has been quantitatively successful at producing upper- and middle-income housing units in high-rise blocks, also continuously expanding the market towards areas of large rent gaps. This real estate market depends largely on local master-plan rezoning enforced by independent municipal governments alongside a national-state subsidy aimed at the acquisition of these residential units. However, it has also been recognized that Santiago's market of renewal produces ground rent dispossession and large-scale socio-spatial displacement, i.e., gentrification by ground-rent dispossession (or GGRD) as the land price paid to owner-occupants is usually below market prices (López-Morales 2011). Since 1990, with the sole exception of Pedro Aguirre Cerda (PAC) *comuna*, ten of the eleven inner-city municipalities of Santiago have redrafted their local urban master plan, specifically amplifying floor-area ratios (FAR), in order to attract the high-rise real estate market into their areas.

PAC is currently not only a territory well connected to the metropolitan center but also a historic working-class area located in the southern inner-city area of Santiago. It contains the emblematic La Victoria *población*,[2] which emerged in 1957 as the first of the many areas involved in the politically structured mass squatting movement in Chile and Latin America. During the military dictatorship (1973–90) there was political obliteration and military repression of its hitherto organized social movements, in addition to deliberate policies of increased social segregation (Finn 2006a, 2006b). The combination of insufficient state-led social housing, low social access to the rental market and iron-handed government opposition to land seizures (Gilbert 1993, 2002) dramatically increased overcrowding in the inner city, which had already been high. Currently, Santiago's inner city shows more than 50 percent of families sharing residence, with the housing shortage in La Victoria *población* is estimated at approximately 600 units, or 18 percent of a total of 3,200 dwellings. The high rate of overcrowding and the lack of green areas are problems occurring not only in La Victoria but also in the whole Pedro Aguirre Cerda *comuna*, which is the municipality of Santiago containing the fewest green areas, with a rate of 1.2 m^2 per capita, yet with a current rate of owner-residence at approximately 80 percent.

Starting in 2003, there has been an aggressive institutional attempt from PAC local government to make radical modifications to its local urban master plan. A first step was to hire a private consultancy firm (PCF) with this precise aim. Should PAC's new plan have been legally approved, it would have cleansed several traditional low-income areas, enlarging the rent gap aimed to attract middle- and high-rise renewal and thus increasing the average social composition of the PAC district. However, PAC's local community realized that by accepting the changes in the urban and building codes contained in the new master plan, larger rent gaps would not necessarily mean that higher land prices would be paid to traditional inner-city, low-income owner-residents as a form of ground rent. In fact, the opposite was the case. The issue of disparity between the ground rent potentially achieved by real estate operators and the actually much lower ground rent cashed in by owner-occupants is core in Santiago. This is because 80 percent of the urban population of Santiago is currently owner-resident and most households in PAC are multi-nuclear, making the actual low capitalized ground rent far insufficient in use value terms for those numerous, agglomerated households to find replacement accommodation elsewhere in a highly segregated and expanding city of 65,000 hectares, leading to large-scale displacement (López-Morales 2009, 2010b, 2011)

This chapter describes and analyzes the collectivized and institutionalized insurgency by the people of PAC, which was not only able to stop the top-down redrafting of the master plan attempted by PAC's local-level municipal government but was also able to produce a bottom-up alternative-draft master plan. This plan was able to control the potential ground rent increase in four key areas of the *comuna* and produce more adequate policies for the area, forcing the municipal authority's external technical apparatus to reach a compromise and draw up a different proposal.

The chapter begins with an illustration of the official definition of social participation in Chilean urban local planning codes, and the actual involvement the people were allowed to have in PAC. The following sections illustrate the political capacity of PAC residents to overcome those structural limitations and decisively intervene at technical and political levels, the political and technical support by external actors that enhanced the PAC community's operative knowledge, the means that PAC grassroots could count on for their struggle (essentially, independent local mass media) and the evaluation of the process carried out by the PCF. Matters of social participation are vital in this analysis because, as this chapter shows, Chilean national law decisively reduces social rights to participation in urban planning processes, disfranchising local communities from effective participation. In the case of PAC, the top-down first draft developed by the PCF and the municipality involved very limited levels of popular participation. A working hypothesis is that the local government of Pedro Aguirre Cerda (PAC) was implementing a managerial-elitist model of policy delivery, pursuing state-entrepreneurial goals (Harvey 1989; López-Morales 2010b; Shin 2009; Macleod 2002) and seeking to attract large-scale, privately led renewal, generate ground rent dispossession and therefore spatial exclusion (i.e., gentrification), with great power concentration in the hands of municipal officials.

The chapter is based on a mixed-method case-study analysis that draws on ethnographic analysis at neighborhood level and a secondary dataset of land value and property records, including qualitative information collected in the field, documents and the main actors' narratives about the process, focusing on the often contradictive discourses from national- and local-level government, and private sector officers. Results indicate that the case of PAC, along with many others in Santiago, shows how the city is being contested by increasingly organized and empowered social forms of local-level urban activism that claim an expanded and more particularized right to the city (Holston 2009; Marcuse 2009). This activism is capable of dealing with structural factors such as the uneven accumulation of ground rent, the loss of economic and cultural heritage as a form of built environment and the negative environmental effects and high levels of displacement caused by large-scale high-rise construction.

Figure 10.1 View of La Victoria *población*.

An exclusionary approach to local planning

Legal definitions of social participation in master-plan elaboration

The levels of participation in the elaboration of district master plans in Chile are currently addressed by the national government in two contradictive ways: on the one hand, in order to produce socially inclusive and comprehensive plans, through technical "recommendations" to planners and local governments; on the other, through "laws" that specifically define when and how citizen participation must be addressed by the local authorities.

Regarding the recommendations, in 1999 the national-level Ministry of Housing and Planning (MINVU) published "Circular-55", an instrument that gives general and specific guidance to Chilean planners for the elaboration of local master plans (MINVU 1999). This document advocates that citizen participation should be iterative throughout the process, because

> the success of a plan largely lies on the ability of the municipality to receive input from those who have the knowledge about the history and lifestyles of the community and its ability to achieve coordinated economic investments, both public and private, leading to better alternatives for community development and the consolidation towards solidarity in their realization.
>
> (MINVU 1999: 11)

This is in accordance with the requirements and expectations generated in the metropolitan area as a whole.

The same document defines that the elaboration of a local-level master plan should proceed through the following steps:

1 Diagnosis, comprising a quantitative and qualitative general approach to the local territory, its social and economic base, its natural and built environment, etc.
2 Proposals and agreements of long-term development alternatives, based on possible forms of growth envisaged.
3 Draft, comprising definitions of local macro-zones or homogeneous areas, land uses and land use intensities assigned.
4 Approval of the plan, granted by all the relevant actors of the *comuna* represented by the Local Economic and Social Council, CESCO (*Consejo Económico y Social Comunal*) and, ultimately, by regional MINVU.

However, depending on every *comuna*'s particularities, the general process might become more or less iterative, depending on the level of success and capacity of agreement the plan generates among the several social and private sector actors involved. Nevertheless, social participation is recommended as a strategic step for all the stages, it being important that municipalities promote

> a broad discussion, either through interaction with the [local] agencies and institutions that [could give] relevant information to the study and/or through interviews and workshops, and provides a schedule that includes number and type of meetings, and identification of participant informants.
>
> (MINVU 1999: 25)

Yet the indications contained in Circular-55 are only recommendations, and thus not mandatory to local-level planners. By contrast, the national Law and By-Law of Planning and Construction (*Ley General de Urbanismo y Construcciones*, LGUC and *Ordenanza General de Urbanismo y Construcciones*, OGUC, respectively) and the Law of Municipalities (*Ley Orgánica Constitucional de Municipalidades*, LOCM) are compulsory over the whole Chilean territory, and they state specifically how social participation must be performed during the elaboration of a new (or redrafting of an existing) local master plan. Nevertheless, as previously noted, the principles established in the LGUC and the LOCM are considerably less participative than those proposed by Circular-55. For example, the LGUC states that social participation comes only after the draft master plan is prepared (see step 3, above). Therefore, before its final discussion and eventual approval by the municipal technical apparatus, the local council must:

1 Inform residents, especially those to be affected by the new plan, about its main features and effects, following what is stated in the General By-Law of Planning and Construction.
2 Conduct one or more public hearings in the most affected neighborhoods in order to present the draft to the community during thirty days, after the public hearings. However, this law does not indicate how many hearings must be conducted, so this can be decided unilaterally by the local administration.
3 After the presentations, the municipality should call the community and constitute the Local Economic and Social Council, CESCO, for a new public hearing specifically aimed at the master plan. The local government must also present a report synthesizing the observations from the community received hitherto. Further enquiries or observations can be received during the fifteen days after the last public hearing.
4 The place for the exhibition of the project, and the venue and time for the public hearings, must be advertised over two days, during two different weeks, either in one of the most popular newspapers of the local territory or through radio advertisement or other local mass media. Projects of master plans are usually exhibited in town councils.
5 From then on, and for the following thirty days, the local council should analyze the community's observations and decide on their inclusion (or not) in the plan. If important changes are to be made in the project, stage 2 must be conducted again. Once the draft is approved at local level, it is sent to the regional office of the Ministry of Housing and Planning (MINVU), which must analyze it and decide whether the plan is in accordance with national

legislation and the metropolitan or inter-municipal master plan (*Plan Regulador Metropolitano de Santiago*, or PRMS), in application since 1994. This latter process must take a maximum of sixty days.

I found agreement among the eleven scholars and fifteen central and regional government officials interviewed that real citizen participation in the elaboration of local master plans is insufficient in Chile. As many of them reckoned, municipalities can even interpret, with a high degree of freedom, what the LGUC, the OGUC and the Law of Municipalities define about the inclusion of citizens' criteria in the final results of these processes (Rajevic 2001). But higher government levels may also share responsibility in promoting a less participatory production of master plans, more specifically separating/isolating adjacent territories. In the specific case of PAC, the regional office of MINVU decided in 2003 to bestow the elaboration of the PAC and Lo Espejo master plans upon two different, isolated external consultants, despite the fact that they are adjacent territories and their plans were conducted at the same time, between 2003 and 2005 (Figure 10.2).

The process of redrafting the PAC master plan began in late 2003.[3] Following current legislation, the regional office of the Ministry of Housing and Planning (MINVU) invited private consulting firms to bid for this project. On this occasion, as usual, two pairs of master plans were offered: PAC was coupled with Curacaví *comuna*, while Lo Espejo (PAC's southern neighboring *comuna*, with which it shares *población* José María Caro) was coupled with Til-Til *comuna*. As Figure 10.2 illustrates, this strategy coupled one inner-city and one peripheral *comuna* in two separate bids. The contracts for Pedro Aguirre Cerda and Curacaví *comunas* were awarded to the private consultancy firm (PCF), while Lo Espejo and Til-Til were awarded to another firm attached to a private university.

Grouping several *comunas* in one single tender is a regular modus operandi for regional offices of MINVU across the country. However, from the beginning the separation of PAC and Lo Espejo seemed awkward for key local social PAC actors. Two inner-city municipalities that share similar problems could have been considered together for the elaboration of two integrated local master plans, or one single inter-*comuna* master plan. Instead, MINVU ran both processes in parallel, with virtually no communication between the two technical teams.

Officially, the government claimed this was a purely technical decision. PAC community leaders were far more skeptical, as most of them thought this was an attempt to divide and disenfranchise, in advance, potential reactions from the local people, especially from the two "conflictive" *poblaciones*, La Victoria and José María Caro, which also belongs to Lo Espejo (see Figure 10.3). Such a suspicion was particularly vivid because these two neighborhoods have been historically coordinated in previous political mobilizations, especially during the military dictatorship (1973–90).

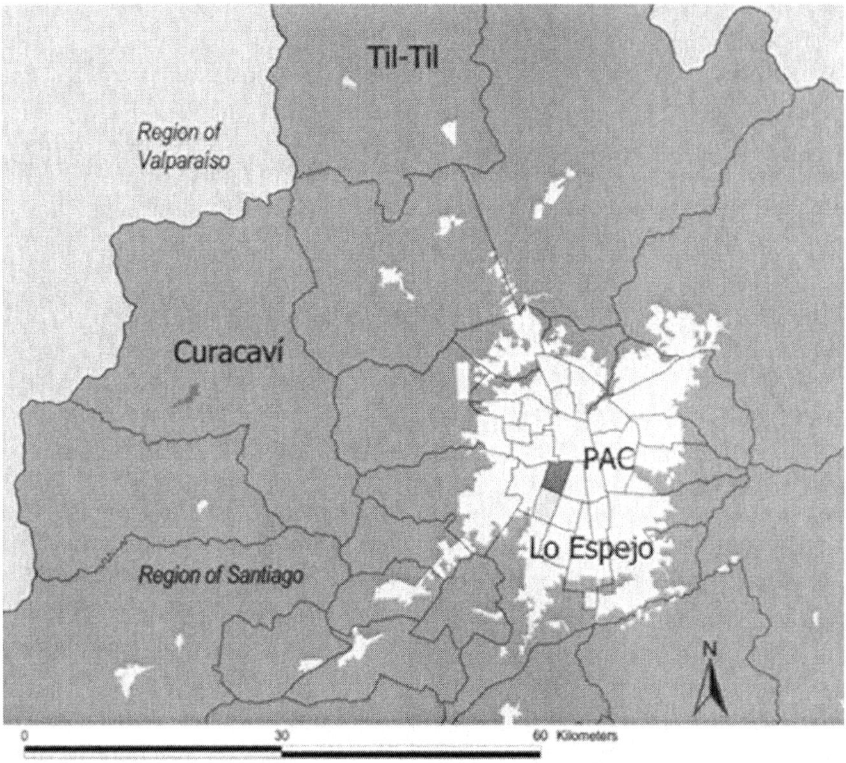

Figure 10.2 The two pairs of *comunas* on bid in the Metropolitan Region of Santiago, 2003.

PAC municipality's interpretation of participation in the master plan

As established by the national Law of Municipalities (LOCM), it is local government that has to select the best proponent for the elaboration of its master plan. PAC's mayor explained in an interview (25 April 2007) that the municipality selected the private consultancy firm (PCF) because it proved more experienced in spatial and development planning. Furthermore, the PAC municipal urban advisor and, legally, the official technical municipal counterpart for the plan, claims the intention of PAC local administration was to be as inclusive as possible. However, the PAC was soon disappointed by the PCF's modus operandi.

> Theoretically speaking, the PCF's planning theory was fantastic. [...] However, they [later] said they would work with representative actors only, not with the whole community. That seemed fine to us again, until they

> started to work here. Then we realized that the participatory process during the diagnosis stage would be really bad, because they didn't actually call the people.
>
> (Interview, PAC municipal urban advisor, 25 April 2007)

This position is not supported by the Regional Office of the Ministry of Housing and Planning chair for Local Master Plans, which claims that the principal reason for the struggle over the new master plan in Pedro Aguirre Cerda was precisely the lack of well-informed social participation during the initial stages of the plan. According to this officer's opinion, PAC municipality had a major responsibility over this, because "it is they who have to say in which direction PAC should develop. But they didn't take on that role, and they let the consultancy firm go alone at working with the residents. And finally, the municipality didn't like the outcome [and] they won't accept it [so] now we're at a stage in

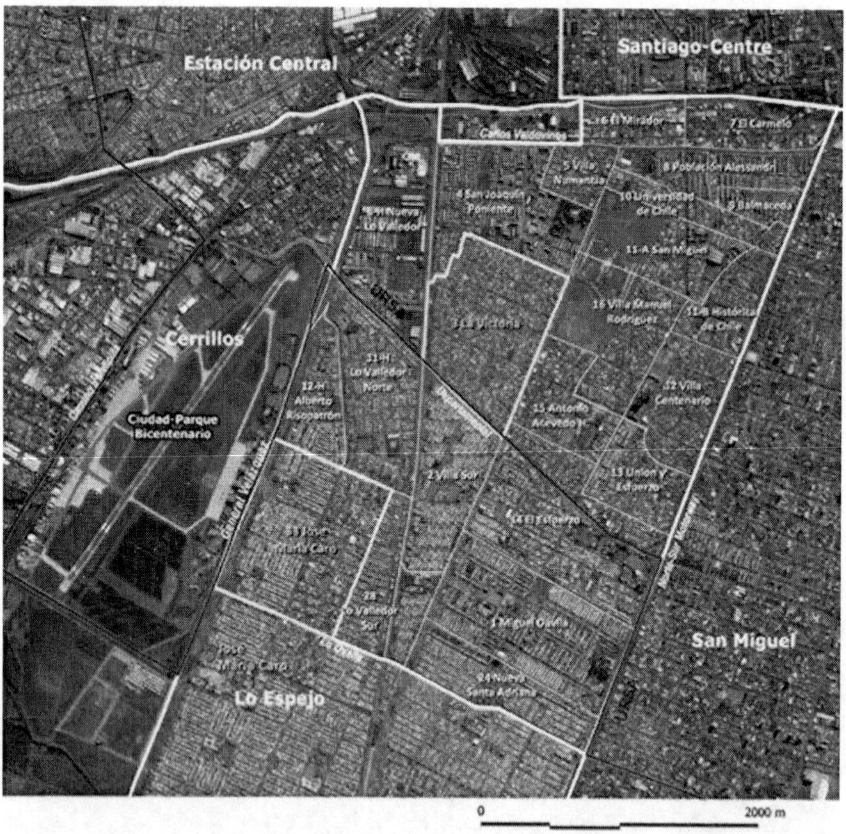

Figure 10.3 Main neighborhood units in PAC (source: author's elaboration based on Microsoft Virtual Earth).

were the job is stalled, unfinished" (interview, regional officer, MINVU, 10 April 2007).

As a report by Núñez (2006) indicates, in mid-2004, PAC's local community realized that the PCF's initial agenda of participation was insufficient and possibly biased to include only specific neighborhood units. These units seemed to have been selected ex-ante by the municipal apparatus because they were identified as the "friendliest" ones, whilst excluding leaders from the more combative *poblaciones*, like La Victoria, José María Caro, El Esfuerzo and Nueva Lo Valledor (see Figure 10.3).

In a report released in August 2005, the PCF accepted its own initial reluctance to be more inclusive, clearly justified, as they claimed, by the thorny municipal electoral panorama of 2004 and even by the characteristics of PAC residents. According to this firm: "a) there is people's ignorance about what a master plan is and about the process of approval; b) the contents of a master plan are highly technical, therefore they should have been translated into a simpler language; c) a weak process of participation and information because it coincided with the local electoral process; d) the latter tended to politicise any activity in the *comuna*" (PCF, 2005: A4–1)

Further to this apparently technical decision, there is evidence to infer that the inclusion of PAC people in the preparation of the new master plan was not totally transparent. In a nighttime chat program for the local La Victoria TV station (see Figure 10.7), a neighborhood representative complained that the address of the venue for the hearing to assess the PCF's first draft was incorrectly printed in official invitations. In consequence, many attendees had been directed to non-existent venues. Other leaders attending those hearings (having received correct invitations) confirmed that there were not enough chairs, and that "other essential facilities for conducting the meeting professionally were lacking". These should not be taken as irrelevant points. Social representatives invited to the program agreed that those two elements were attempts, on the part of the local authority, to demoralize them and exclude them from the process.[4] However, these elements triggered – in advance – even higher levels of unrest among PAC community leaders. The evidence suggests that PAC municipality, acting through the PCF, was able to steer the process, including the selective exclusion of community representatives.

The rough way to produce a bottom-up alternative draft

A historical community mobilization reactivated

By the end of 2004, the PCF had been working for over a year on the first-draft master plan, and the rezoning of the *comuna* had largely taken shape. However, at this time, this proposal was unknown to most grassroots representatives. This led to higher levels of suspicion regarding the imminence of land expropriations and, consequently, evictions. Several neighborhood leaders demanded an official response to these fears from the mayor of PAC. The mayor, in an improvised meeting with these representatives held in front of the council house on 18 May 2005, gave them assurances that, in PAC, there would be "no expropriations whatsoever". However, PAC leaders saw blurriness and inconsistency in the mayor's discourse.

The reasons for their skepticism lay in the fact that the use of expropriations, as defined by the norm contained in the national Law of Planning and Construction (LGUC), is warranted in cases of public interest, such as when a new master plan includes a new open space, or the widening of a road. Article 59 of this law states that urban plots may become "frozen" from the moment a new master plan is approved if the plot is

> affected by declaration of public interest and awaiting public expropriation or acquisition proceedings, it will be not permitted to increase the volume of the buildings existing at the date of approval of the respective master plan, in the affected part of the property in case the expropriation declaratory [partially covers the plot].
>
> (Gobierno de Chile, 2007)

Conscious of this legislation, Claudina Núñez, social leader from La Victoria and then local councillor in PAC, opines that:

> [The municipality] said we [PAC social leaders] were just scaring people, as no plot in PAC would be expropriated. But obviously there would be lots of expropriations! They were widening streets: 30 de Octubre street would be joined with its parallel, Paseo Groher, thenceforth a whole row of houses would disappear, from Galo González street to here [see *ensanche* axes in Figure 10.4]. And in Histórica de Chile neighborhood unit, Club Hípico street was to be widened towards the east up to Manuela Errázuriz street. Therefore, two passages were also disappearing. Other streets were disappearing too in Villa Sur *población*. And in San Joaquín *población*, some residential areas were being rezoned to commercial use. That meant that those social dwellings would be easily subject to expropriation. Urban density considerably increased everywhere. All of that was the consultants' first draft.
>
> (Interview, Claudina Núñez, 31 March 2007)

The initial statement by the mayor, that there would be no expropriation, had soon to be rectified through a municipal official statement. The contradiction with the first draft, which was starting to become known to the community, was evident. Complying with the law, a meeting was organized by the municipality, but restricted to no more than thirteen community leaders, to present publicly the first-draft master plan. The meeting was scheduled for 28 May 2005, in the facilities of Risopatrón School, located in the southwestern corner of PAC. The social representative from neighborhood unit 8-H, Nueva Lo Valledor – one of the excluded ones – recalls what happened in that meeting:

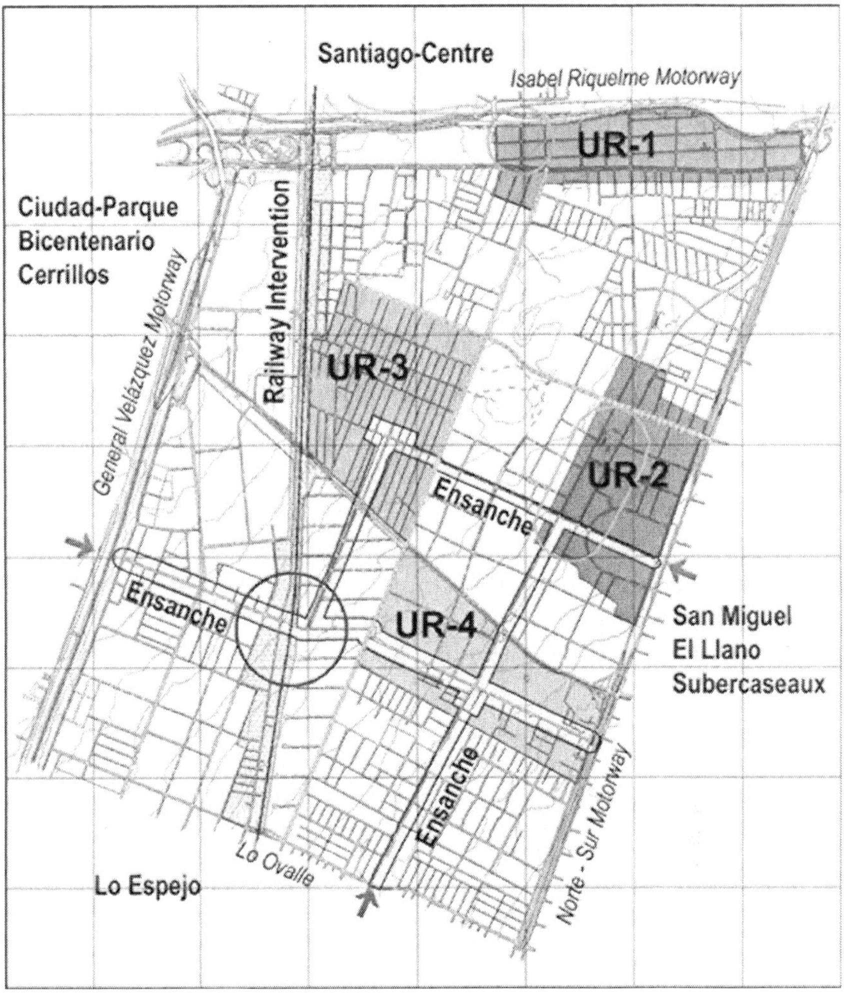

Figure 10.4 First draft master plan proposed by the PCF.

> When we knew they would be having a workshop about the master plan, aimed for twelve or thirteen people, we spread the word amongst leaders and we rushed to take possession of the place, bringing along residents from all of PAC's corners. [...] The municipality was outraged, given that their logic has always been to work by stealth.
>
> (Interview, social representative, Nueva Lo Valledor, 11 May 2007)

The massive attendance at Risopatrón School that evening was broadcast in detail by Channel 3 TV La Victoria (see Figure 10.5). In a report released in

Figure 10.5 Images from the meeting in Risopatrón School in 2005. Top: general view of attendees. Bottom left: people asking about the feasibility of expropriations. Bottom right: Claudina Núñez demanding explanations from the municipality, with the PCF technicians standing at her back (source: local TV station, Channel 3 La Victoria).

2005, the PCF declared that there were between 500 and 600 attendees that evening.

That was the first time the PCF would face, en masse and together, a group of people from almost all the twenty-six neighbourhood units of PAC. Just after the meeting, most people rejected the PCF's first draft, since it was understood to be conceived from "a perspective exclusively technical and arising out of external objectives which neither considered the local identity nor the needs of the community" (Núñez 2006). As observed in Figure 10.4, the proposed rezoning would intervene heavily in four areas of the *comuna* (from UR-1 to UR-4), yet "without a global vision of the territory" (ibid.).

Nonetheless, the evidence suggests that, at this point, the PAC community was unprepared, disorganized and had little technical knowledge to face the authority. But soon after, the leaders from La Victoria requested political advice from the head of the Communist Party, and technical support from the Chilean Architects Union, a professional NGO called Sur and a private architect. These external supporters were instrumental in technically enhancing PAC people's capacities towards the constitution of a collective counter-proposal, which would be partially synthesized in the PCF's final proposal.

What do you mean by a "politicized" movement?

When I asked PAC Municipal Director of Construction, a highly experienced and influential officer within the municipal apparatus, why the PCF's first draft was rejected by the people of PAC, she explained:

> We had an idealistic project that we thought might work well, but when the time of citizen participation came, this project was taken quite politically by the people, and they dismantled it. [...] Since La Victoria *población* doesn't have a single square meter to breathe in, the PCF proposed that there should be a square there, but that idea was not accepted by people from La Victoria. And all that the PCF had thought for the rest of the areas of PAC didn't work either, because La Victoria's people resolved to meddle in every assembly where the PCF was attempting to explain what the plan was about. [...] When you want to transmit an idea to people, you need valid counterparts, but that wasn't done here. [...] People took the issue of expropriations very badly. Nobody thought that they would receive, maybe, an apartment [in exchange for the land [...] Maybe the process [of expropriations] would never exist and everything will end just as a plan, but there was not such a clarifying instance. And there was nothing more to do because the issue had already been politicized. Everything technical that might be said afterwards would be taken politically. [...]
>
> I'm not trying to be pejorative, but if you want to explain something to these people, you have to do it three times. You have to expect they will not understand all you want to say the first time.
>
> (Interview with PAC municipal officer, 2 April 2007)

Throughout the whole interview, this officer claims essentially two things. First, that it is relatively improbable that low-income people can grasp all the implications arising from implementing a master plan, given their limited technical preparation. Second, that the grassroots did not have the capacity to understand the conceptual difference between a territorial master plan and a simple "architectural project". The officer explains that a plan implies a certain set of potential situations that could be turned into reality or not, whilst a project is the representation of a built structure to be materialized. The mistake PAC people apparently fell into was the misinterpretation of the first draft's structure as a factual situation, an imminent process of expropriations, and not a potentiality.

According to this informant, a third reason for the failure of the first draft was that PAC people let themselves take a highly politicized and negative approach, closing the door to further objective discussions about possible and necessary expropriations even though they might have benefited from them. The influence of La Victoria was key in leading the rest of the grassroots' reactions, and this situation conspired against the (otherwise) normal development of the further instances in the production of the plan. The officer summarizes PAC municipality's official explanation of why the process, at a given point, escaped from their hands.

In fact, most municipal and regional MINVU informants agree with the idea that the process of production of the master plan in PAC was "politicized", as illustrated by the unbalanced relation between leaders of La Victoria *población* and the other communal representatives, and the occasional involvement of the Chilean Communist Party as a way to destabilize the process (see Figure 10.6). The evidence shows that both elements were effective and could be observed during the fieldwork. Yet, according to the official interviewees, this destabilization was supposed to have been conducted through the appointment by the Party of Communist professionals, who sought to help local Communist leaders instrumentalize the rest of PAC grassroots' reactions. This is a matter of some speculation that deserves further analysis.

An external architect was directly blamed by the municipality as one of the links between the Communist Party and the grassroots of PAC. In an interview given to me, this person confirms that the grassroots of Pedro Aguirre Cerda could have become, at some point, a mere instrument used by the Communist Party to take over the municipal government, but that this did not happen. In fact, there was disagreement inside the Communist Party regarding this point.

> A meeting was organized in the Victor Jara cultural center, only for Communist social leaders. Once there, a strange situation occurred: a conflict between communists. PAC councillor Ernesto Araneda, a guy with a respectable trajectory because he had once been mayor of San Miguel *comuna* who is also a communist, tried to use the master plan struggle as a political-electoral juncture, in a more subtle way, for seizing by stealth the municipality's direction. But Claudina Núñez firmly demanded that this process should be opened to the whole PAC community. This meant that such information should be released to the public and be disseminated. It was a meeting of twenty or thirty people, and [...] when the time to vote came, Claudina's proposition won.
>
> (Interview, architect, 11 April 2007)

What this architect claims is that there may have been higher levels of autonomy and political independence among PAC grassroots than those the public sector was willing to accept. Of course there was political influence exerted by the Communists, especially those from La Victoria, and to a lesser extent from *población* San Joaquín, over the rest of PAC leaders and the people. This influence sought to oppose the PCF's first draft and take control of the process. In fact, some PAC neighborhood representatives (from Villa Numancia and Histórica de Chile neighborhood units) even withdrew from the process of participation, arguing that they had been quietened by the radicalized mainstream. Yet these were only two leaders, from a total of sixteen social representatives interviewed. Nonetheless, all in all, the political influence in PAC grassroots came from legitimately elected local representatives, not decisively from the Communist Party apparatus. Moreover, it seemed logical to many interviewees that Mrs Núñez, who was possibly running for the mayoral elections in 2008, still needed to legitimize herself as an effective grassroots political leader with a chance of winning the next election (she actually won the 2008 election and is currently mayor of PAC). She needed to prove she was representing the whole PAC community and not only her closest left-wing supporters from La Victoria. Besides, her high level of technical knowledge regarding this specific case has been largely accepted by most PAC grassroots informants interviewed, and even by some governmental officials.

Moreover, the difficult relation between the municipal technical apparatus and mobilized local communities should not be exclusively attributed to this single case. Much less is it a matter of conflict between the social-democratic party the mayor belonged to (the PPD, part of Concertación (Concerted Action), the social-democrat governing coalition of the country until 2010) and the Communist Party. The great majority of the grassroots interviewed made evident their rejection of the mayor's and his technical officers' modus operandi. Some even accused certain municipal officials (including the mayor) of corruption, bribery and even alliance with PAC local drug mafias.[5] In many senses, the conflict experienced in the process of rezoning was to some extent a climax in an already long process of political/social deterioration that has taken place since the mid-1990s.

As an example of bad practice from the municipality, Figure 10.6 shows an anonymous pamphlet that circulated in PAC in early 2005, blaming local leader Claudina Núñez and Councillor Araneda for terrorizing the local population, especially old people, with the threat of expropriations. The pamphlet clearly states there would be no expropriations whatsoever. It further accuses Mrs Núñez and Mr Araneda of taking political advantage of this ploy. Although it is anonymous, La Victoria community leaders believe this pamphlet might have been drafted by someone in the municipal government, because of the level of information demonstrated and the fact that it focuses only on these two actors.

¡No se deje engañar, no hay expropiaciones!

HAY UNA CAMPAÑA DEL TERROR PARA SACAR PROVECHO POLITICO

Los Dramas de la Mentira

Una señora de 80 años de la Población Risopatrón pregunta llorando si es cierto que la van a sacar de su casa porque ahora van a construir la casa del Presidente en Cerrillos y que ella lo único que quiere es morir en su casa.
Una Mentira que está creando angustia en cientos de abuelitos.

Un señor de la Población Dávila tuvo un pre-infarto porque le dijeron que le iban a expropiar su casa.
¡Mentira!, ¿quién se va a hacer responsable de la muerte de alguien?

Un sinvergüenza fue a ofrecer 5 millones de pesos por una casa, diciéndole a la familia que se la van a expropiar por 4 millones de pesos.
Otra Mentira que está siendo usada por los especuladores.

En Los Maitenes dijeron que iban a hacer expropiaciones para hacer una estación del Metro.
Otra Mentira, porque no existe ese plan del Metro.

¡CLAUDINA NUÑEZ MIENTE!
¡ERNESTO ARANEDA MIENTE!

Mienten, porque inventan un peligro que no existe y después aparecen diciendo que lo solucionaron.

Figure 10.6 Text of an anonymous pamphlet that circulated in PAC in early 2005 (source: Claudina Núñez' archives).

Box 10.1 TRANSLATION:

Don't let them fool you, there are no expropriations!

THERE IS A TERROR CAMPAIGN SEARCHING FOR POLITICAL GAIN

The tragedies of this lie:
An 80 year old lady from Risopatrón *Población* asks, in tears, if it's true that she's going to be evicted because the President's House is going to be built in Cerrillos [adjacent *comuna*] and she just wants to die in her house. This lie is creating distress among hundreds of our old residents.

A man from Dávila *Población* had a pre-heart attack because someone said his house would be expropriated. This is a lie! Who is going to be responsible for all these deaths?

A crook offered 5 million pesos [240 UF, or US$9,000] to a family for their house, claiming that it would be otherwise expropriated for $4 million. Another lie being used by speculators.

In Los Maitenes *población* someone said there would be expropriations in order to build a Metro station. Another lie, because such Metro plan does not exist.

CLAUDINA NUÑEZ LIES!
ERNESTO ARANEDA LIES!

They lie, because they invent a danger that does not exist, so later they can claim they fixed it.

The resources of the grassroots: legalizing the illegal

After the event in Risopatrón School had taken place, and had led to a technical counter-proposal, three resources were utilized by the grassroots of PAC to foster citizen participation. First, regular workshops organized, and later just propelled by La Victoria leaders and the PCF technicians. These helped ordinary people keep up to date with the plan's aims, and form their own opinion (highly influenced by La Victoria leaders, though). A second resource was La Victoria local TV station, Channel 3, and its uninterrupted, though illegal broadcasting[6] that kept PAC residents informed about the development of the draft master plan. In 2005, in at least three long evening talk program, technical and political topics related to the rezoning were discussed (see Figure 10.7). The third resource was the support from the following external professional actors:

1 SUR Profesionales, one of the most consolidated NGOs in Santiago focusing on urban development issues of social-democratic orientation. Until 2010, it advised central government and MINVU on housing policy issues. The longstanding connection between SUR and some community leaders from Pedro Aguirre Cerda, especially Claudina Núñez, was vital, since she is a graduate of SUR's Program in Community Leadership. At least two

other grassroots leaders attended the program during 2003 and 2005. Their higher level of preparation, obtained through these studies, is one of the reasons these three actors were the most influential among grassroots leaders in PAC at the time.

2 The architect advising the community, also a member of the Communist Party. This advisor was the most deeply involved during the elaboration of the bottom-up proposal. Under his guidance, the community's alternative proposals took more concrete shape in a draft text, and he was assisted by his third-year undergraduate students from the School of Architecture, Metropolitan University. The latter were instrumental in guiding the community workshops in almost every neighborhood unit.

3 Colegio de Arquitectos de Chile. PAC leaders received initial advice from another architect who first oriented La Victoria leaders about the meaning of a master plan, also indicating crucial aspects to solve, such as the north–south railway that divides the *comuna*.

Ten joint workshops were agreed between the PCF and PAC leaders, to be held between June and July 2005. These meetings sought to disseminate and debate the draft master plan, and were organized in a much more participatory fashion than the two hearings established by the law. They ran in parallel with the PCF's own development of the final version of the plan, and aimed to provide the social feedback they lacked in their initial attempt. But hitherto, leaders from La Victoria had produced several copies of a CD containing the PCF's first draft, with additional information provided by Councillor Araneda, and distributed the information among the other neighborhood units. With this information in their hands, additional workshops were developed in almost all PAC neighborhood units. In these meetings, everyone interested in participating could analyze the consultants' proposal, supported by professionals and a team of students.

In early July 2005, the global proposal was ratified in a communal general assembly by PAC residents. In late July, the communal alternative draft was submitted to the PCF, regional MINVU, municipal government and local councillors, in order to have these community proposals considered, before readjusting and approving the plan by the local and national authority (PAC Community Leaders 2005).

The outline of an alternative plan

Thirteen different micro-proposals of urban regulation produced by nineteen neighborhood units were assembled together in a single alternative-draft master plan, which despite not having graphic representation, clearly showed the local people's perspectives about urban development. The PCF, after an initial reluctance to include the grassroots' direct participation in the process, collaborated and partially assisted them in the production of their proposal. The communal unified draft was structured as follows:

Figure 10.7 Two different night talk programs in the local La Victoria TV station, focused on the master plan redrafting (source: Channel 3 La Victoria).

1 The community expanded the four macro-areas proposed by the PCF into nineteen sub-zones, closer to the number of existing neighborhoods in PAC.
2 A different local road structure, better adapted to the existing local urban fabric and including new open spaces. No new streets would be traced over existing residential space, and the initially proposed system of structural squares and roads (*ensanche* in the first draft) should be eliminated.
3 Definition of small-scale urban renewal zones in the most decayed areas of the *comuna*. Such a proposal should run together with a municipal policy that should allow: a) the right of dwelling extension; b) collective applications to public funds for rebuilding, extending or upgrading existing dwellings; c) any form or morphology of dwellings; and d) increase of the maximum building length in sites (i.e., the percentage of the plot's length occupied by a dwelling) up to 80 percent.
4 Zoning proposal for community facilities and services, and a rezoning of industrial land use. The location of the new municipal council house should also be reconsidered (from its current location in the northeast corner) to a more geometrically central position within the *comuna*.
5 Proposal for open spaces and recreational facilities, comprising community parks across the *comuna*'s space; a pedestrian park parallel to Clotario Blest road (converting the existing high-voltage towers into an underground electric pipeline); a pedestrian path in Dávila *población*; to open Manuela

Errázuriz street. The municipality should consider using part of the municipal budget for converting small vacant plots into open spaces in the most densely populated quarters of PAC.

6 Risk assessment and environmental protection, especially analyzing the effects of petrol stations, transport terminal plots, warehouses, gas supply storages, garbage and rubble collection, surfacing streets and passages, besides other currently lacking basic infrastructure.

Although the grassroots of PAC could not translate their unified proposal graphically into a map, being in fact a list of factors (detailed for almost every neighborhood unit), they clearly established building capacities for the nineteen sub-zones of the *comuna* to be included in a future local plan. The PCF's second and final proposal (see Figure 10.8) is substantially different from the first draft shown in Figure 10.4, as it includes many requirements of PAC people, though not all of them. Thus this final draft is not entirely coincident with the community's ideas, e.g., the potential ground rent proposed by either party differ considerably, especially in the northeast of the *comuna*, in the area surrounding the existing council house. By and large, the PCF resisted reducing their original potential building capacities in the northeast quarter of PAC.

However, in almost 70 percent of PAC's territory, the building capacity proposed is considerably more adequate to the existing spatial structure, and the levels of potential ground rent produced tend to be similar between the local community's and the PCF's positions. The final proposal developed by the PCF is, in fact, a consolidation of the existing PAC road system, only with added changes of direction of traffic flow and the conversion of some local roads into metropolitan-scale avenues. This point, especially, was seen by the community as an achievement because the risk of expropriations was effectively reduced. The potential ground rent planned for these areas is not high enough to attract redevelopment on a large scale, while zones of expropriation are practically nonexistent. Notwithstanding the many differences between PAC grassroots' guidelines and the PCF's final draft, this was seen by the local community as a considerable advance over the previous situation.

Nevertheless, by 2008, the municipality had not made the outcome of the process available to the public, so the plan was stalled. PAC municipality was in fact breaking the National Law of Planning and Construction and the Law of Municipalities due to its reluctance to release the final outcome of the project to final approval and to constitute the Local Economic and Social Council (CESCO). Furthermore, there was considerable pressure from the central and regional government on PAC municipality to release a final resolution regarding the Master Plan. The $76 million Chilean pesos invested by regional MINVU are being compromised by this stalled situation, besides the fact that the municipality has not yet approved the release of final payments to the PCF. My personal impression in the field was that this failed process of rezoning was perceived by regional MINVU as a stain, a regretful pitfall, in their self-confessed 'efficient' performance in local planning issues.

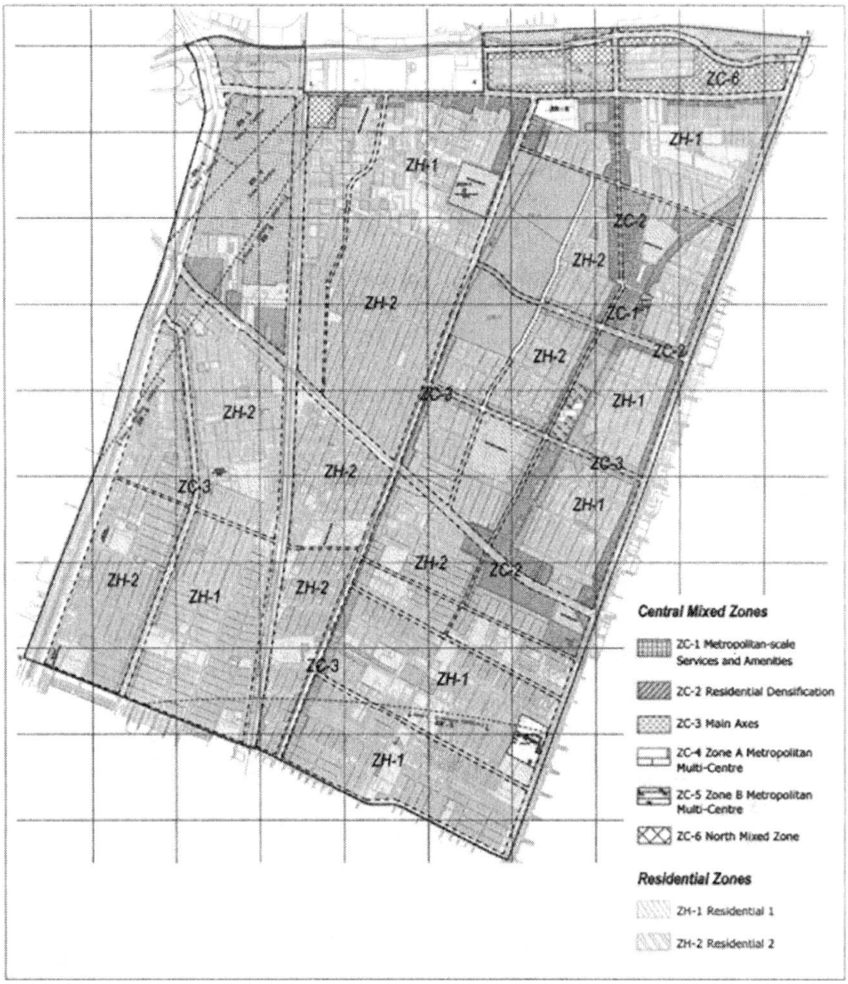

Figure 10.8 Second draft master plan delivered by the PCF (source: López-Morales 2009).

A normative perspective from the private sector

The architect in charge of PAC rezoning process, and also the PCF's CEO, gives a general evaluation of the process seen from the perspective of the consultants. The emphases in the quotation are my additions:

> [La Victoria has] a whole history that I much respect but I also think they have gone against any possible evolution they might have had in the *comuna*. [...] La Victoria is a territory where nobody enters, it is a territory

> that *does not want a change* and also there is a political issue here. [...] But you enter La Victoria and it is, from the point of the urban development, *a total failure*. We are talking about dwellings spontaneously generated with some state support; they might have consolidated some spaces but *there isn't a single public space there*, passages are tiny, they have null connection with the rest of the *comuna*, they don't have urban facilities. It is *a very perverse system* that has been somehow also supported by the authorities. There some strange things happen. [...] Because neither have they managed to generate an integrated *comuna*, which was our goal. It was ultra necessary this integration, otherwise *no alternative way [to development] is possible* and that integration had to be produced by generating a social process. Unfortunately, there are forces there that started to oppose these changes. Why? Because there is too much suspicion, because [grassroots] think external consultants are property developers who will take out their spaces and so they will be expropriated.

This experienced architect and planner categorically delivers a series of statements about the *población*, such as: it "does not want a change", it is "a total failure", "there isn't a single public space there", it is "a very perverse system", no "alternative way [to development] is possible" there. It would be strange if the use of these concepts were involuntary. Instead, to some extent, it might reveal her interpretation of PAC local space as a blank canvas whose urban structure must be radically recomposed and re-imagined. This stance is close to several aspects of "entrepreneurial" planning rationale.

I asked a community leader why La Victoria people were reluctant to accept the proposed "renovation square" planned for the middle of the *población* by the PCF, and she answered: "What do we want more squares for? For our children to have another place to take drugs? And at the cost of sacrificing houses?" This suggests that the whole rationale of urban regeneration – underpinned by the broader official discourse of securitization and eradication of drug dealing in the inner city – considerably contradicts the fact that what La Victoria least needs at present is more visually uncontrolled, large-scale open spaces. Narrow streets and small squares might be criticized by most architects, but at least they seem to have proven means for social control and invigilation by the *población*'s own community.

Conclusions

The PAC case shows that to pretend to conduct purely technical, depoliticized processes of urban planning in current inner-city territories may be inadequate. PAC municipality, from the beginning, imprinted an entrepreneurial agenda for large-scale urban renewal in the process of master-plan elaboration, seeking to enlarge the rent gap. Also, municipal and hired private sector officers defend the privately led perspective on urban redevelopment. They also seem to depict the working-class environment almost purely as a series of highly deprived spaces, sustaining a sort of moral appreciation in regard to the local working-class inhabitants, thus revealing some degree of class bias. Processes of rezoning in socially sensitive spaces like PAC, in times of democracy, cannot be simply left by the central authority to the criteria of external private consultants, many of whom seem to have limited resources, including time, to afford them a deeper insight into the historical contradictions in the local spaces they are supposed to help plan.

However, from the perspective of the local leaders interviewed, the outcome in PAC was in many ways an achievement. Their counter-proposal, despite its technical limitations and the fact that it is not entirely accepted by the consultants, could be highly influential in the elaboration of the PCF's definitive version. Local people could understand what happens when local spatial regulations are abruptly liberalized, and how much ground rent they could be dispossessed of if some of the speculative logics of the market of urban renewal enter into PAC. This is a central historical-political point, and pertains to PAC people's right to land occupation. Given that many working-class urban *poblaciones* were politically and militarily targeted in times of the dictatorship (1973–90) – La Victoria *población* has a special memory of this (Finn 2006b) – those who have stayed there consider their location as part of their political heritage, handed down through almost two decades of fighting military state power. In consequence, most grassroots I interviewed believed the current democratic state – either at municipal or national level – has the obligation to preserve that right.

The evidence suggests that the ability of the PAC community to effect change comes from the technical knowledge their leaders have received from external parties. But this knowledge was also strategically filtered down towards the grassroots. Had this not happened, and had the negotiations been restricted to technicians and local leaders, it would have been extremely difficult for the PCF to include the demands coming from the various neighborhood units and to validate its final proposal. At some point, the radicalization of the process fostered by the leaders of La Victoria *población* (especially Claudina Núñez, in her supposed search for electoral gain) may have helped the PCF to arrive at a relatively consensual outcome, because the communist grassroots leaders used already existing levels of politicization to increase social participation and inform the residents about what was happening with the master plan.

Although PAC people's power has many limitations – including the their incapacity to overcome PAC municipality's internal bureaucracy and their lack of influence over the higher administrative spheres of MINVU apparatus – they do have a high degree of organization and also the political and technical effectiveness to keep within legal boundaries, as shown during their struggle. The current confrontation is clearly different from the long periods of fierce protest and repression experienced during the military dictatorship (1973–90). It is an example of effective social mobilization undertaken by working-class actors within the institutional system of planning. To borrow Holston's words (2009), the methods used in 2005 could in fact be defined as "legalized": no violence was exerted, and practically no organized demonstration was conducted. Most of this should be explained by the high rates of home ownership in PAC. Most of the time, the process of mobilization in PAC stayed within legal boundaries, and, instead of simply opposing the rezoning, the grassroots forced the PCF to deliver a final draft much closer to their own assessments and convenience as owner-occupiers. From the perspective of the grassroots, this is a considerable improvement on the situation twenty years ago, and should inform the development of urban social mobilizations in the future.

It might be paradoxical, but even considering its political relevance during the democratic period since 1990, the case of PAC 2003–5 is still little known – less understood – among scholars and government actors in Chile. Nevertheless, six general lessons can be learned from it, namely:

1 Inner-city urban conflicts are basically class-related conflicts that need to be resolved with that perspective in mind – however, the class-related nature of the inner city is usually forgotten by the national and local government planning apparatus (informed by "normative" depoliticized stances) and this can lead to serious pitfalls in the planning process.
2 The grassroots' participation should be seen not only as a "moral imperative" but as an opportunity for a fluid feedback of information about local problems, thereby focusing public resources investment into the key problems.
3 Technical planning knowledge acquired by social leaders can be successfully (skillfully) filtered down to the grassroots, politically validating the process of planning and enriching the final outcome (i.e., the plan).
4 An effectively politicized leadership by no means disregards the opportunity of involving national-level political spheres into local-level conflicts for the conclusion of its local-level ends.
5 It is possible to achieve real/effective social participation on planning issues, effectively overcoming the boundaries set by national legislation.
6 It is possible to convert and use illegal resources (e.g., the banned community TV channel, or the political pressure exerted on the local government) into legal outputs (e.g., the draft and the final plan).

Ideally, lessons learned here have the potential to contribute to the successful resolution of similar conflicts of this nature in the future, within Chile and throughout the Latin American region.

Notes

1 This paper was written with the generous support of Chile's National Fund for Research and Technology (FONDECYT), Project Code No. 11100337, "Rent Gap, Social Agents and Planning Systems: a Case-Study Analysis of the Property-Led Gentrification of Santiago de Chile's Inner City Area".
2 In Chile, the term *población* refers to any low-income, consolidated, relatively low-density residential estate.
3 The cost of this project (Project Code No. 20180023–0 in SERPLAC's 2005 Regional

Investment Programme) was eighty million Chilean pesos in total (approximately US$143,000), of which 95 percent came from regional sources and the remaining 4 million pesos from the PAC municipality (SERPLAC 2005).

4 Source: night talk program on local TV, Channel 3 La Victoria.

5 Although corroboration of these grave accusations lies beyond this research, it seems that they originate in negative personal perceptions of the mayor, reflecting the considerable level of detachment between PAC governmental apparatus and the people.

6 The Chilean law of telecommunications bans independent local TV and radio broadcasting (Villagra 2009).

References

Castells, M. (1985) *The City and the Grassroots: a Cross-Cultural Theory of Urban Social Movements*, Berkeley and Los Angeles: University of California Press.

Cortés, A. (2009) " 'Nada por Caridad', Toma de Terrenos y Dictadura: La Identidad Territorial de La Población La Victoria", unpublished Masters thesis, IUPERJ, Rio de Janeiro.

Espinoza, V. (1988) *Una Historia de los Pobres de la Ciudad*, Santiago: Ediciones SUR.

Finn, J. L. (2006a) "La Victoria Comprometida: Reflections on Neoliberalism from a Santiago Población", in N. Dannhaeuser and C. Werner (eds) *Markets and Market Liberalization: Ethnographic Reflections*, London: Elsevier.

Finn, J. L. (2006b) "La Victoria: Claiming Memory, History, and Justice in a Santiago Población", *Journal of Community Practice*, 13: 9–31.

Gilbert, A. (1993) *In Search of a Home: Rental and Shared Housing in Latin America*, London: UCL Press.

Gilbert, A. (2002) "Power, Ideology and the Washington Consensus: the Development and Spread of Chilean Housing Policy", *Housing Studies*, 17: 305–24.

Gobierno de Chile (2007) "Ley General de Urbanismo y Construcciones". Online: www.minvu.cl/opensite_20070404173759.aspx (accessed 19 March 2007).

Harvey, D. (1989) "From Managerialism to Entrepreneurialism: the Transformation in Urban Governance in Late Capitalism", GeografiskaAnnaler, Series B, *Human Geography*, 71: 3–17.

Holston, J. (2009) "Insurgent Citizenship in an Era of Global Urban Peripheries", *City & Society*, 21: 245–67.

López-Morales, E. (2011) "Gentrification by Ground Rent Dispossession: the Shadows Cast by Large Scale Urban Renewal in Santiago de Chile", *International Journal of Urban and Regional Research*, 35: 330–57.

López-Morales, E. (2010a) "Santiago de Chile: Examining State Roles in the Transformation of a Postindustrial Inner City Space", paper presented at AAG 2010 Annual Meeting, Washington, DC, April.

López-Morales, E. (2010b) "Real Estate Market, Urban Policy and Entrepreneurial Ideology in the 'Gentrification by Ground Rent Dispossession' of Santiago de Chile", *Journal of Latin American Geography*, 9: 145–73.

López-Morales, E. (2009) "Urban Entrepreneurialism and Creative Destruction: a Case-Study of the Urban Renewal Strategy in the Peri-Centre of Santiago de Chile, 1990–2005", unpublished PhD thesis, University College London.

Macleod, G. (2002) "From Urban Entrepreneurialism to a 'Revanchist City'? On the Spatial Injustices of Glasgow's Renaissance", *Antipode*, 34: 602–24.

Marcuse, P. (2009) "From Critical Urban Theory to the Right to the City", *City*, 13: 185–97.

MINVU – Ministry of Housing and Planning (2008) Actualización Plan Regulador Metropolitano de Santiago. Online: www.minvu.cl/opensite_20080421111026.aspx (accessed 19 March 2012).

MINVU – Ministry of Housing and Planning (1999) Circular DDU No. 55: Plan Regulador Comunal, Santiago: MINVU.

Núñez, C. (2006) "Plan Regulador Comunal: Una Propuesta desde la Comunidad de Pedro Aguirre Cerda", Santiago.

PAC Community Leaders (2005) "Proyecto de Memoria de Alternativa de la Comunidad al Plan Regulador de Pedro Aguirre Cerda", Santiago.

PCF (2005) "Proyecto Plan Regulador Comunal – Comuna de Pedro Aguirre Cerda", Santiago: PAC Municipality.

Rajevic, E. P. (2001) "La Planificación Urbana en Chile. Revista de Derecho, 1". Online: www.cde.cl/wps/wcm/connect/37036b804fbf7edd81b3ab46ce4e7365/5.pdf?MOD=AJPERES (accessed 19 March 2012).

SERPLAC – Regional secretary of planning (2005) "Programa Regional de Inversiones Región Metropolitana de Santiago 2005", Seremi de Planificación y Cooperación-Región Metropolitana de Santiago.

Shin, H. (2009) "Residential Redevelopment and the Entrepreneurial Local State: the Implications of Beijing's Shifting Emphasis on Urban Redevelopment Policies", *Urban Studies*, 46: 2815–40.

Slater, T. (2009) "Missing Marcuse: on Gentrification and Displacement", *City – Analysis of Urban Trends Culture Theory Policy and Action*, 13: 292–312.

Villagra, P. (2009) "Nueva Violación a la Libertad de Expresión: Allanamiento a Umbrales TV", *Clarín* electronic newspaper, Santiago. Online: www.elclarin.cl/index2.php?option=com_content&do_pdf=1&id=16683 (accessed 19 March 2012).

11 Distinguishing the right kind of city

Contentious urban middle classes in Argentina, Brazil and Turkey

Ryan Centner

At the center of the redeveloped waterfront in Buenos Aires, a new addition to the landscape in 2006 seemed to support the right to the city in quite radical fashion. In this privileged quarter of the city, the proprietor of a failed kiosk decided to gift the space to famous local activist Raúl Castells in order to open a soup kitchen for children in need, and to stage graphic posters decrying deadly inequalities across Argentina (Lipcovich 2006). Access to free sustenance and grounds for protesting injustice in this particular location, however, were more remarkable because they were an affront to *the right kind of city*. Indeed, indignation in Puerto Madero over this aberrant tenant was fiery (Murphy 2006). The property holder, while perhaps genuinely supporting these causes, knew that giving ground to Castells would spur an uproar because his presence was so antithetical to the predominantly elite and middle-class character of the neighborhood, which was exactly the point, as this disgruntled impresario sought to express his own ire over local certification disputes with creative flair. While the case of Castells in Puerto Madero is just one illustrative example, this pitting of privileged actors against each other represents an especially contentious yet common urban dynamic among the burgeoning middle classes across the Global South in quests that actually undermine the right to the city by attempting to create the right kind of city, which is always subjective and largely self-referential in class terms.

This chapter examines conflicts over the use and signification of space in the largest cities of three middle-income countries that have attracted global attention for their expanding urban middle classes: Argentina, Brazil and Turkey. The definition of "middle class", however, is imprecise and malleable – features that are important in shaping the nature of these urban conflicts. These groups articulate a right to the city, which Henri Lefebvre (1996: 158) highlights as not "a simple visiting right" but a "right to urban life", which David Harvey elaborates as "a right to change [the city] after our heart's desire" (Harvey 2003: 939). These are meant to be inclusive definitions that prize all those who "inhabit" the city (Lefebvre 1996: 158). Yet what we witness in the cases I survey is an exclusive right to the city, crafted essentially as the circumscribed property of a certain group. Ultimately undermining the right to the city, in the broad sense, different groups that invoke middle-class labels attempt to distinguish and mobilize around visions of what they see as the right kind of city, often with very specific discourses of livability, morality and citizenship. These struggles are clearly rooted in class, yet they also exceed its economic content, as there are often different middle-class groups who militate against each other around disparate notions of appropriate tastes, beliefs and everyday practices in order to stake their claims to the right kind of city.

To illuminate the nature of these struggles, Pierre Bourdieu's scholarship on *distinction* is especially useful. The creation of class-based distinction in the urban realm is, of course, quite counter to the right to the city. Yet it is important to understand how the nature of challenges to spatial justice continues to change. This is particularly the case in these kinds of cities, where growth processes are dynamic but highly inegalitarian, and intersect with longstanding yet unsettled boundaries of right/wrong, us/them, modern/traditional, European/other and so forth. Rather than offer an exactingly comparative case-by-case analysis, this chapter uses excerpts and extrapolations from fieldwork in Buenos Aires, Rio de Janeiro, São Paulo and Istanbul over the last decade, to provide a conceptual toolkit for grasping how contentious middle classes assert exclusionary visions of the right kind of city.

First, in contrast to work that explores whether new middle classes in middle-income countries can be an important subject of politics, it is necessary to understand that the category of "middle class" is blurry and used strategically. In other words, *middleclassness is an important object of politics*: which groups count as middle class, where, when and according to what metrics? Second, it is useful to see space itself as a resource that groups attempt to control by deploying different forms of capital (cultural, social, economic) to make claims on the city. This involves procuring the power literally to take place, a kind of *spatial capital.* Third, above and beyond claims that cities have become increasingly fragmented along class-based lines under the influences of globalized neoliberalism, this chapter illustrates *fractious forms of belonging* that center on rights to urban space but involve quarrelsome, exclusionary claims linked to other scales of legitimization (e.g., nationalism, transnational religions, prestigious foreign citizenships, etc.). By identifying and analyzing the shifting maneuvers that obstruct real inclusiveness even while claiming to promote the right to the city, it is possible to envision what kinds of strategies might be most effective in countering these efforts to impose singular, inegalitarian visions of the right kind of city in these contexts of ongoing urban dynamism in the Global South.

From the right to the city to the right kind of city

Attention to the growing middle classes of middle-income countries, and especially their novel urban milieux, remains strong despite recent global economic turmoil (Davis 2010). Indeed, these groups symbolize the righteous stalwarts of advancement that ostensibly anchor continued progress even in difficult times. Yet there is no agreement on exactly who comprises these much discussed middle classes, ranging sometimes from the nearly elite to the barely subsistent. Bourdieu's insights help us understand practices of marking class distinctions in ways that go beyond economic means. Although his own work centered largely on French society, Bourdieu's research explored disparate social contexts – from Algerian villages to Parisian *arrondissements* – yielding a remarkably transposable framework that pertains to matters of significant upheaval and far-reaching change in a range of locations (Wacquant 2002). Across the great social variation that comprises the category of cities of the Global South, a consistently pivotal feature of middle-class situations is the effort to create some kind of distinction – within city spaces, on individual bodies, in ways of behaving – that renders respectability, itself heavily tied to specific understandings of morality and judgments of what constitutes a desirable way of life. In other words, this is about distinguishing the right kind of city, according to the engrained judgments of middle-class actors.

Bourdieu writes that "[s]ocial subjects, classified by their classifications, distinguish themselves by the distinctions they make, between the beautiful and the ugly, the distinguished and the vulgar, in which their position in the objective classifications is expressed or betrayed" (Bourdieu 1984: 6). There are thus always efforts to discern and enact markers of social difference related to one's own group and others, and that very marking of difference illuminates the social whereabouts of those who seek to distinguish themselves. Emergent middle classes are certainly not the only ones who aim for distinction, but theirs are some of the most fervent and perhaps anxious efforts to do so because of their relatively novel and often precarious social positioning in the cities of the Global South.

These middle classes articulate what is fundamentally a right to the city – for themselves. Echoing Lefebvre's classic formulation, Harvey writes that

> [t]he right to the city is not merely a right of access to what already exists, but a right to change it after our heart's desire. We need to be sure we can live with our own creations (a problem for every planner, architect and utopian thinker). But the right to remake ourselves by creating a qualitatively different kind of urban sociality is one of the most precious of all human rights.
>
> (Harvey 2003: 939)

Rather perversely, however, instead of a city for all its inhabitants, we see struggles over the right kind of city, appropriate for a certain class of urban residents. Yet the identity of the "middle class" itself is an important initial quarrel. While some authors have emphasized these middle classes as new political subjects (Davis 2010; Villegas 2010), we must also consider how middleclassness is politicized in ways that affect how various middle-class groups shape the city in their own image.

Politicizing urban middleclassness

Postulation about the subjecthood of the new middle class is appropriate, as there is a wide variety of definitions given to this group across different national and local contexts. Diane Davis underlines the importance of rejecting "essentialist arguments about so-called middle class culture and its role in economic development, seeking instead to identify differences *among* the middle classes" (Davis 2010: 246, emphasis in original), further asserting that there is not "some essential cultural or political disposition about class politics or class discourses associated with middle 'classness'" (ibid.: 258). But beyond attempting to enumerate exactly who counts as middle class in each setting, and determining whether they are best described as "old" or "new" in their character as political subjects, we must recognize that there are indeed social and cultural attributes ascribed to the middle class that are also a matter of contention, and that there is no single proprietor of these features. In other words, *middleclassness* is a contested ensemble of characteristics, endowed with variable political valences, that different groups seek to own, manipulate, and deploy to a range of ends.

In line with authors writing on other middle-income contexts (e.g., Baviskar and Ray 2011; Kracker Selzer and Heller 2010; Stillerman 2010; Davis 2010), I concur that there is no essence of the middle class that transcends all settings to create a single type of new subject. Instead, what I am suggesting is that the idea of the middle class is a powerful tool that can mean many, sometimes contradictory, things as an object of politics. We can think of this ensemble of traits, in Bourdieu's terms as a field of struggle (Bourdieu and Wacquant 1992: 97) in and through which various actors attempt to craft and control a falsely essentialized sense of middle-class morality. All fields have a particular relation to the field of power, and certainly squabbles over middleclassness are fundamentally about defining the proximity of those with a kind of middling power to those who dominate society. Within fields of struggle, the actions of social groups depend on how they see the field, and make judgments about it, through a particular *habitus*. According to Bourdieu (1990: 53), habitus is a set of accumulated and lasting predispositions that are transposable across distinct social settings. The core features of habitus are shared by large class fractions, as socioeconomic characteristics give a broad – but far from absolute – unity to the experiences of each group in any given social setting. Habitus can, however, be shaped by other dimensions of social life, such as culture, religion, sexuality, regional background, etc. For that reason, there could be many kinds of middle-class habitus in a single setting, and thus a range of contesting views on what are the proper characteristics of the middle class.

Middleclassness, in some sites, might entail the ability to maintain savings, the priority of sending children to private schools, the penchant and wherewithal to consume certain goods from clothing brands to food genres; in others, it could be associated with speaking English crisply, or being in tune with fashion from abroad; in still others, it might be about expressions of outward political disinterest, possession of some kind of professional certification, or adherence to particular religious norms – and all of these might manifest as the *-ness* of the varied middle class in a single city of the Global South. What is consequential, though, is that these disparate characteristics can yield more than just some minor aspect of individual or class identity: they can render legitimacy to specific claims, often in the form of a moral assertion of righteousness in politics. It appears that jockeying over middleclassness as a moralistic trove is a fundamental feature of urban and national politics in a range of developing-world settings.

To associate one's political actions with the interests of the "middle class" seems to be a key way to signal, ostensibly, both innocence and upstanding intentions that selflessly benefit society at large – as especially well documented in studies of several Indian cities (Anjaria 2009; Ellis 2012; Ghertner 2012). Innocence here is rooted in a conscious recognition of middle-class distinction from the situational dispositions of elites and poorer masses alike: according to powerful, widely circulating tropes, the wealthiest segments of society are transparently greedy, corrupt, even abusive, while the poor, purportedly, are unfortunately myopic, easily hoodwinked and quite possibly criminal. Upstandingness, then, derives from middle-class orientations towards hard work, honesty and fairness, which are supposedly also exactly the features that have enabled middle-class people to achieve, if often by tooth and nail, whatever level of success and security they have within an inhospitably bifurcated political economy.

Both "old" and "new" middle classes appeal to these same flexible attributes as a buttressing narrative of moral legitimacy, and indeed people who range from rather wealthy to barely exceeding subsistence might also turn to this ensemble for far-ranging political ends. By stressing that middleclassness is an object of politics quite aside from the question of whether the middle class is a new political subject, I want to emphasize that myriad alliances (e.g., between different middle-class fractions, between old and new middle classes, between middle-class and elite groups) can obtain because the content of that middleclassness is so manipulable as a set of discourses that are relatively vacuous but morally powerful and emotionally loaded. The nature of these alliances, then, will depend on other aspects of local society, culture and politics beyond class positioning itself. Middleclassness must be linked with other mobilizing discourses and deeply felt issues to tip the balance of allegiance among the class fractions involved.

In Argentina, a middle-income country that has long defined its identity in terms of having a more substantial middle class than elsewhere in Latin America (Germani 1962; Germani 1987 [1955]), despite frequent, catastrophic and class-polarizing developmental setbacks (Waisman 1987), there is a strong if contested sense of middleclassness today. Buenos Aires – the capital with a metropolitan population of more than twelve million residents – exhibits particularly well some of the different claims on middleclassness that can manifest with divergent political aims. First, there are vocal articulations of middle-class morality in the city by modest residents who commit to the daily grind of disciplined work and dutiful improvement of both themselves and their community, and who can afford adequate but not luxurious living conditions: a small multiroom condominium in a low-rise building, or perhaps a duplex, with all the basic utilities. In neighborhoods of the city with a longstanding middle-class identity, such as Palermo and Caballito, residents who live in these kinds circumstances are embroiled in conflicts with developers who are erecting massive towers with sumptuously appointed units (see Ciccolella and Mignaqui 2008; Cosacov 2009; Kanai 2010). The terms of these confrontations revolve around the righteousness of middle-class residents and their entitlement to appropriate living conditions as a matter of morality. However, both sides are mobilizing middleclassness as a discourse about themselves, battling over whose claims are more legitimate, more accurately middle class. Some utilize middleclassness as a sign of modesty and precisely the inability to afford skyscraper living, while others demand that the ability to opt for such lifestyles is a matter of middle-class achievement, saving and freedom.

In some ways, this kind of conflict – widespread in Argentina during its recovery from deep neoliberal crisis after 2001–3, but also common in many middle-income countries – represents old versus new middle classes. Yet its cast of characters betrays another feature of this kind of debate: only the *clases populares*, or the poorest Argentines, as well as the most dizzyingly wealthy, except perhaps pertinent investors, are absent from this scenario. Therefore, middleclassness is a field of struggle that involves many, but excludes those who lack the power or desire to evoke its mantle for some particular end. In other conflicts over living space and lifestyles in Buenos Aires, quite different sectors of a very broadly defined middle class all grab onto middleclassness as a flexible but righteous self-identifier in order to scuttle claims of legitimate residence by poorer Argentines in adjacent informal settlements. Whether the employees of the lucrative expanded global service sector who live in the redeveloped glamour of waterfront Puerto Madero's unbeatably central location, or the hardscrabble but mortgage-bearing residents of dilapidated former state housing in Villa Soldati on the city's rusting southern edge, these dramatically disparate groups, at strategic moments, situate themselves as "middle class" in order to delegitimize *villa* (i.e., informal settlement) dwellers – and other precarious urban denizens – whose ramshackle but growing shelters abut their communities (Centner 2012; see also Girola 2007; Lederman 2012). In attempts to eradicate or at least limit these poorer neighbors, both the groups claim middleclassness of some kind – amongst other, shifting bases of legitimacy – in order to manifest a moral claim of greater deservingness over others.

Turkey displays, in some ways, an utterly dissimilar set of issues over middleclassness and morality, tied in significant ways to a vastly assorted Islam (see Turam 2004; Secor 2011). Yet striking political-economic similarities also characterize the comparative recent history of Turkey, Argentina and many other middle-income countries and their largest cities. In these settings, formerly rather elite urban enclaves are now sites of intertwined middle-class and poor settlement, with both ranks growing as rural migrants with meager means continue to seek opportunities in the city while more educated, privileged residents ascend the burgeoning professional heights of expanding global services and other lucrative, high-skilled occupations. That part of the story could apply to almost any major city in the Global South, but in Istanbul, an issue causing division within the middle classes is the rather more specific matter of religion and its expression. More than 90 percent of the Turkish population professes to be Muslim, but the manifestation of this faith in public, and as a political input, demonstrates profound variation across the populace. Merely two decades ago, it would have been rare to see a woman from an economically successful family wearing any kind of *türban* (the Turkish term for headscarves in line with *hijab* – Arabic for the generic, gendered Islamic dress code) in the largest city of this officially secular country. Today, in this same stridently cutting-edge Istanbul, the *türban* is abundantly visible in all areas of the city, including every quarter that might be characterized as any shade of middle class.

While many Istanbullus do not fret over this emergence, there are others, particularly those asserting their middleclassness as a moral justification, who see the marked increase in women's Islamic garb within their neighborhoods and daily rounds as an unwanted infiltration. This shift, from opponents' perspective, treacherously undermines the secular tradition established by the founders of the Turkish Republic, upending the country's orientation towards European civilization while drifting in the direction of ostensibly less civilized Arab peoples, fancifully cast as some kind of peripheral rabble (Navaro-Yashin 1999; Refiğ 2008) – paralleling divisive discourses about the place of public Islam amongst immigrants and their descendants in France and elsewhere (see Gökarıksel and Mitchell 2005). This disagreement surges in areas that have long borne a diffuse middle-class character tied to the city's unique identity as a crossroads of disparate cultures, such as Beyoğlu and other neighborhoods extending from Taksim Square (Arat-Koç 2007; Sözen 2010; Gökarıksel 2012), rather than in districts where the headscarves are actually abundant, many of which are significantly poorer (see Tuğal 2009a). Also in areas where the landscape is defined by sprawling new middle-class construction, such as Levent, this conflict rears up as a struggle over what middleclassness means. *Hijab*-wearing women and their supporters argue there is no dissonance between being an outwardly observant Muslim and a Turkish patriot (Karahan Uslu 2008; Turam 2011): indeed Turkey has become the center of an "Islamic chic" garment industry that recasts femininity and Islam for global markets of the fashion-forward faithful (White 1999; Navaro-Yashin 2002; Gökarıksel and Secor 2009). These women assert that, in fact, to be a morally upstanding member of the middle class means to embrace and defend religious values, including *hijab*, but to do so in a self-styled modern way, expressed with vibrant colors and in line with staunch professionalism in one's career.

The rise of middle-class *hijab* has coincided with the success of the currently ruling Development and Justice Party, or AKP (*Adalet ve Kalkınma Partisi*), a quasi-Islamist yet aggressively neoliberal movement that has vigorously sought to boost economic growth in Turkey (Kalaycıoğlu 2009; Secor 2011). This party has also attempted to speed Turkish accession to the European Union, demonstrating a clear affinity for many legal precedents in Europe, especially regarding religious freedom and its own ability to lead a constitutionally secular country while articulating open adherence to Islam. In several ways, then, the purported "Europeanness" of the AKP and the women who dress in high-fashion *hijab* across Istanbul is somewhat obvious. Less blatantly "European," however, are the masses of poorer Anatolian migrants who have come to Istanbul in recent decades, making homes in interstitial *gecekondular* or other precarious settlements, and who also generally endorse *hijab* (Secor 2002). These urban newcomers and their rural brethren are the mainstay of support for the AKP on a national scale, yet in Istanbul this population makes few gestures at middleclassness (Tuğal 2009b: 430–3). And indeed, the position of these poorer city residents in debates over legitimate presence is rather different. In some cases the religious observance and political leanings of these groups do rouse ire amongst other Istanbullus, but such characteristics typically conflate with an otherness that is mostly about class in a city where gentrification is far-reaching and ongoing (İslam and Behar 2006).[1] The crux of dispute within the middle-income population, then, is what aspects of daily life and appearance, linked to deeper sentiments and potentially to politics, are appropriate for middleclassness. Moreover, the debate revolves precisely around where these expressions are acceptable, and particularly how middle-class places beyond the private realm of the home are inscribed (see Gökarıksel 2012; Turam 2011). This is, therefore, a question of what kind of middleclassness will prevail – will it be "European" or "Arab", "faithful" or "faithless"? These characterizations are patently oversimplified, but such are the frequently invoked terms of the conflict, which points to different idealized visions grounded in disparate forms of ostensibly middle-class habitus. This contentiousness is manifest also in the city's current interventions in cultural landscapes for the European Capital of Culture 2010 mega-event, for which there have been complicated efforts to cast Istanbul as indelibly European and yet of Ottoman – and thus inevitably Muslim – origins (Centner 2010; Göktürk *et al.* 2010). In all of these debates, the Islam of the poor barely registers, and indeed their homes and quotidian venues tend to be the quick victims of unflinchingly stark efforts at urban beautification (see, e.g., Kuyucu and Ünsal 2010), typically aimed at serving those caught up in quarrels over the cultural and aesthetic inflections of middleclassness in remade urban space, the socioeconomic content of which is already clear.

These brief examples from Istanbul and Buenos Aires highlight the great manipulability of middleclassness, and expose how some of the most significant contention related to the middle classes can actually occur among them in battles of habitus that assert differing moral claims. They also show that middleclassness is a political object, and in fact often a tool for jousting with other middle-income groups or for trouncing groups considered to be socioeconomically lesser and, indirectly, less righteous. But these examples also point to the importance of space in these inter- and intra-class conflicts. This is not merely space as location – e.g., urban versus suburban versus rural, and how these coordinates can affect views and actions. Rather, this is space as a field of struggle, specifically over the ability to impose visions of "livability".

The power to take place

Peter Evans (2002: 1–3) asserts that livability entails, by definition, the combination of both livelihood and sustainability, yet this is a normative goal, often rather distant from the actual endeavors of urban dwellers who pursue their own visions of livability that can be much less altruistic, and may have little to do with collective consumption on a societal scale – such as the right to the city. For example, livability today might well include the provision of air conditioning in residences and offices where the new middle classes spend most of their time in steamy cities; this could also extend to the favoring of speedy, comfortable transportation in the form of new, user-pay highways for private cars with routes that strategically link middle-class and elite sites. Clearly these represent strivings at odds with the collective good in many ways, but they betray a certain vision of the right kind of city.

How is this kind of relatively lavish livability legitimized? One way is through recourse to that very middleclassness as a moral compass: it is necessary to live in modest comfort in order to work productively and support the developmental ascent of the national economy, which is an honorable goal that recursively justifies the measure of wellbeing that the middle class ought to enjoy in these working and living quarters, despite the squalor that might surround them in other settlements throughout cities of the Global South. In contexts where new city users have emerged on the urban scene – be they comprised of a particularly sizable occupational cohort, large ranks of immigrants, or members of the new middle class in cities of the Global South – there is likely to be some noticeable effect on the character and use of space, and a recasting of already existing places within the city. When those new city users assemble a range of privileges from social to cultural to economic, as is often the case with new middle classes, then the remaking of place in a quotidian sense is likely to occur on their own terms, often to the exclusion of others. This is, literally, a process of taking place, one that demonstrates an exertion of *spatial capital* in a field of struggle where space is at stake.

In Bourdieu's sense, there are many fields in any given city that exhibit struggles over material space, such as who can use and control different sites that are ostensibly open to everyone, from public spaces to the quasi-public but ultimately private spaces of shopping malls. If we think of these sites as fields, then Bourdieu would suggest that efforts to dominate them are tied to the strategic deployment of cultural, social or economic capital, or some combination thereof. Symbolic capital, in turn, could be any of these other three forms when it gains a greater meaning than its actual value, such as when money is seen also to indicate morality or some kind of right that far exceeds its currency. In a field where space is contested, symbolic capital is connected to spatial power: in turn, spatial capital is a form of symbolic capital in a field where material space is at stake, as I have elaborated elsewhere in relation to privileged new city users (Centner 2008).

Spatial capital, then, is the power to take place, but in a literal sense. It is about controlling and manipulating a place in ways that fit a particular habitus –

those accumulated and transposable dispositions tied to class and class fractions. In this framework, middle-class groups might indeed strive for their own version of livability, but this is likely to be grounded in the visions of a particular habitus of middleclassness, which might itself be contested but which we have little reason to expect would either turn to the state to ensure collective consumption, or be overly concerned with the implications for other classes in a spirit of sustainability, especially when so much effort pours into distinguishing middleclassness. Even if this distinction is somewhat clumsy and convoluted, it is still a matter of marking difference, and doing so with spatial designations that set off middle-class places.

In the deeply unequal landscapes of Rio de Janeiro, we witness competing forms of spatial capital at work in recent conflicts over the use, control and character of various places that have long underpinned notions of middleclassness. A series of increasingly spectacular conflicts have pitted the middle class and various authorities, but especially the police, against diverse criminal gangs, the latter often tied to drug trafficking and rooted in the numerous *favelas* (informal settlements) that riddle the city, with many innocent *favelados* caught all too literally in the crossfire (Alves and Evanson 2011). The conflict is not merely about the regulation of illicit trade; rather, there have been brazen efforts to demonstrate the authority of clandestine power over key sites, highlighting the impotence of the formal rule of law. Certain large, very profitable gangs have effectively demanded the shutdown of key areas of Rio on particular days in recent years – from the central business district to parts of the plainly privileged Zona Sul, which encompasses the famous neighborhoods of Ipanema and Copacabana. On fear of death, businesses and everyday residents have shuttered their doors and resigned themselves to hours or days at a time of forced standstill (see, e.g., Penglase 2005), resulting in eerily silenced places inscribed with the power of gang networks to exert their will on urban space. Counterstrategies have involved massive armed responses by the police and military, including aerial assaults on *favelas* within Rio that have been greeted by anti-aircraft artillery from within the shantytowns, as well as a growing "shock of order" that includes long-term militarized "occupation" of *favelas* by special "pacifying" police units (de Carvalho and de Fátima e Silva 2011: 65), and the tight regulation of spaces of the city that are public but nonetheless associated with middleclassness – especially its beaches (Godfrey and Arguinzoni 2012). In these situations, forms of middle-class morality confront an entirely different habitus with its own justifications and versions of livability. Both sides of this conflict actually envision their own livability in ways that quash the lives of others – perhaps the most ironic manifestation of this endeavor to produce a specific kind of city. These exertions of spatial capital demonstrate vividly how there is little collectively shared striving for improved urban situations across the yawning gulf that separates the predispositions of these disparate forms of habitus.

In Brazil's largest city, São Paulo, there are violent clashes following some of the dynamics outlined above (see Alves 2011). But the transformative use of spatial capital is visible in more peaceable if nonetheless consequential ways within neighborhoods such as Baixo Augusta as its character shifts towards a particular vision of middleclassness. Parts of this area in the blocks downhill from Avenida Paulista bear a somewhat unkempt or even outright dilapidated appearance in some corners. But the character of the area is undergoing major changes, which could be described as gentrification. More analytically illuminating, however, are the group-specific, intra-class spatialized practices through which transformation has been progressing. The imprint of artists, literati and a quickly growing entrepreneurial gay and lesbian presence are marking the area as a particular kind of middle-class place in conscious contradistinction to neighboring, markedly middle-class Jardins with its high-rise towers and demure shopping venues. Baixo Augusta is undergoing an incipient change that refurbishes older structures in a kind of preservationism that is very unusual for São Paulo, while marrying this trend with the insertion of new, edgier, small-scale constructions for a mix of commercial and residential uses that underline the artistic and alternative if still blatantly middle-class character of neighborhood. Despite the presence of one large development – Shopping Frei Caneca – this is mostly a property-by-property shift, with each transformation signaling a small but important taking of place,[2] although this may soon change (Brancatelli and Saldaña 2010). Spatial capital is therefore at play in the slow encroachment of this kind of character on the more rugged working-class district that had formed there in the late twentieth century, and which persists on the margins of Baixo Augusta – indeed, a place name that had not existed in the *paulistano* lexicon until a handful of years ago. So far, violence in this area has been moderate by citywide standards, and the creation of middle-class Baixo Augusta is ongoing.

The point is that we see struggles to assert some sense of middleclassness in particular places – sometimes much more embattled than others – that align with some vision of "livability" but that might well quell the aspirations of different groups who also seek to imprint place with their habitus but have competing forms or lesser quantities of spatial capital. This particular notion of competing bases of justification points to fractious forms of citizenship, wherein different groups mobilize claims of legitimate presence in the city via various, shifting geographic scales.

Fractious citizenships

Differently privileged fractions of the middle class often resort to disparate resources to assert their rights as citizens according to their means rather than their citizenship status. Some, for example, secure expensive private solutions to injustices whereas others turn directly to the state to realize their demands as citizens. This differentiation of citizenship within the urban theater is essential to understanding how new middle classes seek to craft the right kind of city among competing claims from fellow citizens of diverse economic and social backgrounds.

James Holston (2008), writing about the quickly urbanizing peripheries of São Paulo, conjectures that a willingness amongst average Brazilians to familiarize themselves with applicable laws and then realize their rights as citizens through the judiciary signals a new, insurgent form of citizenship – one that in itself grants the institution of citizenship a worth that it never enjoyed in Brazil before the late twentieth century. For much of Brazilian history, to be a citizen, Holston shows, was to be "an anonymous other, a John Doe – a person, in fact, without rights … 'a nobody.'" (Holston 2008: 4). Yet in his research, Holston finds that a rapid nationwide shift to city dwelling in the postwar years, coupled with overcoming the hardships of authoritarian government from the 1960s to the 1980s, has created a more robust sense of rights that can indeed be activated for and by citizens, with constitutional guarantees and the courts providing a means to level social differences in one of the most inegalitarian countries in the world. Insurgent citizenship, then, is about disadvantaged groups improving their own lot as they, essentially, become more immersed in middleclassness – as new small landowners in the city with vested property they seek to protect – and thus seek to trump older elitist hierarchies that had excluded them from protection, as citizens who were mere "nobodies."

Holston's research on insurgent citizenship chronicles important transformations in Brazil, and marks the rising importance of citizenship discourses there and elsewhere in cities of the Global South (see also Davis 2010: 256–7). Yet there is a much more varied picture related to citizenship and its mobilization by differently positioned urban populations in their quest to make a range of demands on city space. There may indeed be the kind of insurgent citizenship that Holston delineates among less advantaged but increasingly middle-class-oriented groups, but there are also what I capture elsewhere as *microcitizenships* (Centner 2012). These go beyond the ambit of the poor and the powerful – i.e., the main groups of focus for Holston as well as other citizenship scholars working in the Global South (e.g., Ong 2006) – to encompass the differentiated citizenship assertions of all manner of urban dwellers, including quite variegated middle classes, who turn to disparate bases of legitimacy to make claims on the city.

To return to Buenos Aires, in an arc of middle-class claims from the rather elite Puerto Madero to quite comfortable Palermo and Belgrano to much more modest San Blas and Villa Devoto, it has become common over the last decade for Argentines with immigrant ancestry to formalize a second nationality by obtaining passports from Italy or Spain – the two most common genealogical backgrounds in Argentina, which also have some of the most accessible qualifications for citizenship by descent (via one grandparent) amongst European Union countries. This can be a means of easing travel or becoming eligible for some kind of overseas retirement benefit, but it technically also opens up work opportunities anywhere in the twenty-seven EU countries. There are no figures on exactly how many Argentines have availed themselves of this dual citizenship and the ability to work abroad in much higher-paying professional settings, but it appears several thousand are indeed living in Spain (Margheritis 2007), while there is a very high rate of return to Argentina by dual nationals of all ages (Novick 2007). Awareness of Europe, or even an imaginary conception of "Europeanness", has had profound effects on middle-class residents of Buenos

Aires, their understandings of citizenship and how they make demands to shape the city.

In their efforts to seek a certain kind of livability in Buenos Aires, many of these middle-class groups seek not only to improve their immediate circumstances through personal initiative or by appealing to local authorities or social movements. Instead, they are animated by visions of key elsewheres – especially Barcelona, Madrid, Milan and Rome. These are places where they may have spent some time, but at the very least in which they have some vision of functionality and order, and due to their secondary citizenship there, they aim to replicate those often highly fantasized characteristics in Argentina, ostensibly as a matter of birthright. If Barcelona can manage to have impeccable and safe public spaces, then Buenos Aires ought to pursue the same basic achievement: a sense of beauty and security in Spain should be transportable by these dual nationals anywhere they live, so the logic goes. In this way, middle-class residents of Buenos Aires are staking citizenship claims in an entirely different scale, that of the Italian or Spanish transnation, in order to make demands on urban space in Argentina, and often at the expense of other Argentine citizens who do not share this same imagination or basis for legitimacy in how they seek to use or improve the same terrains of the city. Those others, however, might have very different bases for claiming legitimate presence in urban space, guided by different forms of habitus (see Centner 2012). Differentiation of claims by habitus is evident in this case as there is a roughly even distribution of Italian and Spanish descent across all income levels in contemporary Argentine society, but not all groups have the interest, wherewithal and funds to seek EU citizenship. It is the middle classes – familiarized with images of Europeanness and keen to the benefits of middle-class societies – who see value in seeking out a second citizenship, and also appeal to the urban conditions that they imagine prevailing there as justification for the kind of environment they demand in Buenos Aires.

A particular feature of microcitizenships is that there are appeals to distinct geographic scales (e.g., the city, the nation, the transnation) even as the object of contention remains city space itself. The overall result is multiple and competing citizenships that articulate fractious forms of belonging in urban space due to their basis in differently scaled legitimacies, each with its own vision of territorialized rights. When successful, *microcitizenships* are thus habitus-specific negotiations with some level of the state that entail both recognition and service provision in order to grant exclusive but temporary rights to particularized legitimate uses of urban space. Although middle classes are only one group that might seek to make such claims, they are often some of the most effective, partly because of their ability to resort to middleclassness as a hallowed if flexible political tool in buttressing their other-scaled legitimacies.

Conclusion

All of these features of middle classes in cities of the Global South point to different ways in which distinction is not only crafted, but spatialized. Distinction is a thorny matter, whether between recognizable classes or, especially, within broad classes wherein groups seek to differentiate themselves from each other. Of course these assertions of distinction become rather more crucial when they involve the remaking of space and the taking of place. This is especially so in cities – such as many in the Global South – where dense conditions mean that space is at a premium, making the wielding of spatial power absolutely momentous. Certainly there are many actors apart from the middle classes who have the ability to transform urban space, including not least the state and more elite echelons, including private developers. The expansion of urban middle classes in the Global South, however, marks some important shifts in how the fate of cities is being shaped.

Quarrels over middleclassness might seem, on the surface, to be debates of self-absorption, but they are one key aspect of urban transformation in the Global South, as different groups lay claim to some kind of middle-class identity and mobilize it as a weapon of moral exclusion. Moves to exclude, whether openly aggressive or more subtly transformative, point to the use of spatial capital as a key mechanism through which the varied middle class is engaged in urban change in the Global South – sometimes as a spatialization of those conflicts over middleclassness, sometimes pitting a fraction of the middle class against poorer or wealthier groups. In addition to political appeals to morality that map onto space, we also now see inventive, multiscalar use of different citizenship claims to ensconce actions on urban space within disparate legitimacies that may very well conflict each other due to these distinct geographic bases of justification. These are not the only struggles to create distinction by middle-class groups in cities of the Global South, but they represent particular modes of spatialized struggles. Primarily, these concepts and examples show that class is a key feature of urban struggles but that it is almost always coupled with other dimensions of social distinction that have significant rooting in locally specific contingencies.

To conclude, all of these efforts to spatialize distinction revolve in significant ways around the habitus of middle-class groups. But from Istanbul to Buenos Aires, it is clear there are many different kinds of middle-class habitus at work in shaping what each group sees as the right kind of city. Nonetheless, we can discern loose similarities across the vast diversity of the urban Global South – and the Global North, for that matter – in such struggles over urban distinction: it is inherently conflictive, inegalatarian and self-referential. In sum, the quest to imprint the city with one's own image of righteousness is deeply antithetical to the encompassing sense of the right to the city propounded by Lefebvre and Harvey, even though it is about changing the city "after our heart's desire". The problem centers on defining "we" and "our", which is why a critical focus on the self-referencing of middleclassness is so crucial to overcoming these conflicts. Successful strategies to bolster the right to the city for all will need to push these visions of the right kind of city, to reconcile themselves with the fact that cities are always shared, collective achievements that can only thrive when space – literally – is made for everyone, including those who are on the "wrong" side of distinction.

Notes

1 In addition to class, but often tied to it, sometimes issues of ethnic origin are pivotal in processes of othering in Istanbul, particularly but not only regarding Kurds (Secor 2004; Mills 2010).
2 This particular kind of new middle-class place is emerging with little involvement by any level of the state, other than the most basic infrastructural provision and regulation, which had already been in place for more than a half-century anyway.

References

Alves, J. A. (2011) "Topografias da Violência: Necropoder e Governmentalidade Espacial em São Paulo", *Revista do Departamento de Geografia*, 22: 108–34.

Alves, M. H. M. and Evanson, P. (2011) *Living in the Crossfire: Favela Residents, Drug Dealers, and Police Violence in Rio de Janeiro*, Philadelphia, PA: Temple University Press.

Anjaria, J. S. (2009) "Guardians of the Bourgeois City: Citizenship, Public Space, and Middle-Class Activism in Mumbai", *City & Community*, 8: 391–406.

Arat-Koç, S. (2007) "(Some) Turkish Transnationalism(s) in an Age of Capitalist Globalization and Empire: "White Turk" Discourse, the New Geopolitics, and Implications for Feminist Transnationalism", *Journal of Middle East Women's Studies*, 3: 35–57.

Baviskar, A. and Ray, R. (eds) (2011) *Elite and Everyman: the Cultural Politics of the Indian Middle Classes*, New Delhi: Routledge India.

Bourdieu, P. (1984) *Distinction: a Social Critique of the Judgement of Taste*, Cambridge, MA: Harvard University Press.

Bourdieu, P. (1990) *The Logic of Practice*, Stanford, CA: Stanford University Press.

Bourdieu, P. and Wacquant, L. J. D. (1992) *An Invitation to Reflexive Sociology*, Chicago, IL: University of Chicago Press.

Brancatelli, R. and Saldaña, P. (2010) "Antes Degradada, Região do Baixo Augusta vira Alvo de Construtoras". Online: www.estadao.com.br/noticias/geral,antes-degradada-regiao-do-baixo-augusta-vira-alvo-de-construtoras,623320,0.htm (accessed 20 March 2012).

Centner, R. (2008) "Places of Privileged Consumption Practices: Spatial Capital, the Dot-Com Habitus, and San Francisco's Internet Boom", *City & Community*, 7: 193–223.

Centner, R. (2010) "Cities and Strategic Elsewheres: Developments in the Transnational Politics of Remaking Cities", *New Global Studies*, 4: 1–7.

Centner, R. (2012) "Microcitizenships: Fractious Forms of Urban Belonging after Argentine Neoliberalism", *International Journal of Urban and Regional Research*, 36: 336–62.

Ciccolella, P. and Mignaqui, I. (2008) "Metrópolis Latinoamericanas: Fragilidad del Estado, Proyecto Hegemónico y Demandas Ciudadanas. Algunas Reflexiones a Partir del Caso de Buenos Aires", *Cuadernos del Centro de Estudios del Desarrollo*, 69: 47–68.

Cosacov, N. (2009) "Dinámica del Capital y Movilización de Vecinos: Aproximaciones a un Análisis Microespacial de un Conflito Urbano en un Barrio de la Ciudad de Buenos Aires", *Intersticios*, 3: 193–204.

Davis, D. (2010) "The Sociospatial Reconfiguration of Middle Classes and their Impact on Politics and Development in the Global South: Preliminary Ideas for Future Research", *Political Power & Social Theory*, 21: 241–67.

de Carvalho, V. A. and de Fátima e Silva, M. d. R. (2011) "Política de Segurança Pública no Brasil: Avanços, Limites e Desafios", *Revista Katálysis*, 14: 59–67.

Ellis, R. (2012) "'A World Class City of Your Own!': Civic Governmentality in Chennai, India", *Antipode*, 44 (4): 1143–60.

Evans, P. (2002) "Looking for Agents of Urban Livability in a Globalized Political Economy", in P. Evans (ed.) *Livable Cities?: Urban Struggles for Livelihood and Sustainability*, Berkeley, CA: University of California Press.

Germani, G. (1962) *Política y Sociedad en una Epoca de Transición: De la Sociedad Tradicional a la Sociedad de Masas*, Buenos Aires: Editorial Paidos.

Germani, G. (1987 [1955]) *Estructura Social de la Argentina: Análisis Estadístico*, Buenos Aires: Ediciones Solar.

Ghertner, A. (2012) "Nuisance Talk and the Propriety of Property: Middle Class Discourses of a Slum-Free Delhi", *Antipode*, 44 (4): 1161–87.

Girola, M. F. (2007) "Procesos de Apropiación del Espacio y Sociabilidad Vecinal en un Gran Conjunto Urbano Situado en la Ciudad de Buenos Aires", *Anthropologica*, XXV: 131–55.

Godfrey, B. and Arguinzoni, O. (2012) "Regulating Public Space on the Beachfronts of Rio de Janeiro", *Geographical Review*, 102: 17–34.

Gökarıksel, B. (2012) "The Intimate Politics of Securalism and the Headscarf: the Mall, the Neighborhood, and the Public Square in Istanbul", *Gender, Place & Culture*, 19: 1–20.

Gökarıksel, B. and Mitchell, K. (2005) "Veiling, Secularism, and the Neoliberal Subject: National Narratives and Supranational Desires in Turkey and France", *Global Networks*, 5: 147–65.

Gökarıksel, B. and Secor, A. (2009) "New Transnational Geographies of Islamism, Capitalism and Subjectivity: the Veiling-Fashion Industry in Turkey", *Area*, 41: 6–18.

Göktürk, D., Soysal, L. and Türeli, İ. (eds) (2010) *Orienting Istanbul: Cultural Capital of Europe?*, London: Routledge.

Harvey, D. (2003) "The Right to the City", *International Journal of Urban & Regional Research*, 27: 939–41.

Holston, J. (2008) *Insurgent Citizenship: Disjunctions of Democracy and Modernity in Brazil*, Princeton, NJ: Princeton University Press.

İslam, T. and Behar, D. (eds) (2006) *İstanbul'da Soylulaştırma: Eski Kentin Yeni Sakinleri*, Istanbul: İstanbul Bilgi Üniversitesi Yayınları.

Kalaycıoğlu, E. (2009) *Hicab, Türban, and Democracy: Religious Freedom versus Political Protest*, Aberdeen, United Kingdom: Centre for the Study of Public Policy, University of Aberdeen.

Kanai, M. (2010) "The Politics of Inequality in Globalizing Cities: How the Middle Classes Matter in the Governing of Buenos Aires", *Environment and Planning A*, 42: 1887–901.

Karahan Uslu, Z. (2008) "Beneath the Headscarf: a Modern Individual", *New Perspectives Quarterly*, 25: 43–5.

Kracker Selzer, A. and Heller, P. (2010) "The Spatial Dynamics of Middle-Class Formation in Postapartheid South Africa: Enclavization and Fragmentation in Johannesburg", *Political Power & Social Theory*, 21: 171–208.

Kuyucu, T. and Ünsal, Ö. (2010) "'Urban Transformation' as State-led Property Transfer: an Analysis of Two Cases of Urban Renewal in Istanbul", *Urban Studies*, 47: 1479–99.

Lederman, J. (2013) "'But this is a Park!' The Paradox of Public Space in a Buenos Aires 'No Man's Land'", *Citizenship Studies*, 17 (1), February.

Lefebvre, H. (1996) *Writings on Cities*, Cambridge, MA: Blackwell.

Lipcovich, P. (2006) "Un Debate entre Mate y Tortas Fritas". Online: www.pagina12.com.ar/diario/sociedad/3–63826–2006–03–03.html (accessed 3 December 2011).

Margheritis, A. (2007) "State-Led Transnationalism and Migration: Reaching out to the Argentine Community in Spain", *Global Networks*, 7: 87–106.

Mills, A. (2010) *Streets of Memory: Landscape, Tolerance, and National Identity in Istanbul*, Athens, GA: University of Georgia Press.

Murphy, M. (2006) "Argentina, entre el Hambre y el Lujo – 10 de marzo de 2006". Online: http://news.bbc.co.uk/hi/spanish/latin_america/newsid_4791000/4791972.stm (accessed 3 December 2011).

Navaro-Yashin, Y. (1999) "The Historical Construction of Local Culture: Gender and Identity in the Politics of Secularism versus Islam", in C. Keyder (ed.) *Istanbul: Between the Global and the Local*, Lanham, MD: Rowman & Littlefield.

Navaro-Yashin, Y. (2002) *Faces of the State: Secularism and Public Life in Turkey*, Princeton, NJ: Princeton University Press.

Novick, S. (ed.) (2007) *Sur-Norte: Estudios sobre la Emigración Reciente de Argentinos*, Buenos Aires: Catálogos Editora.

Ong, A. (2006) *Neoliberalism as Exception: Mutations in Citizenship and Sovereignty*, Durham, NC: Duke University Press.

Penglase, R. B. (2005) "The Shutdown of Rio de Janeiro: the Poetics of Drug Trafficker Violence", *Anthropology Today*, 21: 3–6.

Refiğ, H. (2008) "Woman's *Türban* is Sign of Rebellion Against Atatürk", *New Perspectives Quarterly*, 25: 40–2.

Secor, A. (2002) "The Veil and Urban Space in Istanbul: Women's Dress, Mobility and Islamic Knowledge", *Gender, Place and Culture*, 9: 5–22.

Secor, A. (2004) "'There is an Istanbul that Belongs to Me': Citizenship, Space, and Identity in the City", *Annals of the Association of American Geographers*, 94: 352–68.

Secor, A. (2011) "Turkey's Democracy: a Model for the Troubled Middle East?", *Eurasian Geography and Economics*, 52: 157–72.

Sözen, Ü. (2010) "Gentrification, Cosmopolitanism, and Consumption in the Beyoğlu District: a Case Study of Metropolitan Habitus", unpublished thesis, Koç University.

Stillerman, J. (2010) "The Contested Spaces of Chile's Middle Classes", *Political Power & Social Theory*, 21: 209–38.

Tuğal, C. (2009a) *Passive Revolution: Absorbing the Islamic Challenge to Capitalism*, Stanford, CA: Stanford University Press.

Tuğal, C. (2009b) "The Urban Dynamism of Islamic Hegemony: Absorbing Squatter Creativity in Istanbul", *Comparative Studies of South Asia, Africa and the Middle East*, 29: 423–37.

Turam, B. (2004) "The Politics of Engagement between Islam and the Secular State: Ambivalences of 'Civil Society'", *British Journal of Sociology*, 55: 259–81.

Turam, B. (2011) "Ordinary Muslims: Power and Space in Everyday Life", *International Journal of Middle East Studies*, 43: 144–6.

Villegas, C. (2010) "Revolution 'from the Middle': Historical Events, Narrative, and the Making of the Middle Class in the Contemporary Developing World", *Political Power & Social Theory*, 21: 299–312.

Wacquant, L. J. D. (2002) "An Inventive and Iconoclastic Scientist", *Berkeley Journal of Sociology*, 46: 177–86.

Waisman, C. (1987) *Reversal of Development in Argentina: Postwar Counterrevolutionary Policies and their Structural Consequences*, Princeton, NJ: Princeton University Press.

White, J. (1999) "Islamic chic", in C. Keyder (ed.) *Istanbul: Between the Global and the Local*, Lanham, MD: Rowman & Littlefield.

12 Bloggers' right to Cairo's real and virtual spaces of protest

Wael Salah Fahmi

Introduction

This chapter examines Cairo's contested real and virtual public spaces as urban youth and bloggers protest for democratic reforms. The study attempts to answer the question: does virtual space affect and modify real space, and give a new dimension to human communication interactions? The research for this study used a qualitative ethnographic analysis of Egyptian bloggers' real and virtual experiences of political activism and civil participation (Fahmi 2009). The author's discussions with various bloggers focused on their opinions regarding cyberactivism and the blogosphere as spaces of protest. Anecdotal records are based on participant observation during bloggers' sit-ins and street demonstrations for democratic reforms over a period of six months (March–August 2006), and highlight the social dimension of the blogosphere as spaces of protest (Fahmi 2009).

Through the study of the Egyptian bloggers' right to real and virtual spaces of protest, the empirical survey attempts to answer the following questions.

- How does the virtual space of the Internet relate to geographically definable, "real" locations? Can they still be clearly distinguished and how do they merge?
- How is the understanding of space and communication changing within the relatively small, relatively privileged group of those active in alternative media with the rapid appropriation of information technology?

In spite of security surveillance and restrictions on its activities within public spaces, the emerging street protest movement continues a process of urban resistance under the slogan "the street is ours", as blogger activists contest power in spheres beyond their community level, appealing to grassroots networks and to local practices. Bloggers' street tactics aim to stretch across real and virtual spaces, constructing relational social and spatial networks, while (re)appropriating and (re)claiming Cairo's contested spaces of protest in downtown Unions Street and Midan al Tahrir (Tahrir Square).

Faced with formidable challenges to expression in Cairo's public spaces, urban blogger activists have developed new means of articulating dissent with spatial tactics from boycott campaigns, cyberactivism and protest art to innovations in mobilization, means of communication and organizational flexibility. This is particularly evident in the way urban blogger activists have employed urban installations and street graffiti, prompting the construction of a significant site of urban resistance and spatial contestation. The emergence of grassroots street activism unravels hidden urban spaces of protest in the Global South, whilst opening up a new public sphere through which the role of urban governance might be contested to accommodate cultural identities within various forms of spatiality and popular democracy.

This chapter is organized into three sections. The first section will explore the theoretical discussion on networked social movements in terms of the active role of street blogging as a new form of mobile interaction within public spaces, encouraging civil participation and online political activism. The right to the city and the significance of the cityscape for urban mobilizations are then related to Cairo's European Quarter and its public spaces, which witnessed the Spring 2006 pro-democracy street rallies and sit-ins organized by urban youth and blogger activists.

The second section will provide the historical review and qualitative ethnographic analysis of Egyptian bloggers' real and virtual experiences of political activism and civil participation. The historical chronology of different street rallies and political events during 2005 will emphasize the significant role played by the Egyptian Movement for Change (Kefaya)[1] in political mobilization activities, which were posted and distributed by bloggers on the Egyptian blogosphere. The empirical survey will highlight various issues raised during the author's informal discussions with bloggers, focusing on their opinions regarding cyberactivism and the blogosphere as spaces of protest. The survey also includes anecdotal records of bloggers' sit-ins and street demonstrations during April 2006, within Cairo's downtown Tahrir Square and Unions Street (in front of the Press Syndicate and Judges Club). These street actions were organized in solidarity with Judges' demanding judiciary reforms and independence from the state-appointed Ministry of Justice.

The third section will first cover the period between 2006 and 2008, exploring the emergence of new civil society mobilization of various social groups who organized an urban protest movement. This movement, which is different from the networked social movements, was inspired by Kefaya's earlier street actions and encouraged by bloggers' online and offline political activism. This section will mention the 6 April 2008 general strike, which gained local popularity and international exposure through SMS, emails, blogs, Facebook, YouTube and Twitter. Simultaneously, riots erupted in the Textile Company in the northern Mehalla al Kobra City, as a result of confrontations between workers and police security. The 6 April events are considered an example of a national collective action, where bloggers collaborated with various non-politicized groups who started gaining mass protest experience. This was a fundamental development, as noted during the 2011 winter uprising. This section of the chapter will also examine the emerging hybrid spaces of protest in 2011, where more organized mass street protests were mobilized by "Facebookers" in cyberspace, as protestors reclaimed and contested main public squares and streets. The 2011 uprising was characterized by hybrid spaces of protest in terms of street rallies and confrontations in public spaces, online streaming of images, news updates on Twitter and twenty-four-hour live TV broadcasting of the eighteen-day sit-in in Tahrir Square. The current study indicates that online activism was very much intertwined with bloggers' offline political actions, as virtual space affected and modified real space, giving a new dimension to Egypt's mass protest movement.

Networked social movements and geographies of protest

This section explores the active role of the blogosphere within networked social movements, as bloggers become part of a virtual transnational community, encouraging civil participation and online political activism.

Spontaneous urban social movements demanding a right to the city are related to material reality in the city and to spatially fixed groups, such as residents protesting against rehabilitation, renovation and gentrification, and demanding infrastructure provision. From the "renovation-deportation" movement in Paris, opposing planned gentrification and residential squatters in London and Amsterdam, to suburban popular squatting in many cities of the Global South, people everywhere have been claiming the city. Rapid urbanization and the colonization of urban space have provoked an urban social movement against authority for the control of peripheral urban land. This has been explored as a spontaneous subordinate culture and a massive urban social movement, defying the state and contravening its planning regulations (Castells 1997; Leontidou 2006).

However, Leontidou (2006) notes that the recent marginalization of grassroots urban social movements as a theoretical and material spontaneous occurrence is attributed to various factors. First, there is the recent shift in collective cultures around urban social movements as a result of the "digital revolution", as well as the contested concept of "globalization" – a spatial globalization discourse, within which urban social movements appear somewhat parochial in the face of transnational spatiality. Second, there is more emphasis on organization theory, especially the "third sector" of non-governmental organizations (NGOs) and voluntary organizations in general (McAdam and Scott 2005), as well as concepts of social capital, resource mobilization and political opportunity structures. This resulted in informal popular mobilizations claiming the city, with the "right to the city" being underplayed in favor of more formal transnational organizations, and various environmental, political, social and cultural movements converging in anti-globalization rallies that have escalated since those massive demonstrations against the World Trade Organization Summit in Seattle in 1999 (Leontidou 2006). Third, the digital divide, knowledge gaps and control over information resources during political protest have led to the exclusion of large masses of the urban poor from online activism (Van Dijk 2005; Norris 2004). Such a digital divide is noted by Della Porta and Tarrow:

> [e]very new form of communication both heightens ties between those who already know one another, and raises the walls of exclusion for those lacking access to the new medium of communication.
>
> (Della Porta and Tarrow 2005: 4)

This new medium of communication is related to Rheingold's notion of smart mobs, which refers to a group of people who cooperate with each other by using information and communication technologies to organize activism; smart mobs include "[p]eer-to-peer collectives, pervasive computing, social networks, and mobile communications" (Rheingold 2002: 66). Accordingly, public spaces are transformed into interfaces for new smart mobs with ubiquitous access to mobile communication networks during social actions and street rallies. Such moblogging or mobile blogging, also known as street blogging (ibid.), is a social practice that enables bloggers to post and publish text, images (photoblogging) and video clips (video blogging or vlogs) during public rallies and street mobilizations.

Melucci (1996) and Castells (1997) describe such new social movements (NSMs) in terms of loosely organized and open networks. Castells regards "resistant networks" of social movements as a defense

> against the placeless logic of the space of flows characterizing social domination in the Information Age. Such "global fluids" consist of flows or waves of people, information, images, easily crossing borders with a greater degree of flexibility than ever before.
>
> (Castells 1997: 358)

The age of communications, with its associated transnational public spheres, has contributed to the emergence of these new social movements, waging discursive resistance, and challenging meanings, identity constructions and media stereotypes (Leontidou 2006). Characterized by "Notopia" (no topos, no place, no location, no maps), the Internet has become inhabited by global cyber "hacktivists" (hackers and activists) who have contributed to development of the new digital social movements (Papadimitriou 2006). These itinerant *flâneur* activists mobilize and organize massive anti-globalization rallies via the Internet (Leontidou 2006), constantly traveling and reassembling for street demonstrations in various cities around the globe. However, new social movements, with their do-it-yourself approach to information and communication technologies, have mixed old and new technologies, merging virtual and physical spaces into "networks of alternative communication" (Hamm 2003). These networks are linked to "social centers" that shelter transnational networks in unoccupied buildings and abandoned urban spaces, where innumerable media groups work with WiFi, video streaming, satellite wireless connections and the open source operating system Linux (Notes from Nowhere 2003). From the national "print publics" that informed movements to the "electronic sit-in" by activists blocking the World Trade Organization's website in 1999, information tends to outrun institutional networks, flowing along more invisible spaces and "submerged networks" (Anderson 1988). The spaces of resistance constituted within these hybrid physical and virtual worlds have created new geographies of protest. On the one hand, due to their geographic mobility, loose organizational models and access to communications, global networks have shifted their campaigns and resources to alternative virtual venues (Tarrow 1998; Della Porta *et al.* 1999). On the other hand, as events are reported through websites, blogs and streams in a collaborative social process, a means of navigation is provided for street protestors (Hamm 2006). Media activism uses the Internet as a workspace, social center and project workshop, so that virtual and physical spaces are experienced almost as a single space of communication. The expansion of the communication space and creation of public spheres for political organization and network formation have contributed to holding together the diversity of the anti-globalization movement (Hamm 2003), mainly through public access points regarded as hubs for the exchange of information (Notes from Nowhere 2003: 241).

Since the 1999 protests in Seattle against the World Trade Organization talks, the process of reclaiming the streets and producing emancipatory globalized public spheres has taken different forms of expression and practice, with the launching of IndyMedia (the Internet-based network of independent media centers or IMCs), which emerged as the backbone of communication for the broad coalition of anti-capitalism movements. In the spring of 2003, IndyMedia, which attracted anarchists, libertarians and techno-utopians hoping to act out frontier freedoms, was affiliated with 600 to 700 electronic mailing lists and sixty IRC (Internet relay chat) chatrooms, in addition to collaborative web pages (wikis) used by social forums. IndyMedia has linked decentralized actions and networks with familiar forms of struggle such as mass marches, rallies and reclamations of space, challenging the boundaries between cyberspace and real space (Notes from Nowhere 2003: 241).

It has become common to think not of a single movement (Pickerill and Chatterton 2006), but of a "movement of movements" (Mertes 2004), lacking a center, difficult to control, monitor and police, and not conducive to the formation of hierarchies, leaders and centralization of power. This use of affinity groups sharing common goals and areas at a human scale has been common with alternative and independent media, such as IndyMedia, allowing them to develop non-hierarchical techniques for communication (Kidd 2003).

The next section will explore the relationship between real and virtual spaces of protest in the context of urban activists' right to the city and Cairo's contested public spaces.

The right to the city and Cairo's contested public spaces

At the 2006 World Social Forum meeting, the anti-globalization movement prepared a Charter for the Right to the City (Portaliou 2007), challenging the neoliberal attack on public spaces (Castells 1983). According to its preamble, the

Charter "is an instrument intended as a contribution to the urban struggle and as an aid in the process of recognition of the right to the city in the international human rights system" (World Social Forum 2006). Social movements' street actions are related to the concept of the right to the city, which implies the right "to freedom, to individualization in socialization ... to participation and appropriation", and also includes "the right to the use of the [city] center" (Lefebvre 1996). This has highlighted people's right to participate in the public sphere, spatializing this right and associating it with everyday urban life. In this case, the urban experience, the social encounter in the street, becomes a political event.

Within the domain of contested city spaces, many young people consider the street the stage for performance, where they construct their social identity in relation to other members of society (Malone 2002). Urban youth's consumption of public spaces regards the street as a venue for social encounters and political protest, as a site of domination and resistance, and as a place for constructing youth culture (Fyfe 1998). This indicates the image of an alternative city as an object of contestation from the point of view of the grassroots and in juxtaposition to dominant social and political forces (Portaliou 2007). This is an alternative city, which comes out of urban struggles and the action of urban social movements, against the neoliberal attack on urban space and on the rights to the city (Castells 1983).

Social movement activities engender new "public spaces of representation", which provide opportunities for movement actors to recognize themselves and be recognized as alternative constituencies or counterpublics (Melucci 1996; Alvarez *et al.* 1998; Warren 1998). These processes create critical openings for injecting messages and demands into mainstream decision-making processes and translating them into potential foundations for change (Melucci 1985). White (1995) suggests that "publics" are special social spaces, designed to hold maximum participants with minimum friction, that allow for an opening of communication where participants switch from everyday interaction rituals into collective action. The long-term pressure of mobilization is part of what keeps democratic public spaces vital and open (Giugni 1999). Sheller (2003) suggests that "social flow" can be used as a new way of theorizing how social movements actually occur, emphasizing the very micro-level interest in the physical movement of bodies and "collective actors" occurring in named places. It could be argued that social movements reconfigure time and space through collective action, by selectively linking elements of past, present and future through mobilizing narratives, and by transforming everyday spatial boundaries. One kind "social movement" is intrinsically connected to actual movements of people and information through geographical space. A classic repertoire of contention used by social-movement actors involves flooding into public spaces, filling them with a special kind of active presence (Melucci 1985), and stopping other kinds of flows (Sheller 2003). Literal flows of people (and blockages) have always been crucial to social movements (Jordan 1998), from the barricades in Paris in 1848 (Gould 1995) to the "Reclaim the Streets" (RTS) events organized by anti-road protesters in the 1990s. These actions of appropriating space have made visible the latent power of state authority to exercise coercion at the most personal level by stopping freedom of movement and by incarcerating people.

Leontidou (2006) emphasized the significance of the cityscape for urban mobilizations, and the ways governments introduce planning measures to control crowd behavior. This could be noted in the case of Haussmann, who rebuilt central Paris between the 1848 and 1871 insurrections, surrounding the Arc de Triomphe with monumental boulevards as a means to decentralize discontent (Harvey 1996). Nevertheless Haussmann's wide boulevards and grand squares provided a perfect urban venue for massive demonstrations in Paris. This could be as noticed in the city barricades of the Paris Commune in 1871 and the student uprisings in May 1968, which took place within Boulevard St Germain in the Quartier Latin.

A similar relationship between space and politics was evident during the spring 2006 pro-democracy street rallies and sit-ins organized by urban youth and blogger activists within Cairo's European Quarter, this part of the city being regarded as a contested site for collective action and as a symbolic space for urban youth's political participation and spatial appropriation (Malone 2002; Batuman 2003). This can be seend in Figure 12.1, which shows the main public spaces that were venues for street rallies and the routes of major demonstrations from Midan al Tahrir (Tahrir Square), through Talaat Harb Square, Mostafa Kamel Square to Union Street, where protestors usually gathered on the stairs of the Press Syndicate and sometimes opposite the Judges' Club.

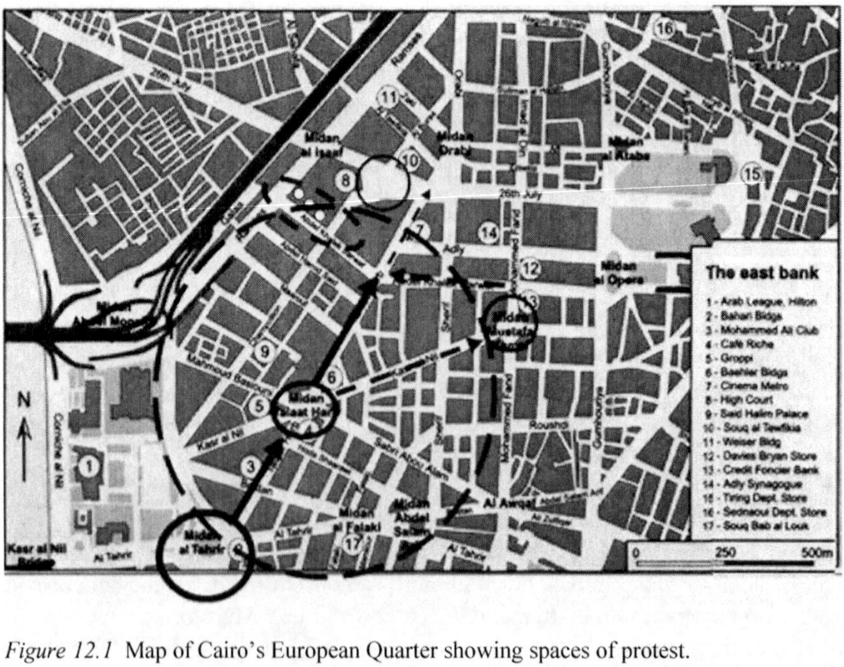

Figure 12.1 Map of Cairo's European Quarter showing spaces of protest.

The following section will provide an ethnographic analysis of the Egyptian bloggers' real and virtual experiences in political mobilizations and street blogging during the spring 2006 demonstrations within Cairo's European Quarter and downtown public spaces.

The bloggers' movement and the Egyptian blogosphere

Egyptian bloggers[2] are part of a loose network not affiliated with a particular political group, although some are active members of the Egyptian Movement for Change (Kefaya). Many bloggers who were active members of Kefaya decided to organize themselves into a proper collective, Youth for Change, in 2005. Previously, they had mainly been involved in the 2003 spring anti-Iraq war movement (known as 20 March) and Palestine solidarity campaigns. While the government's massive round-up and imprisonment of hundreds of protesters in the weeks following the 2003 anti-war demonstrations did impede the movement, the immediate war period constituted an important step in opening doors for critics of domestic policy. Initially regarded as the semi-official youth section of Kefaya, Youth for Change's five rotating committees (media, communications, culture, art and outreach) were run by a five-person steering committee. Active members of the group formed temporary social centers, as they moved with their laptops and mobile phones between coffee shops, university halls, unions and syndicates throughout the city. The group was technologically savvy, spreading their message through emails, SMS messages and blogs. Youth for Change's strategy is almost guerrilla style in relation to some of the more conservative tactics of Kefaya.

However, there are marked ideological differences between the youth and older members of Kefaya, both about the nature of outreach tactics and about Kefaya's failure to reach beyond an exclusively intellectual, Cairo-based crowd. This has been a reaction to Kefaya's general discourse of human rights and democracy, which fails to address the more fundamental, daily concerns of Egyptians, its focus on regime change seen as impeding the elaboration of a practical agenda. As Youth for Change distanced its activities from Kefaya, the bloggers discussed the everyday problems of the excluded: the poor, the homeless and the unemployed. In this way the bloggers' movement, demanding elementary rights to the city for everyone, underlined and clarified the content of contemporary urban struggles.

On 25 May 2005, after government supporters assaulted protesters demonstrating against the presidential referendum, bloggers started posting reports, with several photo and video blogs of police attacks and arrests spreading throughout the Egyptian press, on Arabic satellite television stations, and on the Internet. As a blogger posted in May 2007:

> Bloggers' voices are clearly heard worldwide. They are independent and free. They act individually, campaign collectively. A police state would not accept bloggers in Egypt who shape public opinion and raise topics about

> torture, corruption, and election falsification on mainstream media. As bloggers are summoned to State Security headquarter, some stop blogging, fearing the consequences of speaking out, while others lose their jobs. The regime is labeling bloggers a "bunch of kids" who go to jail.
>
> (Interview with Blogger A, 2007)

Confronted by an emergency law enacted in 1981, which restricted the organization of public rallies and distribution of posters in the streets, the Egyptian blogosphere has developed into a virtual platform for socio-political expression. Blogger activists have created an Internet hub, encouraging "citizen journalism" by sending accounts of demonstrations, including pictures of police abuses, and posting them on the web as a contemporary archive of Egypt's battle for democracy. Such bloggers are becoming increasingly employed in directing political campaigns and mobilizing people for rallies and demonstrations.

Under Egypt's emergency laws, which have been in place for nearly twenty-five years, these bloggers could be jailed indefinitely. The State Security Court reviews such detentions every fifteen days, applying article 80(d) of the penal code, which criminalizes "harming Egypt's image". An elected official declared,

> They could be punished for what they are writing according to the law, because it is libel. If they think that what they are doing is an expression of their freedom, they should remember who gave them this chance, and who is insisting on its continuity.
>
> (Interview with official, April 2006)

Such official reactions have nevertheless strengthened the networks of alternative communication, resulting in technical improvements and new connections between bloggers, advocacy groups and civil liberty organizations. Since 2005 Egyptian bloggers have created spaces of protest that are a hybrid of the physical and virtual worlds. Their extensive use of information and communication technologies was appropriated through collaborative online content management systems (wikis) and media streams, in convergence with various forms of street protest. By June 2005 they had taken their message directly to the street, targeting working-class Cairo communities like those of Shoubra and Sayeda Zeinab (Table 12.1). Groups of two to four activists visited public squares and distributed flyers as they engaged people on the street in impromptu discussions. This was described by a blogger:

> Youth for Change's central strategy is to explain to people that they have certain rights and someone has a responsibility to listen to their demands. In seeking to link daily concerns with failures of the political process, activists broach issues that touch all sectors of life – from transport costs to healthcare access to unemployment.
>
> (Interview with Blogger B, April 2006)

Table 12.1 Chronology of Cairo's street movements during 2005

March–December 2005
• 25 May – Demonstrations against the referendum on the Constitutional article 76. • 1 June – Women in black protest against security harassment organized by the Association for Egyptian Mothers at the Press Syndicate. • 8 June – A candle vigil at Dareeh Saad (shrine of the leader of the 1919 revolution). • 9 June – The "Street is Ours" movement launched to defend women's political rights. • 15 June – Youth for Change and bloggers' demonstration at Sayeda Zeinab Mausoleum. • 22 June – Demonstration organized in Shubra by the Popular Campaign for Change under the slogan "Freedom now". • 26 June – Demonstration in front of the Ministry of Interior, celebrating the International Day against Torture. • 29 June – Protest at the Virgin Mary Church in Zeiton. • 14 July – Kefaya's unemployment protest in Abdeen Square. • 20 July – The National Coalition for Change and Reform rally of 5,000 protestors to support the judges. • 31 July – Protest against presidential elections. • 1 August – Sit-ins for the release of detainees of 31 July protest at the General Prosecutor's Office in the High Court and at the Press Syndicate. • 2 August – Artists and Writers for Change protest in Talaat Harb Street chanting, "Big brother is watching you". • 3 August – Demonstration against corruption in Opera Square. • 8 August – Demonstration at Dar el Hikma (Doctors' Syndicate) for the release of detained doctors. • 14 August – The National Coalition for Change and Reform street demonstration, with 7,000 protestors calling, "No to the regime, Yes to Judiciary Independence". • 16 August – The Movement for Popular Monitoring of the Elections "Shayfeenkom" ("We are watching you") was officially launched. • 21 August – Youth for Change flash demonstrations, each around 30 minutes, with thousands of flyers distributed in the low-income and densely populated Giza village of Nahia. • 23 August – Artists and Writers for Change pro-reform demonstration in Talaat Harb Square. • 26 August – Youth for Change protest in Rod el Farag district, with thousands of leaflets publicizing blog URL addresses. • 1 September – The National Coalition for Change and Reform rally to support judges' independence and the end of corruption. • 8 September – Friends and families of victims of Beni Suef Theatre fire, "5 September Group", protest at the state-sponsored funeral and demand inquiries regarding the death of more than thirty artists. • 27 September – Demonstration organized by Kefaya, bloggers, 5 September Group and detainees' families. • 12 December – Kefaya marking its first anniversary with a protest against the newly elected members of the parliament.

Source: extracts from various blogs, with modifications by author (accessed March 2008).

The empirical survey – 2006 pro-judiciary reform spring demonstrations

Many bloggers use pseudonyms, as they feel this gives them more freedom to write about politics without being detained. However, the recent establishment of the General Administration for Information and Documentation and the Department for Confronting Computer and Internet Crime, two security units within the Ministry of Interior in charge of online surveillance and monitoring, has led to the detention of some bloggers and the blocking of their blogs by the authorities. The qualitative study reported here explored the Egyptian bloggers' movement, looking at their experiences with cyberpolitical activism and street actions for democratic reforms. In addition to drawing on alternative media online publications posted on the Egyptian blogosphere, the author conducted informal discussions and unstructured interviews with a number of bloggers, and was a participant observer at sit-ins and street demonstrations over a period of six months (March–August 2006). Narratives and digital images were employed to record bloggers' navigation between the blogosphere and Cairo's public spaces.

The bloggers were selected based on their political involvement in street demonstrations and history of blogging. Given the recent detainment of some bloggers and the blocking of their blogs by security authorities, the author agreed not to reveal the names of those bloggers who do not use pseudonyms, nor the names of blogs referred to in the chapter.

> Where the press is tightly controlled, pro-democracy human rights activists and journalists, who are shut out from the mainstream media, have taken to the web to disseminate information as they set up their own blogs free of charge.
>
> (Interview with blogger A, 2006)

> Most bloggers have been infused by politics rather than having driven the political movement. Maybe in few years when we see thousands of blogs we can claim that blogs affected politics. Right now I think it's the other way around: the political climate has affected bloggers.
>
> (Interview with blogger B, 2006)

> Some bloggers have been politicized by the blogs and have become part of the Egyptian Movement for Change (Kefaya). They interact with the movement both intellectually (through analysis and criticism) and practically (by physically joining demonstrations). This is why they are attracted to Kefaya, which is more of a flexible coordinating framework than an actual organization with a single ideology.
>
> (Interview with blogger C, 2006)

> Blogs are read by those who have access to computers and high-speed Internet and have the luxury of time to sit around and talk about politics.

> Although I am convinced that blogging will eventually influence pro-democracy reform initiatives, this does not necessarily mean taking to the streets. Despite being fueled by bloggers, political activism remains a personal choice.
>
> (Interview with blogger D, 2006)

> I think activist bloggers are forming an interactive popular electronic journalism, which is more truthful, reliable and revealing than the coverage of any other form of mainstream media of recent political events in Egypt. This kind of effective activism is informative whilst creating a new kind of awareness amongst Internet users.
>
> (Interview with blogger E, 2006)

> Internet activism contributes substantially to initiatives of reform. I believe that detaining activist bloggers exposed security measures used against Egyptians in general if they tried to express themselves. The detentions drew the attention of most human rights organizations in Egypt and abroad and raised so many questions about the seriousness of the reform process.
>
> (Interview with blogger F, 2006)

> Whilst mainstream media would report a protest through a short clip on TV or a photo in a newspaper, bloggers give a very detailed description of street protests, with many photos and anecdotal observations compared to the official media.
>
> (Interview with blogger G, 2006)

> Blogs have been a good means in providing accounts of recent street protests, since they are not regulated by censorship or by any political affiliations. Mailing lists, online forums, emails and SMS messages could provide forms of citizen journalism and are effective ways of delivering information and mobilizing people.
>
> (Interview with blogger H, 2006)

Bloggers' narratives revealed that their alternative news websites were probably the most important sites through which networks of critical and informed constituencies were formed; yet the extent and efficacy of these new virtual spaces of contention remained limited. Accordingly, the street is considered the most vital locus for the audible expression of collective identities (Bayat 2003), as urban youth activists and bloggers gather in coffee shops, unions and syndicates throughout the city on a daily basis.

During the spring of 2006 a group of young activists and bloggers (calling themselves 30 February Group) organized two sit-ins in solidarity with judges demanding judiciary reforms and independence from the state-appointed Ministry of Justice, which directly influences any elections. During the first sit-in, in support of two judges facing charges for having reported September 2005 presidential election violations, Tahrir Square was occupied for two days with protestors calling for democracy, political changes and judiciary reforms (Figures 12.2 and 12.3).

A month later, there was another sit-in, in front of the Press Syndicate and Judges' Club (Figure 12.4, Table 12.2). For five days urban youth activists transformed the street into zones for public protest, employing urban installations and street graffiti, prompting the construction of spaces of protest, a significant site of urban resistance. This event stunned the government and major precautions were implemented to suppress the pro-judiciary reform demonstrations organized by urban youth and bloggers. During the second sit-in, the government planned to overpower the demonstrators by arranging a counter-demonstration of their own supporters. Although anti-riot police forces cordoned off the protest and were able to divide the masses, this did not stop demonstrators from trying to break through cordons, resulting in more confrontations. Thousands of demonstrators made it to central Cairo, where police chased them down and reportedly arrested 300 protesters. After five days, Unions Street was completely empty as anti-riot security forces and police vehicles occupied the city center.

Since the 2006 spring demonstrations, street cafés within the Stock Exchange district have turned into venues for political public dissent, social centers where urban youth gather and organize future public rallies, even in the face of an emergency law forbidding more than four people from meeting in open spaces

Figure 12.2 Photoblogs of street demonstrations – Midan al Tahrir (Tahrir Square).

Figure 12.3a Video (vlog) stills of street demonstrations – Midan al Tahrir (Tahrir Square).

Figure 12.3b Video (vlog) stills of street demonstrations – Midan al Tahrir (Tahrir Square).

Figure 12.4 Photoblogs of sit-in – Unions Street, Judges' Club.

without government permission. Consequently, the government imposed security measures and police checkpoints to control street rallies, whilst introducing CCTV surveillance and occupying public squares with baton-wielding security forces and plainclothes agents.

In contrast to the period between 2004 and 2006, the Egyptian Movement for Change (Kefaya), with its structured, incremental, protest-centered strategies and orientation towards the vast, excluded and indifferent Egyptian public, has faded from the political scene, devolving into contending factions and ceasing to capture the public's imagination. After 2006, demonstrations became less frequent, attracting a limited number of youth activists. As a consequence of political and ideological conflicts within Kefaya, various factions split off, such as Youth for Change. These groups were mainly composed of bloggers, who accused Kefaya's leaders of being authoritarian and compromising.

The following section will explore the emergence of a civil society mobilization and an urban protest movement, different from the bloggers' networked social actions. This movement, inspired by Kefaya's street actions and bloggers' political activism, will be shown to have been instrumental in the eruption of the 2011 winter uprising that brought down the regime.

Table 12.2 Bloggers' posts of pro-judiciary reform demonstrations – spring 2006

Week one

- Despite nearly 200 activists being detained during the past few weeks, a solidarity demonstration was planned today with judges who face charges for reporting 5 September presidential elections' violations.
- Pro-democracy demonstrators marched in support of the judges, stopping in front of the Judges' Club, where judges organized a silent vigil. Protestors carried huge posters of prominent pro-reform judges. The posters were designed by the bloggers and young activists (30 February Group, who organized the March 2006 Midan al Tahrir sit-in)
- At 10 a.m. around 700 activists started chanting, "*judges put your strength together – rescue us from the tyrants, freedom, freedom where are you – emergency status between us and you, one by one … tomorrow the hundred becomes a thousand, revolution, revolution till victory – revolution in all Egypt's streets and alleys*".
- The demonstration failed to take place at the High Court, as streets were occupied by thousands of uniformed and plainclothes police agents. The protest was forced to divert to a side street before being violently dispersed: it did not last more than fifteen minutes. While thousands of State Security agents with batons, sticks and shields sealed most downtown streets, plainclothes police agents attacked protestors before arresting them. I started taking photos from the other side of the street. People were running in different directions. Wherever I looked there were people I thought to be secret police.
- As hundreds of disguised police occupied the streets, one protestor told me to go home immediately as violence was extending to all side streets. Despite all this, in Adly Street, I saw fifty people chanting and raising posters: "Egypt's pharmacists support Egypt's judges".
- As protestors entered the Metro movie theater to escape security forces, I went to Excelsior Café; whenever I tried to get out plainclothes police agents would follow me, so I decided to stay inside the café for an hour. There was news of a protest in Ramsis Street and more arrests.
- Judges' hearing was postponed to the following week.

Week two

- While the Ministry of Interior issued a statement two days ago warning that any gathering in Cairo would be illegal, more demonstrations were planned today.
- At 7.30 a.m., as young activists were wandering downtown Cairo in groups of fives, they were attacked by plainclothes police from the side, with state security forces attacking them from the front.
- At 8:30 a.m. the Press Syndicate and Lawyers' Syndicate were sealed. The High Court was also surrounded by security forces.
- A protest, consisting of 500 people, started at Souk el Tawfikia Street (next to Hisham Mubarak Law Center), proceeding from al Gomhoreya Street in front of Sydnawi Shopping Store, to Ataba Square, to El Gabarooni Alley, then to Assem Alley, and finally to Kamel Sedky Street in Faggala. After each police attack, protestors were dispersed and would withdraw, only to gather again in a new location, standing hand in hand as they faced the security forces, who used the "squeezing protestors technique". Few injuries were reported amongst protestors.
- By the end of the day there was news from the High Court that one judge was found innocent, with the other judge receiving a severe reprimand. The Judges' Club refused the verdict and vowed to appeal.
- Opposition sources announced that between 240 and 400 demonstrators were arrested today. I have seen at least fifteen arrests, with beatings of reporters and photographers. My photos feature eleven arrests.

Source: extracts from various blogs, with modifications by author (accessed March 2008).

The new civil society mobilization

Over the period between 2006 and 2008, a new civil society mobilization emerged, making specific demands of the authorities. Inspired by Kefaya and urban bloggers' activism, ordinary people, not belonging to political parties, trade unions, or pro-democracy groups, broke the psychological barrier of fear as they exercised their right to public protest and demanded basic rights (housing, salaries, water and services). Regarded as a new template for political action in Egypt, street demonstrations by groups of ordinary people were less documented and less celebrated compared to similar actions organized by Kefaya and Youth for Change. This was notable in the revival of the labor movement between December 2006 and September 2007, when workers organized rallies and sit-ins inside Mehalla al Kobra Textile Company (Ghazl al-Mehalla) in northern Egypt, protesting against layoffs and demanding better compensation, better working conditions, and democratically elected unions. The Real Estate Tax Collectors' (civil servants') strike during December 2007 lasted nearly ten days, and involved a massive sit-in in front the Prime Ministry Cabinet Office, which attracted media attention and public solidarity.

On 6 April 2008, a broad coalition of civil society representatives and online Facebook activists called for a general strike to demand decent living conditions and to protest against corruption, nepotism, inflation, torture, poverty and police brutality. The plan was to stay at home and not report to work or school, or alternatively to join protestors in street processions converging on main city squares. As the 6 April general strike gained local popularity and international exposure through SMS, emails, blogs, Facebook, YouTube and Twitter, riots erupted in Mehalla al Kobra City, as the result of confrontations between workers and police security. Simultaneously a number of bloggers and Internet activists were detained, particularly those "Facebookers" involved in initiating the online mobilization for the general strike. Although the call for the general strike coincided with Mehalla al Kobra textile workers' strike, no direct coordination was noted between the Facebookers' virtual actions and the workers' physical mobilization. An Egyptian blogger, in a post criticizing what he described as a call "coming from the cyberspace by bloggers, Facebook activists", wrote that,

> We, the Egyptian bloggers, have always prided ourselves on the fact that we have one foot on the ground and the other in the cyberspace. ... But this time, it seems some have thrown both their feet as well as brains in the cyberspace and are living some virtual reality, mistakenly believing (helped by the media sensationalist coverage of the "Facebook activism") that they are the ones behind the events in Mehalla.
>
> (Blogger K, blog post, May 2008)

As a result of the 6 April events, which were considered a national collective action, the government imposed an additional amendment to the emergency law, prohibiting demonstrations without prior permission from the Ministry of

Interior, obliging protestors to define the location and time of street rallies and effectively forbidding any criticism of the regime and any calls for democratic and political reforms. Consequently, bloggers started to experiment with the Internet as an additional space to articulate political dissent, organizing their repertoire of contention through web-based campaigns. This contributed to major street demobilization for the next three years, and was only reversed during the 2011 winter uprising, discussed in the next section.

The 2011 winter uprising

On January 2011 more street events erupted within various areas in Egypt, with Cairo providing more visual media coverage. Unlike the fragmented nature of the 2006 spring street movement, the 2011 winter uprising provided more organized collective action by various groups that had gained experience in protesting against the authorities. In this case, workplace protests organized by industrial laborers, civil servants, students and trade practitioners joined associational protesters from Lawyers' and Doctors' syndicates in reclaiming and contesting main public squares and streets.

During the 2011 uprising events, "technopolitics" was mobilized by Facebookers in cyberspace, highlighting the resistance and subcultures that had emerged in the electronic terrain as street rallies flooded into Cairo's public spaces. Images of demonstrations and streets confrontations were re(packaged) as visual media and presented to a global TV screen audience, while news was "retweeted" across the net among cyberactivists. For eighteen days, striking images of the sit-in in Tahrir Square captured the world's imagination, especially those of the urban youth. Amongst other groups, there were large numbers of bloggers and cyberactivists who occupied main public squares in Cairo and Alexandria.

The "mediatized" nature of the 2011 uprising has drawn local political governance into global scrutiny, through twenty-four-hour live broadcasting of the eighteen-day sit-in and the fact that TV screens are the shape of the contemporary public sphere. The staging of the city was manifest in the way Tahrir Square was transformed into a festival of resistance with mass nonviolent direct actions, marches, street theatre, music and celebrations. The intrusion of the global media into Tahrir Square's theatre of protest created a visuality which gave the bloggers' movement new forms of agency, enabling them to raise and legitimize issues of injustice ranging from poverty to democratic reform.[3]

Emerging hybrid spaces of protest

According to Hamm (2006), the social construction of geographies of protest is characterized by "deterritorialization", where events are reported through websites, blogs and streaming in a collaborative social process, changing the subjective experience of online activists while providing navigation for street protestors. Simultaneously, new boundaries are reterritorializing the emerging hybrid virtual and real public spaces, creating socially constructed and temporary geographies of protest.

The current study indicates that virtual space affects and modifies real space, and gives a new dimension to human communication interactions. Blogging was very much related to and intertwined with bloggers' practical "offline" everyday experiences. They navigated between the virtuality of the blogosphere, where they practiced their online activism, and into the reality of Cairo's (downtown) public spaces, where they regularly expressed political dissent and contention in the form of street rallies, demonstrations and sit-ins (Fahmi 2009). Despite showing similar tactics, the street actions of Egyptian bloggers were not initially connected to anti-globalization movements. As they gained experience in street tactics and popular online journalism, both bloggers and Youth for Change brought a globalized dimension to their local protest movement, linking with international media through vlogs and photoblogs.

Bloggers' cyberactivism extended through spaces of protest to include a broader (virtual) community of "hacktivists" (hackers and activists) who perceived themselves as part of wider online global social movement. Through online and offline discourses and practices, Egyptian bloggers have created temporary geographies of protests that have changed the spatial and temporal construction of space. As they organized their repertoire of contention through street demonstrations and sit-ins, and through web-based campaigns, they were influenced by the practices of transnational movements against neoliberal globalization with their globally synchronized activities. Nowadays the Egyptian blogosphere provides more than a news resource, representing an alternative urban hub and acting as an interface between events in the streets and digital communication network platforms. Therefore, bloggers created an alternative city as a result of their socio-political actions and ubiquitous access to mobile communication networks within public spaces. Their long-term pressure through mobilization, as they sought to realize their right to the city, contributed to the emergence of public spaces that are democratic, vital and open, thus generating modern forms of urbanism in the Global South.

Notes

1 The Egyptian Movement for Change (Kefaya, literally Enough) was formed in 2004, when intellectuals and public figures signed a founding statement demanding real political changes and an end to economic inequality and corruption. Between 2004 and 2005, Kefaya transformed from an umbrella organization binding various groups, ideologies and visions into a protest street movement.

2 According to an official report issued by the IDSC (Information Decision Support Cabinet) in 2009, nearly 16 percent of Egyptians access the Internet. The report highlighted a significant rise in the number of Egyptian blogs since 2007, recording nearly 160,000 (30 percent of the Arab World's blogs). Most bloggers are between the ages of twenty and thirty-five, with 20 percent being considered political; 68 percent use Arabic, and 20 percent use both Arabic and English (Fahmy 2010).

3 Peoples' "mediatized" protest in Cairo and Alexandria's public squares (the center) was fundamental in supporting "real" actions in provincial cities (the periphery), where

workers and urban poor were engaged in guerrilla street confrontations with security forces. Protestors in various cities evacuated and occupied official buildings, police stations and the ruling National Democratic Party headquarters. There was less visual coverage of these events, which constituted the deciding factor in determining the outcome of the uprising with the removal of the former regime.

Bibliography

Alvarez, S. E., Dagnino, E. and Escobar, A. (1998) "Introduction: the Cultural and the Political in Latin American Social Movements", in S. E. Alvarez, E. Dagnino and A. Escobar (eds) *Cultures of Politics, Politics of Cultures: Re-Visioning Latin American Social Movements*, Boulder: Westview Press, pp. 1–23.

Anderson, B. (1988) *Imagined Communities*, London and New York: Verso.

Batuman, B. (2003) "Imagination as Appropriation: Student Riots and the (Re)Claiming of Public Space", *Space and Culture*, 6 (3): 261–75.

Bayat, A. (2003) "'The Street' and the Politics of Dissent in the Arab World", MERIP 226. Online: www.merip.org/mer/mer226/226_bayat.html (accessed 11 October 2007).

Castells, M. (1997) *The Power of Identity*, Oxford: Blackwell.

Castells, M. (1983) *The City and the Grassroots*, London: Edward Arnold.

Della Porta, D., Kriesi, H. and Rucht, D. (eds) (1999) *Social Movements in a Globalizing World*, New York: St Martin's Press.

Della Porta, D. and Tarrow, S. (2005) "Transnational Process and Social Activism: an Introduction", in D. Della Porta and S. Tarrow (eds) *Transnational Protest and Global Activism*, Lanham, Boulder, New York, Toronto, Oxford: Rowman and Littlefield Publishers, pp. 1–19.

Fahmi, W. (2009) "Bloggers' Street Movement and the Right to the City: (Re)Claiming Cairo's Real and Virtual Spaces of Freedom", *Environment and Urbanization*, 21 (1): 89–107.

Fahmy, N. A. S. (2010) "Revealing the 'Agenda-Cutting' through Egyptian Blogs: an Empirical Study", paper presented at the Eleventh International Symposium on Online Journalism, University of Texas, Austin, April. Online: http://online.journalism.utexas.edu/2010/papers/Fahmy10.pdf (accessed 10 April 2011).

Fyfe, N. (1998) *Images of the Street: Planning, Identify and Control in Public Space*, London: Routledge.

Giugni, M. (1999) "Introduction: How Social Movements Matter: Past Research, Present Problems, Future Developments", in M. Giugni, D. McAdam and C. Tilly (eds) *How Social Movements Matter*, Minneapolis: University of Minnesota Press, pp. xiii–xxxiii.

Gould, R. (1995) *Insurgent Identities: Class, Community and Protest in Paris from 1848 to the Commune*, Chicago: Chicago University Press.

Hamm, M. (2006) "Reclaiming Virtual and Physical Spaces. Indymedia London at the Halloween Critical Mass", *Open 11* (Hybrid Space – How Wireless Media Mobilize Public Space, NAi Publishers SKOR): 96–111.

Hamm, M. (2003) "A r/c tivism in Physical and Virtual Spaces", *republicart.net*. Online: http://republicart.net/disc/realpublicspaces/hamm02_en.htm (accessed 10 December 2007).

Harvey, D. (1996) *Justice, Nature and the Geography of Difference*, Oxford: Blackwell.

Jordan, J. (1998) "The Art of Necessity: the Subversive Imagination of Anti-Road Protest and Reclaim the Streets", in G. McKay (ed.) *DiY Culture: Party and Protest in Nineties Britain*, London and New York: Verso, pp. 129–251.

Kidd, D. (2003) "Indymedia.org: a New Communications Commons", in M. McCaughey and M. Ayers (eds) *Cyberactivism: Online Activism in Theory and Practice*, London: Routledge, pp. 47–70.

Lefebvre, H. (1996) *Writings on Cities, Part II: Right to the City*, transl. and ed. E. Kofman and E. Lebas, Oxford: Blackwell.

Leontidou, L. (2006) "Urban Social Movements: from the 'Right to the City' to Transnational Spatialities and Flâneur Activists", *City*, 10 (3): 259–68.

Malone, K. (2002) "Street Life: Youth, Culture and Competing Uses of Public Space", *Environment and Urbanization*, 14 (2): 157–68.

McAdam, D. and Scott, W. R. (2005) "Organizations and Movements", in G. F. Davis, D. McAdam, W. R. Scott and M. N. Zald (eds) *Social Movements and Organizational Theory*, New York: Cambridge University Press, pp. 4–40.

Melucci, A. (1996) *Challenging Codes: Collective Action in the Information Age*, Cambridge: Cambridge University Press.

Melucci, A. (1985) "The Symbolic Challenge of Contemporary Movements", *Social Research*, 52 (4): 789–816.

Mertes, T. (2004) *The Movement of Movements. A Reader*, London and New York: Verso.

Mitchell, W. J. (2003) *Me++: the Cyborg. Self and the Networked City*, Cambridge: MIT Press.

Norris, P. (2004) "The Digital Divide", in F. Webster (ed.) *The Information Society Reader*, New York: Routledge, pp. 273–86.

Notes from Nowhere (2003) "Indymedia's Precursors and Birth", in Notes from Nowhere (eds) *We are Everywhere, the Irresistible Rise of Global Anticapitalism*, London and New York: Verso, pp. 230–44.

Papadimitriou, F. (2006) "A Geography of 'Notopia'. Hackers *et al.*, Hacktivism, Urban Cyber-Groups/Cyber-Cultures and Digital Social Movements", *City*, 10 (3): 317–26.

Pickerill, J. and Chatterton, P. (2006) "Notes towards Autonomous Geographies: Creation, Resistance and Self-Management as Survival Tactics", *Progress in Human Geography*, 30 (6): 730–46.

Portaliou, E. (2007) "Anti-Global Movements Reclaim the City", *City*, 11 (2): 165–75.

Raafat, S. (2003) *Cairo: the Glory Years – Who Built What, When, Why and for Whom*, Alexandria: Harpocrates Publishing.

Rheingold, H. (2002) *Smart Mobs. The Next Social Revolution*, Cambridge, MA: Perseus Books Group.

Sheller, M. (2003) "The Mechanisms of Mobility and Liquidity: Re-Thinking the Movement in Social Movements", working paper, Lancaster University, Lancaster.

Tarrow, S. (1998) *Power in Movement: Social Movements and Contentious Politics*, Cambridge: Cambridge University Press.

Van Dijk, J. (2005) *The Deepening Divide. Inequality in the Information Society*, London and New Delhi: Sage.

Warren, K. (1998) "Indigenous Movements as a Challenge to the Unified Social Movement Paradigm for Guatemala", in S. E. Alvarez, E. Dagnino and A. Escobar (eds) *Cultures of Politics, Politics of Cultures: Re-Visioning Latin American Social Movements*, Boulder: Westview Press, pp. 165–95.

White, H. (1995*) Where Do Languages Come From?, Switching between Networks, Part I*, New York: Center for the Social Sciences at Columbia University Pre-Print Series.

World Social Forum (2006) "World Charter on the Right to the City", discussion paper.

Afterword

Re-engaging with transnational urbanism

Martin J. Murray

In the current phase of globalization, the conventional approaches to understanding transnational urbanism and the contemporary modern metropolis have become unsettled. The unprecedented hyper-growth of the sprawling mega-cities of the (so-called) Global South, coupled with the massive city-building projects of the Asia Pacific Rim, the "instant cities" of southern China and the "spectacular urbanism" of the Persian Gulf, has fundamentally altered both the pace and the scale of worldwide urbanism (Marshall 2003). Yet the dominant theories and methodical approaches used to study cities – constituting what Ananya Roy (2011a) has called the universalizing knowledge-space of "global urbanism" – remain largely tied to the urban experience of the leading world-class cities of Europe, North America and other core areas of the world economy. In the conventional urban studies literature, debate and discussion over the past two decades have largely revolved around three analytic conceptions: the "world-cities hypothesis" (Friedmann 1986), "global cities" (Sassen 2001) and "world city networks" (Taylor 2004). Whatever disagreements exist within this hegemonic discourse, the dominant theories in urban studies remain tied to the premise that so-called "world-class" or "global" cities occupy the apex of an implicit hierarchy of preeminence and functional importance (Therborn 2011).

While the focus on global cities, world-class cities and their spinoffs (creative cities, tourist-entertainment cities, the post-modern metropolis) has captured the lion's share of scholarly attention over the past several decades, an insurgent countercurrent of urban scholarship has called for the rethinking of conventional urban studies in ways that capture a wider and fuller range of diversity, heterogeneity and complexity, characterizing the multiple trajectories of transnational urbanism in the twenty-first century. Recent calls for "provincializing global urbanism" (Ananya Roy and Helga Leitner[1]), for engaging with "urban theory beyond the West" (Edensor and Jayne 2012) and for discovering "new geographies of theory" (Roy 2009a) indicate a growing interest in re-conceptualizing the field of urban studies in ways that move beyond the universalizing and teleological impulses of the prevailing Western theories of urban growth and transformation. What is needed is a rethinking in urban studies, away from readymade, deductive theories seeking to identify single pathways for urban transformation and towards a kind of theoretical openness and flexibility that seeks through comparison and contrast to account for the historical-spatial specificity of those "ordinary cities" (Amin and Graham 1997; Robinson 2006, 2008), which, as Jennifer Robinson (2002) has put it, are cities off the map. Engaging horizontally with a "world of cities" – instead of remaining fixated with the ranked hierarchies of "global cities" – can open a broader conversation, or dialogue, about what it means to talk about transnational urbanism in the twenty-first century (Grant and Nijman 2002; McFarlane 2006).

It is time to write these ordinary cities into urban theory, beginning by moving beyond conventional categories (First World versus Third World; Western and "non-Western"), labels (global cities, world-class cities), and hierarchies (developed and developing) that, by definition, foreclose on other possible approaches or angles of vision. These binary characterizations only contribute to a conceptual blindness that obscures our ability to think outside the constraints of linear time. Rethinking conventional urban studies requires us to decenter the dominant optic that has largely focused on the leading "world-class" cities of North America and Europe as privileged objects of knowledge and primary sources of urban theory, and begin to build alternative analytic frameworks that take seriously the complexity and heterogeneity of global urbanism (Ehrkamp 2011; McFarlane 2010; Robinson 2002, 2006). Bringing ordinary cities back into the mainstream of urban studies enables us to appreciate the diverse and multiple trajectories of urban growth and development, the endless possibilities for leaping ahead or falling backward and the hybrid mixing, all of which characterize transnational urbanism in the twenty-first century (McFarlane 2007; Robinson 2005; Roy 2009a, 2009b).

The steady stream of new image-categories – such as instant cities (Wright 2008), mega-cities (Demissie 2011), slumdog cities (Roy 2011a), shadow cities (Neuwirth 2006), invisible cities (de Boeck 2002), subaltern urbanism (Roy 2011b), pirate towns (Simone 2005), postcolonial cities (Yeoh 2001), postcolonial urbanism (Bishop *et al.* 2003), hybrid urbanism (AlSayyad 2001) and so forth – reflects a growing unease in urban studies scholarship with the current state of urban theory (Abbas 2000; Roy 2011a; Simone 2006; de Boeck 2002; Bishop *et al.* 2003; Yeoh 2001). These efforts to take seriously different kinds of urbanity provides ample testimony for the need for urban studies scholarship to broaden its fields of inquiry and expand its horizons if we wish to account for the diversity and complexity of urban forms and experiences that comprise the transnational world of cities today (Watson 2009; Pieterse 2008).

Reframing the ordinary cities of the (so-called) Global South

In so many ways, the megacities of the Global South are among the most innovative and vibrant places on the planet. The inventive use of pirated technologies (Sundaram 2010a), the experimentation with cutting-edge building design practices (Marshall 2003: 201–4) and the creative adaptation of sustainable systems of low-cost infrastructure (Simone 2001a, 2001b, 2003) has enabled some ordinary cities of the Global South to leapfrog over the older historic cities of the (so-called) Global North and to carve out niches for themselves at the forefront of global innovation. Yet there is no shortage of caricature, hyperbole and unreflective sensationalism about what makes these large, sprawling megalopolises in the so-called developing world such quintessential exemplars of failed or distorted urbanism (see Myers and Murray 2006). Even a cursory review of popular portrayals of everyday urbanism in the megacities of the Global South – whether in journalistic accounts, documentary and commercial films, fiction writing or photographic exhibitions – reveals a clear preoccupation with the chaotic, disorderly and dangerous features of city life (Gberie 2005). Seen within the context of contemporary globalization, these places certainly face a host of seemingly insurmountable challenges, including stalled economic growth and restricted opportunities for wage-paid employment, crumbling infrastructure, massive rural to urban migration, environmental degradation, lingering social conflict over scarce resources and state violence directed at controlling social groups who have asserted their "right to the city". This seemingly endless flow of negative images has prompted many urban theorists to regard so-called "Third World cities of the developing world" as globally insignificant and structurally irrelevant. This stress on their alleged failures and deficiencies has meant that the sprawling megacities of the Global South have been largely viewed through the normative lens of developmentalism, an approach which favors prescriptive policy prescriptions directed at improving capacities for good governance, service delivery and productivity (Robinson 2002, 2006, 2010). The developmentalist stress on good governance, rehabilitation and improvements often overlooks the practices through which these decentered ordinary cities actually work (Appadurai 2001; Simone and Rao 2012).

The title of this collection of essays – *Locating Right to the City in the Global South* – indicates its geographical focus and its thematic interests. The aim of this edited volume is to provide a fuller picture of the complicated dynamics of urban life in the so-called megacities of the Global South. These cities can neither be understood apart from the pressures of globalization, which has exposed them to the unpredictable vagaries of the world market, nor from a retreating public administration subjected to a triumphant neoliberal ideology. The essays take as their point of departure the premise that these large megalopolises operate in accordance with distinctive logics, norms and values, producing highly complex social worlds that cannot be simply dismissed as hopelessly unmanageable and chaotic.

The growing recognition for the need to theorize the diversity of urban trajectories cannot be accomplished without unsettling and disrupting conventional approaches to understanding transnational urbanism. The contributors to this volume resist the idea that cities of the Global South can be studied on the basis of established knowledge about universalizing processes of urban transformation taking place in linear time. Their essays address a number of important questions that can assist us in rethinking conventional urban theory and in helping to shape current debates in the scholarly literature. At the risk of oversimplification, it is possible to identify seven features of contemporary transnational urbanism that require more attention in order to broaden the horizons of the field of urban studies:

1 theorizing the so-called "slum cities" of the Global South;
2 instant urbanism, or the rapid growth of cities outside North America, Europe, and other core areas of the world economy;
3 "bubble urbanism", or the spatial fragmentation of urban landscapes;
4 splintering urbanism and impermanent infrastructure;
5 new modes of urban governance that have emerged under the shadow of neoliberalism;
6 the right to the city and social justice; and
7 the (re)appropriation of public space.

The seemingly dysfunctional landscapes of the megacities of the Global South – for which "slum" and "informality" constitute the twin foundational tropes – form the empirical and analytical points of departure for much of the research work and writing that takes these cites as their subject (Kudva 2009: Rao 2006: 228). As a general rule, the conventional urban studies literature locates the so-called "slum cities" of the "Third World" on a "continuum from nascent/informal to developed/post-industrial", where the latter represents the apex of achievement in city-building (Pieterse 2008: 108). Implicit in this linear trajectory of urban development is the normative prescription that calls for progressively replacing "the chaotic, malfunctioning city of informality" with an orderly, coherent, optimally functioning modern metropolis. As Edgar Pieterse (2008: 108) has put it, "This narrative tends to function on the basis of an assumption that modern, gleaming, skyscraper-filled cities, with adequate networked infrastructures in place to support them, are the only and ineluctable way into the urban future."

Rather than objects of knowledge, the megacities of the Global South are typically classified as sites of intervention – dystopian places in need of remedial action to address failed municipal management, broken-down infrastructure, overcrowded streetscapes and derelict housing (Robinson 2002, 2006). The developmentalist framework is unable to accommodate the complexity and diversity exemplified by contemporary cities of the Global South. The essays in this volume offer an opportunity to revise our understanding of development, not as a staged trajectory moving inexorably towards a common endpoint but as a complex process with multiple pathways towards indeterminate outcomes, and bring into view new practices and experiences suppressed by the developmentalist discourse of remedial intervention. Viewed horizontally rather than vertically, these cities of the Global South provide the occasion to rethink "the contours of modernity in a global age" (Rao 2006: 227).

Slum cities and the Global South

In *Planet of Slums*, Mike Davis draws our attention to the accelerated growth of urban slums, informal settlements and shantytowns in the hyper-growth cities of the Global South as the "radical new face of inequality" on a world scale (Davis 2006b: 201). Following in the footsteps of a well-publicized United Nations Human Settlement Program (UN-HABITAT), Davis argues that, instead of functioning as a focal point for growth and prosperity, cities in the (mis-named) "developing world" have become vast dumping grounds for a surplus population eking out a bare existence in unskilled, unprotected and low-wage informal service activities and trade (Davis 2006b: 175). Huge, sprawling slums have become "gigantic concentrations of poverty", which house an "outcast proletariat" – "the fastest-growing and most novel social class on the planet" (Davis 2004: 11). Trapped in a vicious cycle of poverty and with no options but to generate income through access to informal economies, this growing underclass is almost completely delinked, or disconnected, from industrial-manufacturing growth and from the dwindling supply of formal-sector employment (Davis 2006; Breman 2006; Huchzermeyer 2004, 2011; Gilbert 2007; Benjamin 2000; Simon 2011).

Because he paints such a grim picture of the apocalyptic future of these "slum cities" of the Global South, critics have charged that Davis unnecessarily flattens out differences in both history and social-spatial form, homogenizes the experience of slum-dwellers (Cunningham 2007; Rao 2006: 227), and thereby conjures up images of "huge undifferentiated neighborhoods filled with hopeless underemployed masses" (Angotti 2006: 961). While much has been written about broken-down infrastructure, overcrowded streetscapes, and lack of municipal oversight in the so-called "slum-cities" of the Global South, much less is known about the everyday lives of urban residents. In focusing on the common lot of misery and deprivation, scholars often overlook the dynamics of cooperation and conflict that characterize the daily existence of the urban poor (Harding 2007: 25–7). Asef Bayat (1997, 2000, 2009) – following in the footsteps of James Scott (1985), who coined the phrase "weapons of the weak" to capture everyday forms of peasant resistance – refers to the "quiet encroachment of the ordinary" to capture the stealthy, opaque tactics through which marginalized urban residents carve out usable spaces for themselves in uninviting urban landscapes.

Conventional urban studies literature tends to treat "informality" as an indicator of socioeconomic backwardness, a sign of underdevelopment, and the manifestation of the persistence and durability of pre-capitalist forms of artisanal production and circulation. In contrast, scholars like AbdouMaliq Simone (2004, 2010) and others uncover how informality is the creative wellspring for collaborations, networks and associational life that enables poor urban residents to "get by" with very little by making linkages with wider social worlds. This scholarship inverts the largely a priori assumptions of the conventional urban studies literature by suggesting a different reading of dysfunctionality and modernity in the contemporary era of globalization.

In Indian cities like Mumbai, Delhi and elsewhere, a shadowy tenure-by-occupancy – rather than formal legal title – generally governs access to informal housing. Similarly, cash, rather than reliance on the legal banking system and informal credit, lubricates commodity circulation and drives production cycles

(Sundaram 2010a). For example, Benjamin (1991) has examined the development of a vast, decentralized manufacturing zone in East Delhi that operates outside of legal restrictions and oversight, cobbled together through complex paralegal strategies where small-scale entrepreneurs make use of unauthorized electricity connections, cultivate political favors from local elites and rely upon informal financing. Similarly, Huchzermeyer (2004, 2011), Fawaz (2008, 2009c) and others have investigated the role of informal social networks and dynamics of self-help housing in the construction of informal settlements. In looking at the seamy side of urban informality, others (Auyero 2000; Weinstein 2008; Hoffman 2007, 2011) offer a more sobering assessment of the actual possibilities by focusing on intense competitiveness, exploitation and power hierarchies that merely reinforce the vulnerability of the "poorest of the poor".

"Instant urbanism": the rapid pace of urban transformation on a global scale

Dubai burst onto the global urban scene as an instant iconic symbol of size and spectacle (Elsheshtawy 2008, 2010). Virtually overnight, an out-of-the-way fishing village was totally transformed into a gigantic metropolis, growing at breathtaking speed and attracting a huge influx of corporate executives, financiers, consumer-driven tourists and guest workers from all corners of the world. Almost immediately, it became a brand, a model that was widely emulated, copied and mimicked (Abbas 2000; Acuto 2010). The rapid-fire urbanism that characterized the building of Dubai exemplifies the emergent pattern of building entirely new cities from scratch. Dubai is the reigning exemplar of an aspiring world-class city that has achieved a great deal of adulation. It has often been asserted and assumed that if Manhattan represents the apotheosis of twentieth-century congested urbanism, then Dubai has become the emergent metropolitan prototype for the twenty-first century – a futuristic oasis floating unanchored above the urbanism of yesteryear (Pithouse 2008: 567). Critics, however, dismiss Dubai as an unsustainable experiment in global excess, an enclosed gated community for the super-rich, rooted in mindless consumption and economic injustice, and sustained by under-paid indentured labor, physically segregated in featureless shack-like accommodation (Davis 2006a; Davidson 2008; Elsheshtawy 2008, 2010; Kanna 2011). It is the seamy side of such city-building efforts – the "grit beneath the glitter", as Hal Rothman and Mike Davis suggest for Las Vegas – that enables us to counter the overly romanticized rhetoric of "successful urbanism" that often accompanies celebratory stories about Dubai and other new cities of the Gulf States (Rothman and Davis 2002).

Yet the emergence of these "instant cities" undermines ideas embedded in the conventional urban studies literature, which suggests that "underdeveloped cities" in "developing areas" proceed through distinct stages of growth along a single pathway in linear time. These new "instant cities" are cobbled together from generic models derived from elsewhere (Prouty 2008; Arkaraprasertkul 2010). The secret key to the success of these city-building efforts is serial repetition. While variation in the assemblage of the parts produces different configurations, real estate developers tend to make use of similar architectural stylistics, functional components, environmental programs, building typologies, legal envelopes, aesthetic formats and construction technologies (Koolhaas, 1998). Working hand in glove with international architects and urban design specialists, real estate developers have borrowed, plagiarized and ransacked planning motifs from generic prototypes. In transferring design specifications from the global to the local, real estate developers "cherry-pick" the signature features of universally acclaimed cities and recycle them as the basic building blocks of these new "instant cities".

This kind of "instant urbanism" that has blossomed around the world has precedents in contemporary city building in southern China, the Asia Pacific Rim, the Persian Gulf, Singapore and India, where large-scale, master-planned cities appear as if out of nowhere (Wright 2008; Marshall 2003; Campanella 2008). They conform to the new urban design politics of bringing together aesthetically pleasing architecture with entertainment spectacles (Elsheshtawy 2008). The serial reproduction of business conference centers, up-to-date transport infrastructure, glitzy airports with global reach, luxury shopping and entertainment districts and towering skyscrapers offer amenities comparable to so-called First World cities. It is not unusual for these copies to surpass the original in terms of grandeur, aesthetics and style.

Bubble urbanism: fragmented urban landscapes

What has been largely overlooked in the scholarly literature on the ordinary cities of the Global South are the changing dynamics of social polarization and spatial fragmentation. From Shanghai to Saigon (Ho Chi Minh City), from Delhi to Dubai, cities from across the rapidly urbanizing world have become experimental sites for improvisational engagements with modernity, cosmopolitanism and globalization. Scholarly research on mega-projects has shown the increased reliance upon large-scale, iconic and expensive office complexes, commercial and entertainment sites, luxury housing and infrastructure undertaken in cities throughout the world. Whether spurred by outside capital investment in real estate or driven by "home-grown" local propertied interests, property developers in so-called "Third World" cities have turned to large-scale redevelopment projects as a way to achieve world-class status (Makdisi 1997; Silver 2008; Shatkin 2008). One current in the scholarly literature argues that large-scale redevelopment is part and parcel of a global strategy of metropolitan gentrification. Neil Smith (2002: 427), for example, has argued that the process of gentrification, "which initially emerged as a sporadic, quaint, and local anomaly in the housing markets of some *command–center* cities", has blossomed into a far-reaching urban revitalization strategy on a global scale, "densely connected into the circuits of global capital and cultural circulation". In a similar vein, Rowland Atkinson and Gary Bridge (2005) have referred to global gentrification as "the new urban colonialism".

Constructing huge mega-projects and inserting them into existing metropolitan fabrics has become a standardized city-building strategy for real estate developers seeking to emulate what a world-class city should look like. Pablo S. Bose (Chapter 6, this volume), for example, shows how the combination of new upscale housing developments, the construction of business parks and the environmental upgrading of threatened wetlands – all designed to integrate Kolkata (Calcutta) into transnational circuits of the world economy – have come at the cost of a seemingly endless cycle of evictions, land expropriations, privatization of municipal services and environmental devastation.

While much has been written about large-scale mega-projects that have transformed the spaces of luxury consumption in such diverse cities as Beirut (Makdisi 1997), Jakarta (Kusno 2000; Silver 2008; Simone and Rao 2012), and Johannesburg (Murray 2008, 2011), very little is understood about such contextual differences as variations in the systems of institutional governance, the balance between private and public financing and the nature of state power (Kanai and Ortega-Alcázar 2009). In contrasting large-scale redevelopment projects in Accra (Ghana) and Mumbai, Richard Grant and Jan Nijman (2002) stress the importance of local context. Historically specific hydrological, climatological and topographical features endow cities with unique assemblages of physical characteristics that undermine any universalizing logics of generic urban form. In contrasting the divergent outcomes of high-profile mega-projects in Shanghai and Delhi, Xuefei Ren and Liza Weinstein (Chapter 5, this volume) make a strong case against a crude convergence hypothesis, an argument that suggests cities are increasingly converging around generic (that is, similarly designed and formatted) mega-projects. Similarly, David Walker (in this volume) echoes this anti-convergence position, arguing that the historically specific regeneration of the Centro Histórico (Mexico City) was the result of a peculiar intersection of the conjoined processes of gentrification, neoliberalism and globalization.

In aspiring world-class cities around the world, real estate developers – acting either alone or in concert with municipal authorities – have launched ambitious schemes design to lure international investments and global tourists (Boano *et al.* 2011). City boosters have stressed public-private partnerships, place-marketing and culture-led development as the key components of competition-based and market-led strategies to jumpstart revitalization projects. This "fight for the global catwalk" (to borrow a useful metaphor from Monica Degen) has typically involved the sponsorship of flagship property-led redevelopment projects, where "cities compete with each other by parading made-up images of different areas of the city which advertise these spaces as favorable and attractive to business and leisure" (Degen 2003: 867–8). Faced with the imperatives of global competitiveness, city builders in so-called "developing countries" have turned to up-to-date mega-projects with modern infrastructure as a way of "securing optimal productive platforms for mobile and local capital" (Pieterse 2008: 37–8). This frantic race to fashion a world-class city on the margins of modernity recasts city-building efforts into a business operation, where privately led strategic initiatives replace holistic planning under the watchful eye of a vigilant public administration, and where "profitability replaces the public good" (Perlman 2010: 249).

The current frenzy for constructing spectacular architecture goes well beyond functional use to creating some sort of trophy building (Fernandez-Galiano 2005: 1–7). For aspiring world-class cities, no building structure quite embodies the heightened importance of symbolic meaning in architecture as the high-rise skyscraper. A high-rise office tower is more than a marvel of engineering prowess; it is also a symbol of urban, national, and corporate identity – a claim to modernity (Angelini 2005: 69). City boosters use spectacular architecture as a way to announce their hoped-for emergence on the world stage (Elsheshtawy 2008: 164). As exemplars of the principle that height is a symbol of power, unusually "tall buildings" like the Burj Al Arab Hotel in Dubai, the Pentronas Towers in Kuala Lumpur and Taipei 101 in Taiwan, along with other equally spectacular architectural monuments, have greatly enhanced the cachet of aspiring world-class cities seeking to secure their place among the leading metropolitan centers (Jacobs 2006; Elsheshtawy 2010: 152–4; King 1996, 1999).

While large-scale mega-projects disrupt the urban fabric in ways that physically separate the urban landscape into sites of luxury and zones of impoverishment (Shatkin 2008; Rodgers 2004; Pírez 2002), small-scale revitalization projects produce a kind of micro-fragmentation (Centner 2012), which creates arenas of social and economic contestation. What often characterizes these micro-spaces – or small additions to the urban landscape – is that they function like oases of tranquility in the midst of chaos, of cleanliness surrounded by environmental degradation, of safety amidst crime, of stability within a world of uncertainty.

Splintering urbanism and impermanent infrastructure

Up-to-date infrastructure, or the installation of new modes of technological interconnectivity, has long been integrally linked to the vision of the modern city, the dream of what a decent urban life offers its inhabitants (Ali and Rieker 2008: 4). The multifaceted process that Stephen Graham and Simon Marvin (2001) refer to as "splintering urbanism" consists of the unbundling of urban infrastructure as a result of the widespread movement towards liberalization, privatization and outsourcing of service delivery (Graham 2000). They begin with the premise that neoliberal modes of urban governance stress economic efficiency rather than social equity, and favor competition, outsourcing and privatization. They argue that the conjoined forces of economic liberalization and the collapse of comprehensive urban planning has resulted in the unbundling of basic urban infrastructure, that is, the segmentation of such infrastructural services as transportation, telecommunications, power and water into different networks and packages (Graham and Marvin 2001: 141). The regulatory regimes which support unbundling of integrated infrastructure have contributed to a growing chasm in urban landscapes between globally connected spaces of cosmopolitan urbanity and "bypassed" areas of blight, abandonment and neglect. No longer saddled with the responsibility to provide comprehensive infrastructure, city builders customize "premium networked spaces" (Graham 2000) in order to meet the needs of "valued" users, enabling these powerful urban elites to retreat into "secessionary" spaces that are disconnected "from the wider urban fabric", yet connected to similar places, thus forming archipelagos of "globalized enclaves" (Graham and Marvin 2001: 249, 268, 389). Splintering urbanism widens the gap between connected and disconnected places and people, and hence both contributes to and helps to "sustain the fragmentation of the social and material fabric of cities" (Graham and Marvin 2001: 33 [source of quotation], 288, 383; Coutard 2002; Zérah 2008; Jaglin 2008; Harvey 1989; Silva 2000).

By employing the metaphorical expression "people as infrastructure", Abdou-Maliq Simone (2005) suggests an underexamined mode of engagement through which poor urban residents turn to improvisational practices to compensate for outmoded, broken-down or absent infrastructural systems. Similarly, Simone speaks of a kind of "pirate urbanism" in Africa, where the urban poor – who have been abandoned by urban planners with their grand schemes appealing to corporate investors and globetrotting tourists – have incrementally reconstructed collapsing infrastructures and forged new networks of informal collaborations that extend far and wide (Simone 2006). Because of the ease with which telecommunications enable the transfer of money and the sharing of information, the current age of globalization actually opens up new possibilities for this "pirate urbanism" by fostering the emergence and development of low-cost informal artisanal work and trading networks spread across regions, and weakening national-state sovereignty (Simone 2006: 357; see also Simone 2001a, 2001b, 2003). Ravi Sundaram (2010a) characterizes "pirate modernity" as a novel set of circumstances that constitute a post-media urbanism in postcolonial cities in India: "pirate modernity connects media piracy, grey market commodities and informal urban infrastructures" (Sundaram 2010b: 5). Subaltern populations are able to attach themselves to existing technological networks (like electricity and media networks) and redeploy them for new mobilities. Pirate modernity typically bypasses formal and intellectual property rights, providing a crucial but illegal gateway for subaltern populations to gain access to new infrastructures. In looking at existing technologies in Kano (Nigeria), Larkin (2004, 2008) reveals how "pirated", informal media economies have become an impromptu substitute for crumbling municipal infrastructure.

Wael Salah Fahmi (Chapter 12, this volume) reveals how the street demonstrations and socio-political events that took place in the public squares and gathering places of Cairo during the tumultuous early months of 2011 created new political spaces for contestation that reflected a deeper power struggle between the security forces and urban youth. Easy access to mobile communication networks enabled youthful protesters to engage in widespread street blogging as a way of gathering and passing along information in real time (Srinivasan and Fish 2011; Edwards 2011). The protest movement used its unconventional army of blogger activists to tap into a new kind of cyberactivism that fundamentally redefined the real and virtual public spaces of the city.

New modes of urban governance under the sign of neoliberalism

The turn towards privatization, deregulation and liberalization – for which the "neoliberal city" (Hackworth 2007) serves as convenient shorthand – figures prominently in new research and writing on transnational urbanism. In responding to the neoliberal challenge, municipal authorities in megacities of the Global South installed new regulatory regimes that have removed institutional constraints, legal barriers and administrative barriers in order to pave the way for making the market function efficiently through downsizing, outsourcing and rightsizing municipal services (Demissie 2011).

Yet as an epistemological category, neoliberalism (and all of its surrogate formulations) often functions in the conventional urban studies literature as a kind of blunt instrument, used to explain the deleterious effects of privatization, urban entrepreneurialism and downsizing. While the term has heuristic value in helping to frame our thinking about urban governance, one must take care not to overextend its use as a one-dimensional, one-sided analytic device that seeks to explain too much (Gupta and Ferguson 2000).

In unpacking the term, Jan Nijman (2008: 74) distinguishes three ways in which slum rehabilitation has undergone modification under the sway of neoliberal policies: first, the shift from government intervention to almost exclusive reliance on the regulative mechanisms of the free market; second, the shift of responsibility from municipal agencies to civil society; and, third, the rescaling of state structure of administration from the central to the local level. In their study of informal housing schemes in Ouagadougou (Burkina Faso), Wouter Bervoets and Maarten Loopmans (Chapter 3, this volume) demonstrate how the introduction of private property markets in informal housing settlements resulted in large-scale property speculation and uncontrolled urban expansion. Similarly, in his investigation of the *Villes sans Bidonvilles* (Cities without Slums) program undertaken in Morocco, Koenraad Bogaert (in this volume) shows how the introduction of new kinds of property taxation, public-private partnerships and streamlined access to credit has lured slum-dwellers into new institutional arrangements where the opportunities, incentives, sanctions and constraints of private property markets have created a new set of circumstances that shape individual decision-making. Similarly, Ernesto López-Morales (Chapter 10, this volume) reveals how popular mobilization at a local, grassroots level has been able to force the municipal administration in Santiago (Chile) to compromise on their ambitious redevelopment schemes. Mona Fawaz (Chapter 1, this volume) argues that, ironically, the weakness and fragility of the public administration in Beirut has meant that those who have settled illegally in informal settlements have been unable to mount a struggle to secure a "right to the city" with municipal officials but instead have turned to private militias for support.

Neoliberal policies and practices operate on the principle that private market mechanisms rather than public authorities can more efficiently and effectively regulate urban space. The shift to neoliberal modes of urban governance does not necessarily mean a decrease in the number of open, accessible places for social congregation and chance encounter, yet it does mean a change in how these post-public spaces are produced and regulated. A great deal of scholarly attention has been directed at understanding how municipalities enter into contracts – especially the increasingly popular type of hybrid ownership called "public-private partnerships" – to provide incentives for the private companies to create or manage public spaces (Banerjee 2001; Németh and Schmidt 2007; Németh 2009; Németh and Hollander 2010). Similar scholarly writing on Buenos Aires, for example, reveals how the shifting modes of municipal governance have converted public space into a sort of visual spectacle, an ethereal space that provides the illusion of genuine social mixing, masking the reality of the space itself as accessible to only a privileged few (Guano 2002; Pírez 2002).

In drawing attention to the ghosts of Durban (racial segregation under apartheid) and the ghosts of Berlin (Nazism, ruination, spatial division), Christine Hentschel (Chapter 9, this volume) weaves together a complicated narrative that create new divisions in urban space. In her view, the tactics of displacement and exclusion – what Foucault (1977: 26) referred to as the "micro-physics of power" – are always small scale and incremental, silent and stealthy. At the end of the day, these softer, indirect and seemingly benign techniques of "social sorting" and gentrification by stealth accomplish their goals of social cleansing just as effectively as large-scale strategies of urban removal and spatial clearances.

The right to the city and social justice

Recent writing on cities has substantially unsettled conventional notions of urban space/place, belonging and citizenship. Once thought of as relatively fixed, stable and durable, these concepts have been subjected to a high degree of bending, blending and stretching. While helpful as points of departure, such binary categorizations as "inclusion" versus "exclusion", "private space" versus "public space", and "citizen" versus "subject" take us only so far in understanding the complex dynamics that produce everyday precariousness. For instance, the theorization of such fuzzy relationships as "insurgent citizenship" (Holston 1995, 1999, 2007, 2009), "flexible citizenship" (Ong 1999, 2006), "active citizenship" (Miraftab and Wills 2000), "differential citizenship" (Lake and Newman 2002) and "micro-citizenship" (Centner 2012) has drawn attention to the malleability of rights and entitlements in the use of urban space. With the retreat of the state administration and the shift from government to governance, many of the functions once reserved for public authorities have been absorbed by a burgeoning not-for-profit sector comprising a "shadow state" increasingly responsible for social service delivery and community development. The dependence of urban residents on the "shadow state" results in differential citizenship or selective disenfranchisement (Lake and Newman 2002; Trudeau 2012). Similarly, in theorizing the rupture of various forms of social belonging, Robert Castel (2000: 534, 520) develops the concept of "disaffiliation" to unpack the "cumulative logic of deprivation", where "economic insecurity becomes destitution and fragility of relationships becomes isolation". Working with this concept of "disaffiliation" – or the dissociation of the conventional social bonds that tie urban residents to the social worlds around them (ibid.: 520) – enables us to theorize the differential inclusion of the urban poor into the mainstream of urban life along the dimensions of access to regular work, to decent housing and to the protective shield of state assistance.

The claim of a "right to the city" provides a powerful normative framework to support non-legal occupation of land, infringement of property rights and the bypassing of building regulations by poor urban dwellers (Fawaz 2009a: 827; 2009b; Parnell and Pieterse 2010; Roy 2009a, 2009b). Assertions of a "right to the city" are entangled with questions of citizenship, belonging and social justice. As David Harvey (2008: 23) has argued, "the right to the city is far more than the individual liberty to access urban resources: it is a right to change ourselves by changing the city". A great deal of urban scholarship has focused attention on how otherwise marginalized and deprived urban residents carve out spaces for themselves in the city through improvisation, stealth, subterfuge and sheer persistence. James Holston (2007, 2009), for example, uses the framework of "insurgent citizenship" to underscore how the urban poor of São Paulo and Brasília engage in a wide variety of clandestine or semi-clandestine tactics that place them outside the protections of the law. Along with others, Holston draws on the groundbreaking work of Henri Lefebvre (1996, 2003), who developed the notion of the "the right to the city". The shaping of insurgent public space suggests a mode of city making that is different from institutionalized approaches to city building and their association with master planning and official policy making. Unlike the conventional practice of urban planning, the making of insurgent public space implies that social groups and individuals can play a distinct role in shaping urban landscapes in defiance of official rules and regulations (Hou 2010: 15).

In *Politics of the Governed*, Partha Chatterjee (2004) distinguishes between claims for a rightful place in the city between citizens, who operate within the domain of what liberal political theorists call "civil society", and subaltern populations, whose demands fall outside the realm of legal subjectivity. Well-to-do urban residents, who enjoy the full rights and benefits of citizenship have a range of state-legitimated, institutionalized channels open to them through which they are able to pursue their interests. In contrast, subaltern populations have no choice but to transgress the strict lines of legality in their everyday struggle to live and work.

This distinction between citizens (as privileged insiders) and populations (as excluded outsiders) enables us to unpack the hierarchical modalities of urban governmentality, that is, to distinguish between those who have a legitimate right to claim rights and, conversely, those who are suspended in a lawless, extra-legal limbo, unable to call upon the law for support yet simultaneously subjected to its rules and regulations (Agamben 2005). In looking at pre-paid water meters in post-apartheid South Africa, Antina von Schnitzler (2008) shows how neoliberal reforms used rationalization to create "calculative citizens" enlisted in their own self-governance. Jia-Ching Chen (Chapter 4, this volume) shows how neoliberal policies operating under the banner of "ecological harmony" use the idea of "green development" actually to entrench social inequalities in rapidly growing cities. Working through the wide-angle lens of governmentality, Rachel Berney (Chapter 7, this volume) charts the emergent geographies of spatial governance in Bogota, Columbia, within which poor and marginalized groups seek to make claims upon the municipality. For Berney, these assertions of a "right to the city" assume a fundamentally different tenor than the claims of "citizens", whose bundle of rights and entitlements is affirmed and empowered through constitutional guarantees. In contrast, poor and marginalized groups assert their "right to the city" through their engagement and active participation in illegal activities, violating the rules which govern civil society and transgressing the laws of private property (Bayat 1997; Appadurai 2000; Chatterjee 2004).

(Re)appropriation of public space

One recurrent theme that characterizes the emerging scholarly literature on ordinary cities of the Global South is contestation over the "proper" and "appropriate" use of public space, that is, easily accessible places for social congregation, interaction, exchange and chance encounter. Under circumstances where dictatorships and autocratic regimes have given way to electoral democracy, political projects seeking to recover the public sphere are linked with notions of expanding access to public space. In this sense, vibrant and robust public spaces of the city have become visible symbols of democratic openings, synonymous with social equality, political liberty and freedom of expression. Yet the transformation of public space into "designer landscapes" that aim to sell the city in the image-conscious world of global neoliberalism has gone hand in hand with new mechanisms of social exclusion and marginalization of "unwanted" Others. While appearing to be open and inviting, these post-public spaces have become sites of conflict, rivalry and contestation.

Painted in broad brushstrokes, the privatization of public space and the accompanying commercialization of places of social congregation and chance encounter have produced new kinds of administered space in cities, variously referred to as "privatized public space", "quasi-public space", "privately owned public space", "pseudo-public space", "publicly accessible spaces", "post-liberal space" or "post-public space". In trying to account for the exclusionary logics which regulate access to (and hence the use of) these packaged built environments, an entirely new vocabulary – domestication, purification, militarization, pacification, Disneyification, revanchism and the like – has come into play.

On balance, conventional urban studies literature tends to coalesce around two opposing currents. On the one side, there are those who stress the growing "militarization of urban space" (Davis 1992; Flusty 2001) and "revanchist urbanism" (Smith 1996; Coafee 2005), and the use of physical force and violence. As John Allen has argued, the exercise of power in these privatized public spaces rests "ultimately on some form of domination, where the choice over who enters is constrained by certain kinds of watchful power or discriminating rules of entry" (Allen 2006: 443). This kind of dominating power is typically equated with the marked presence of physical barriers, visible policing and surveillance technologies (Davis 1992).

In describing the downtown reconstruction of Beirut in the wake of the Lebanese civil war (1975–90), Monica Fawaz and Marwan Ghandour (2010), for example, draw attention to how urban warfare and armed conflict normalizes the presence of visible security measures in the form of barricades, road blocks or army checkpoints in the everyday life of the city. The rehabilitated city center – carried out by the private real estate company *Solidere* – preserves and perpetuates the isolation of the historic core of the city in abiding by the boundaries articulated during the civil war. *Solidere* transformed the battle zone from a space of military violence into a space for the free flow of capital without altering the spatial detachment of the historical core from the rest of the city.

On the other side, there are those who focus on the more subtle, less visible, yet equally effective methods of regulating the use of public space (Atkinson 2003; Degen 2003; Zukin 1995; Allen 2006). The scripted nature of post-public, accessible spaces that depend for their commercial success upon the staging of a certain kind of accessibility, inclusiveness and openness enables anonymous strangers to move freely, yet they "unknowingly remain subject to a form of control that is regularized, predictable, and far from chaotic" (Allen 2006: 443). Closure and exclusion in these multifunctional sites relies upon complex modes of power, less obtrusive and more subtle than those mechanisms of spatial control that monitor and regulate the use of space in functionally specialized enclaves (Atkinson 2003). There is more to the exercise of power in commercial post-public places than the exclusionary logics of physical barriers and visible displays of authority. Closure in the kinds of post-public spaces that rely upon a kind of spatial openness is not just about domination but also about seduction. The power of seduction works "through the *experience* of the space itself", through its *ambient*, sensory qualities of sights, sounds and smells (Allen 2006: 442). Ambient power derives from the character of these post-public spaces, where the spatial layout, aesthetically pleasing architecture and controlled experience produce "a particular atmosphere, a specific mood, a certain feeling" that together induce conformity with expected behavior and action. Somewhat akin to the evocative metaphor of "pacification by cappuccino" coined by Sharon Zukin (1995: 28), ambient power works through the flexible powers of seduction rather than simply brute force, through persuasion and not strict regulation.

The essays by Ryan Centner, Rachel Berney, David Walker and Christine Hentschel (Chapters 11, 7, 8 and 9, this volume) highlight these oscillations and tensions between the use of physical force and quiet persuasion to maintain order in post-public spaces and, in so doing, reveal the complicated nature of civic belonging in class-stratified cities. By using public space as a site for informal trading and petty entrepreneurship, poor and marginalized groups encroach upon the aesthetic expectations of respectable and proper urban citizens. Informality and its shadow legality undermine the proper use of public space, and thereby trigger the reaction of middle-class citizens who have claimed this space as their own. The contemporary city of Bogotá thus becomes a stage-set for enacting new dramas of competition between the urban poor, who seek a place in the city as informal street vendors, and middle-class groups aspiring to comfortable lifestyles in the newly refurbished public spaces of the city (Berney 2011; Donovan 2008; Crossa 2009).

In cities throughout the Global South, mushrooming street economies have engendered ongoing tensions among street vendors, property owners, large retailers and municipal authorities regarding who is entitled to use the public spaces of the city and for what purposes. In efforts to "appropriately develop" and to "modernize" the public spaces of their cities, municipal authorities typically impose strict regulatory regimes that place limitations on the legitimate uses of city streets, public parks and other social gathering places (Bromley 2000; Morales 2007; Cross and Karides 2007: 19; Douglass and Danière 2009; Hunt 2009). Because their efforts of to make a living often fall outside the law, street vendors are vulnerable to harassment and intimidation, and subjected to fines, prosecution and confiscation of their trading goods (Cross 1998, 2007; Cross and Karides 2007; Drummond 2000). Despite the vigilance of municipal authorities, the filtered spaces of the neoliberal city are porous, malleable and pliable (Németh 2009; Law 2002). B. Lynne Milgram (2011) has shown how vendors in Baguio City (Philippines) have developed strategies to evade, circumvent or negotiate restrictions on their trade, thereby challenging municipal regulations privileging regular wage-paying work that corresponds with the city's vision of progress and development.

Yet one must be careful not fall into the trap of treating acts of insurgency in public space as if they easily fall into the binary oppositions of "top down" versus "bottom up". Unconventional efforts to reclaim and remake public space are not always spontaneous, participatory and inclusive. Ryan Centner (Chapter 11, this volume) contrasts the "right to the city" – which Henri Lefebvre (1996: 158) defined as an ontological "right to urban life" – with "the right kind of city", or an imaginary space that reflects a middle-class preoccupation with aesthetically pleasing places. In drawing on examples from Istanbul, Rio de Janeiro and Buenos Aires, he unveils how middle-class urban residents assert exclusionary visions of the right kind of city that, by definition, exclude marginalized groups that are socially beneath them.

Theorizing public space in the Global South

The excessive focus on the narrow range of features that define what constitutes a "global city" – and the attendant fixation on rank ordering cities in accordance with their position in a hierarchy of relative achievements – tends to flatten out our understanding of the contemporary experience of urban modernity. As a general rule, conventional urban studies literature has not been able to break sufficiently away from the regulative fiction that the process of urbanization is a more or less singular pathway that unfolds in stages, where cities are distributed along a continuum that stretches from the "developed" metropolis at one pole to those that lack all the features of modern urbanity at the other, with a lot of "developing" ones somewhere in between, striving to catch up. This failure to dispense with ranked hierarchies – such as the unhelpful distinction between First World and Third World cities, or "developed" and "developing" – locks urban scholars into one-dimensional ways of thinking that conceive of urban transformation only in terms of growth and development.

The point of departure for understanding contemporary transnational urbanism – the emergent cities of the twenty-first century – is recognizing the heterogeneity of spatial form, the diversity of opportunities for social action/political agency and diverse trajectories of transformation (what kinds of growth? what kinds of decline?). Thinking about contemporary transnational urbanism requires a sensitivity to gaps and lags, bypassing and leapfrogging, accelerating and slowing down (Roy 2011a, 2011b).

To a large extent, the discussion around the "right to the city" has taken for granted the convention distinction between public space and private space that characterized the birth of the modern capitalist city in the West. On the one hand, the rights of private proprietorship enable the owners of private property to restrict and monitor access to urban spaces to those deemed worthy of admission. In contrast, public space was inextricably linked to the imagined ideal of the public sphere: the arena of modern urban life defined by civic engagement, democratic debate and popular assembly. The public administration of accessible spaces of the city acted as a counterweight to the powers of private property owners to restrict access and define use of space. The use of public space was governed not by the logic of market tyranny but by the communitarian values of sociability, social interaction and chance encounter.

This kind of analytic framework for understanding "the right to the city" has lost some of its salience. As Aihwa Ong (2006: 7) has argued, the emergence of new modes of urban governance adopted in cities of the Global South have produced new patchwork patterns of "non-contiguous, differentially administered spaces of graduated or variegated sovereignty". The introduction of new zoning practices has created a withering array of regulatory regimes that have undermined the once sovereign power of public authority over the social congregating spaces of the city. Spatial configurations like securitized office parks, gated residential communities and shopping malls are neither "public" nor "private" in the conventional sense.

Existing only in embryonic form, contemporary transnational urbanism proceeds along multiple pathways of growth and development, abandonment and decline, which defy simple classification. Conventional urban studies literature has continued to rely on analytic categories, concepts and descriptive markers that were first developed and gradually refined to account for the meteoric rise in the nineteenth and twentieth centuries of the great modern industrial metropolises that constituted the core of the capitalist world economy. Although urban theorists have always recognized the importance of cities as privileged sites of cultural consumption, theorizing about the modern metropolis has typically taken the production of manufactured goods and services as the key to understanding the dynamics of urban growth and development (Lloyd and Clark 2001: 357). As the new, post-industrial "informational city" of high finance, producer services, information technology and digital media has replaced the prototypical "industrial city" of mass production, city builders have turned to culture-led strategies for downtown revitalization, focusing on entertainment, tourism and leisure services as the lynchpins for restoring urban prosperity. At the same time, the sprawling megacities of hyper-growth – with massive informal settlements, broken-down infrastructure and environmental degradation – have produced new spatial configurations that have never existed before in such exaggerated form. What characterizes contemporary transnational urbanism is the remaking of cities in ways which bypass and leapfrog over the patterns of urban growth and development that defined the historic era of the great modern industrial metropolises of the capitalist West. The production of a multiplicity of different cities that respond to different pressures and logics makes the call for a more thorough, cosmopolitan re-imagining of urban theorizing on a global scale even more urgent (Robinson 2002: 549).

Note

1 This is the title of a 2011 Social Science Research Council (SSRC) dissertation proposal development workshop organized by Ananya Roy and Helga Leitner.

References

Abbas, A. (2000) "Cosmopolitan De-Scriptions: Shanghai and Hong Kong", *Public Culture*, 12 (3): 769–86.

Acuto, M. (2010) "High-Rise Dubai Urban Entrepreneurialism and the Technology of Symbolic Power", *Cities*, 27: 272–84.

Agamben, G. (2005) *State of Exception*, transl. K. Attell, Chicago: University of Chicago Press.

Ali, K. and Rieker, M. (2008) "Introduction: Urban Margins", *Social Text*, 26 (2 95): 1–12.

Allen, J. (2006) "Ambient Power: Berlin's Potsdamer Platz and the Seductive Logic of Public Spaces", *Urban Studies*, 43 (2): 441–55.

AlSayyad, N. (2001) *Hybrid Urbanism: on the Identity Discourse and the Built Environment*, Westport, CT: Praeger.

Amin, A. and Graham, S. (1997) "The Ordinary City", *Transactions of the Institute of British Geographers*, 22: 411–29.

Angelini, A. (2005) "Book Review of *Spaces of Global Cultures*", *Future Anterior*, 2 (1): 67–71.

Angotti, T. (2006) "Apocalyptic Anti-Urbanism: Mike Davis and his *Planet of Slums*", *International Journal of Urban and Regional Research*, 30 (4): 961–7.

Appadurai, A. (2001) "Deep Democracy: Urban Governmentaliy and the Horizon of Politics", *Environment & Urbanization*, 13 (2): 23–43.

Appadurai, A. (2000) "Spectral Housing and Urban Cleansing: Notes on Millenial Mumbai", *Public Culture*, 12 (3): 627–51.

Arkaraprasertkul, N. (2010) "Power, Politics, and the Making of Shanghai", *Journal of Planning History*, 9 (4): 232–59.

Atkinson, R. (2003) "Domestication by *Cappuccino* or a Revenge on Urban Space? Control and Empowerment in the Management of Public Spaces", *Urban Studies*, 40 (9): 1829–43.

Atkinson, R. and Bridge, G. (2005) *Gentrification in a Global Context: the New Urban Colonialism*, New York: Routledge.

Auyero, J. (2000) "The Hyper-Shantytown: Neoliberal Violence(s) in the Argentine Slum", *Ethnography*, 1 (1): 93–116.

Banerjee, T. (2001) "The Future of Public Space: Beyond Invented Streets and Reinvented Places", *Journal of the American Planning Association*, 67 (1): 9–24.

Bayat, A. (2009) *Life as Politics: How Ordinary People Change the Middle East*, Stanford, CA: Stanford University Press.

Bayat, A. (2000) "From 'Dangerous Classes' to 'Quiet Rebels': Politics of Urban Subalterns in the Global South", *International Sociology*, 15: 533–56.

Bayat, A. (1997) 'Uncivil Society: the Politics of the "Informal People"', *Third World Quarterly* 18 (1): 53–72.

Benjamin, S. (2008) "Occupancy Urbanism: Radicalizing Politics and Economy beyond Policy and Programs", *International Journal of Urban and Regional Research*, 32 (3): 719–29.

Benjamin, S. (2000) "Governance, Economic Settings and Poverty in Bangalore", *Environment and Urbanization*, 12: 35–56.

Benjamin, S. (1991) *Jobs, Land and Urban Development: The Success of Small Manufacturers in East Delhi*, Boston: Lincoln Institute of Land Management.

Berney, R. (2011) "Pedagogical Urbanism: Creating Citizen Space in Bogotá Colombia", *Planning Theory*, 10 (16): 16–34.

Bishop, R., Phillips, J. and Yeo, W. (eds) (2003) *Postcolonial Urbanism: Southeast Asian Cities and Global Processes*, New York and London: Routledge.

Boano, C., Lamarca, M. and Hunter, W. (2011) "The Frontlines of Contested Urbanism: Mega-Projects and Mega-Resistances", *Journal of Developing Societies 2011*, 27 (3–4): 295–326.

Breman, J. (2006) "Slumlands", *New Left Review*, 40: 141–8.

Bromley, R. (2000) "Street Vending and Public Policy: a Global Review", *International Journal of Sociology and Social Policy*, 20 (1): 1–29.

Campanella, T. (2008) *The Concrete Dragon: China's Urban Revolution and What it Means for the World*, New York: Princeton Architectural Press.

Castel, R. (2000) "The Roads to Disaffiliation: Insecure Work and Vulnerable Relationships", *International Journal of Urban and Regional Research*, 24 (3): 519–35.

Centner, R. (2012) "Microcitizenships: Fractious Forms of Urban Belonging after Argentine Neoliberalism", *International Journal of Urban and Regional Research*, 36 (2): 336–62.

Chatterjee, P. (2004) *Politics of the Governed*, New York: Columbia University Press.

Coafee, J. (2005) "Urban Renaissance in the Age of Terrorism: Revanchism, Automated Social Control or the End of Reflection?", *International Journal of Urban and Regional Research*, 29 (2): 447–54.

Coutard, O. (2008) "Placing Splintering Urbanism: Introduction", *Geoforum*, 39: 1815–20.
Coutard, O. (2002) "Premium Network Spaces: a Comment", *International Journal of Urban and Regional Research*, 26 (1): 166–74.
Cross, J. (2007) "Pirates on the High Streets: the Street as a Site of Local Resistance of Globalization", in J. Cross and A. Morales (eds) *Street Entrepreneurs: People, Place and Politics in Local and Global Perspectives*, New York: Routledge.
Cross, J. (1998) *Informal Politics: Street Vendors and the State in Mexico City*, Stanford, CA: Stanford University Press.
Cross, J. and Karides, M. (2007) "Capitalism, Modernity, and the 'Appropriate' use of Space", in J. Cross and A. Morales (eds) *Street Entrepreneurs: People, Place and Politics in Local and Global Perspectives*, New York: Routledge.
Crossa, V. (2009) "Resisting the Entrepreneurial City: Street Vendors' Struggle in Mexico City's Historic Center", *International Journal of Urban and Regional Research*, 33 (1): 43–63.
Cunningham, D. (2007) "Slumming It: Mike Davis's Grand Narrative of Urban Revolution", *Radical Philosophy*, 142: 8–18.
Davidson, C. (2008) *Dubai: the Vulnerability of Success*, C. Hurst & Co.
Davis, M. (2006a) "Fear and Money in Dubai", *New Left Review*, 41: 47–68.
Davis, M. (2006b) *Planet of Slums: Urban Involution and the Informal Proletariat*, New York: Verso.
Davis, M. (2004) "The Urbanization of Empire: Megacities and the Laws of Chaos", *Social Text*, 22 (4): 9–15.
Davis, M. (1992) "Fortress Los Angeles: the Militarization of Urban Space", in M. Sorkin (ed.) *Variations on a Theme Park*, New York: Noonday Press.
de Boeck, F. (2002) "Tales of the Invisible City", in O. Enwezor, C. Basualdo, U.-M. Bauer, S. Ghez, S. Majaraj, M. Nash and O. Zaya (eds) *Under Siege: Four African Cities*, Ostfildern-Ruit, Germany: Hatje Cantz Publishers.
de Boeck, F. and Plissard, M.-F. (2006) *Kinshasa: Tales of the Invisible City*, Antwerp: Ludion.
Degen, M. (2003) "Fighting for the Global Catwalk: Formalizing Public Life in Castlefield (Manchester) and Diluting Public Life in el Raval (Barcelona)", *International Journal of Urban and Regional Research*, 27 (4): 867–80.
Demissie, F. (2011) "Global Cities of the South/Urban Subjects", *Journal of Developing Societies*, 27 (3, 4): 217–28.
Donovan, M. (2008) "Informal Cities and the Contestation of Public Space: the Case of Bogota's Street Vendors, 1988–2003", *Urban Studies*, 45 (1): 29–51.
Douglass, M., and Danière, A. (2009) "Urbanization and Civic Space in Asia', in A. Danière and M. Douglass (eds) *The Politics of Civic Space in Asia: Building Urban Communities*, London and New York: Routledge.
Drummond, L. (2000) "Street Scenes: Practices of Public and Private Space in Urban Vietnam", *Urban Studies*, 37 (12): 2377–91.
Edensor, T. and Jayne, M. (eds) (2012) *Urban Theory Beyond the West: a World of Cities*, New York: Routledge.
Edwards, R. (2011) "Tahrir: Ends of Circulation", *Public Culture*, 23 (3): 493–504.
Ehrkamp, P. (2011) "Internationalizing Urban Theory: Toward Collaboration", *Urban Geography*, 32 (8): 1122–8.
Elsheshtawy, Y. (2010) *Dubai: Behind an Urban Spectacle*, New York: Routledge.
Elsheshtawy, Y. (2008) "Navigating the Spectacle: Landscapes of Consumption in Dubai", *Architectural Theory Review*, 13 (2): 164–87.

Fawaz, M. (2009a) "Neo-liberal Urbanity: a View from Beirut's Periphery", *Development and Change*, 40 (5): 827–52.

Fawaz, M. (2009b) "Hezbollah as Urban Planner? Questions to and from Planning Theory", *Planning Theory*, 8 (4): 323–34.

Fawaz, M. (2009c) "The State and the Production of Illegal Housing: Public Practices in Hayy el-Sellom, Beirut-Lebanon", in K. Ali and M. Rieker (eds) *Comparing Cities: the Middle East and South Asia*, London: Oxford University Press.

Fawaz, M. (2008) "An Unusual Clique of City-Makers: Social Networks in the Production of a Neighborhood in Beirut (1950–1975)", *International Journal of Urban and Regional Research*, 32 (3): 565–85.

Fawaz, M. and Ghandou, M. (2010) "Spatial Erasure: Reconstruction Projects in Beirut", *ArteEast Quarterly*, Spring. Online: www.arteeast.org.

Fernandez-Galiano, L. (2005) "Spectacle and its Discontents; or, the Joys of Architainment", in W. Saunders (ed.) *Commodification and Spectacle in Architecture: a Harvard Design Magazine Reader*, Minneapolis, MN: University of Minnesota Press.

Flusty, S. (2001) "The Banality of Interdiction: Surveillance, Control and the Displacement of Diversity", *International Journal of Urban and Regional Research*, 25 (3): 658–64.

Foucault, M. (1977) *Discipline and Punish: the Birth of the Prison*, London: Penguin Books Ltd.

Friedmann, J. (1986) "The World City Hypothesis", *Development and Change*, 17: 69–83.

Gberie, L. (2005) "Review Article – Africa: The Troubled Continent", *African Affairs*, 104 (405): 337–42.

Gilbert, A. (2007) "The Return of the Slum: Does Language Matter?", *International Journal of Urban and Regional Research*, 31 (4): 697–713.

Graham, S. (2000) "Constructing Premium Network Spaces: Reflections on Infrastructure Networks and Contemporary Urban Development", *International Journal of Urban and Regional Research*, 24 (1):183–5.

Graham, S. and Marvin, S. (2001) *Splintering Urbanism: Networked Infrastructures, Technological Mobilities and the Urban Condition*, London: Routledge.

Grant, R. and Nijman, J. (2002) "Globalization and the Corporate Geography of Cities in the Less Developed World", *Annals of the Association of American Geographers*, 92 (2): 320–40.

Guano, E. (2002) "Spectacles of Modernity: Transnational Imagination and Local Hegemonies in Neoliberal Buenos Aires", *Cultural Anthropology*, 17 (2): 181–209.

Gupta, A. and Ferguson, J. (2000) "Spatializing States: Toward an Ethnography of Neoliberal Governmentality", *American Ethnologist*, 29 (4): 981–1002.

Hackworth, J. (2007) *The Neoliberal City: Governance, Ideology, and Development in American Urbanism*, Ithaca, NY: Cornell University Press.

Harding, J. (2007) "It Migrates to Them: The Coming Mega-Slums", *London Review of Books*, 29 (5), 8 March: 25–7.

Harvey, D. (2008) "The Right to the City", *New Left Review*, 53: 23–40.

Harvey, D. (1989) "From Managerialism to Entrepreneurialism: the Transformation in Urban Governance in Late Capitalism", *Geografiska Annaler*, 71: 3–17.

Hoffman, D. (2007) "The City as Barracks: Freetown, Monrovia, and the Organization of Violence in Postcolonial African Cities", *Cultural Anthropology*, 22 (3): 40–48.

Hoffman, D. (2011) "Violence Just in Time: War and Work in Contemporary West Africa", *Cultural Anthropology*, 26 (1): 34–57.

Holston, J. (2009) "Insurgent Citizenship in an Era of Global Urban Peripheries", *City & Society*, 21 (2): 245–67.

Holston, J. (2007) *Insurgent Citizenship: Disjunctions of Democracy and Modernity in Brazil*, Princeton, NJ: Princeton University Press.

Holston, J. (1999) "Spaces of Insurgent Citizenship", in J. Holston (ed.) *Cities and Citizenship*, Durham, NC: Duke University Press.

Holston, J. (1995) "Spaces of Insurgent Citizenship", *Planning Theory*, 13: 35–51.

Hou, J. (2010) "(Not) your Everyday Public Space", in J. Hou (ed.) *Insurgent Public Space: Guerrilla Urbanism and the Remaking of Contemporary Cities*, New York: Routledge.

Huchzermeyer, M. (2011) *Cities with "Slums": from Informal Settlement Eradication to a Right to the City in Africa*, Cape Town: University of Cape Town Press.

Huchzermeyer, M. (2004) *Unlawful Occupation: Informal Settlements and Urban Policy in South Africa and Brazil*, Trenton, NJ: Africa World Press.

Hunt, S. (2009) "Citizenship's Place: the State's Creation of Public Space and Street Vendors' Culture of Informality in Bogota, Columbia", *Environment and Planning D: Society and Space*, 27 (2): 331–51.

Jacobs, J. (2006) "A Geography of Big Things", *Cultural Geographies*, 13 (1): 1–27.

Jaglin, S. (2008) "The Differentiation of Technical Services in Cape Town: Echoing Splintering Urbanism?", *Geoforum*, 39 (6): 1897–906.

Kanai, G. and Ortega-Alcázar, I. (2009) "The Prospects for Culture-Led Urban Regeneration in Latin America: Cases from Mexico City and Buenos Aires", *International Journal of Urban and Regional Research*, 33 (2): 483–501.

Kanna, A. (2011) *The City as Corporation*, Minneapolis, MN: University of Minnesota Press.

Katodrytis, G. (2005) "Metropolitan Dubai and the Rise of Architectural Fantasy", *Bidoun Magazine*, 4: 2–3.

King, A. (1999) "(Post)Colonial Geographies: Material and Symbolic", *Historical Geography*, 27 (1): 99–118.

King, A. (1996) "Worlds in a City: Manhattan Transfer and the Ascendance of Spectacular Space", *Planning Perspectives*, 11: 87–114.

King, R. (2008) "Bangkok Space, and Conditions of Possibility", *Environment and Planning D: Society and Space*, 26 (2): 15–37.

Koolhaas, R. (1998) "The Generic City", in R. Koolhaas and B. Mau (eds) *S,M,L,XL*, New York: Monacelli Press.

Kudva, N. (2009) "The Everyday and the Episodic: the Spatial and Political Impacts of Urban Informality", *Environment and Planning A*, 41 (7): 1614–28.

Kusno, A. (2000) *Behind the Postcolonial: Architecture, Urban space and Political Cultures in Indonesia*, New York: Routledge.

Lake, R. and Newman, K. (2002) "Differential Citizenship in the Shadow State", *GeoJournal*, 58 (2–3): 109–20.

Landry, C. (2000) *The Creative City: a Toolkit for Urban Innovators*, London: Earthscan.

Larkin, B. (2008) *Signal and Noise: Media, Infrastructure, and Urban Culture in Nigeria*, Durham, NC: Duke University Press.

Larkin, B. (2004) "Degraded Images, Distorted Sounds: Nigerian Video and the Infrastructure of Piracy", *Public Culture*, 16 (2): 289–314.

Law, L. (2002) "Defying Disappearance: Cosmopolitan Public Spaces in Hong Kong", *Urban Studies*, 39: 1625–45.

Lefebvre, H. (2003) *The Urban Revolution*, Minneapolis, MN: University of Minnesota Press.

Lefebvre, H. (1996) *Writings on Cities*, Cambridge, MA: Blackwell.

Lloyd, R. and Clark, T. N. (2001) "The City as Entertainment Machine", in K. F. Gotham (ed.) *Critical Perspectives on Urban Redevelopment, Volume 6*, New York and Amsterdam: Elsevier.

Makdisi, S. (1997) "Laying Claim to Beirut: Urban Narrative and Spatial Identity in the Age of *Solidere*", *Critical Inquiry*, 23 (3): 660–705.

Marshall, R. (2003) *Emerging Urbanity: Global Urban Projects in the Asia Pacific Rim*, London and New York: Spon Press.

McFarlane, C. (2010) "The Comparative City: Knowledge, Learning, Urbanism", *International Journal of Urban and Regional Research*, 34 (4): 725–42.

McFarlane, C. (2007) "Urban Shadows: Materiality, the 'Southern City' and Urban Theory", *Geography Compass*, 2: 340–58.

McFarlane, C. (2006) "Crossing Borders: Development, Learning and the North–South Divide", *Third World Quarterly*, 27 (8): 1413–38.

Milgram, B. (2001) "Reconfiguring Space, Mobilizing Livelihood: Street Vending, Legality, and Work in the Philippines", *Journal of Developing Societies*, 27 (3–4): 261–93.

Miraftab, F. and Wills, S. (2000) "Insurgency and the Spaces of Active Citizenship: the Story of the Western Cape Anti-Eviction Campaign in South Africa", *Journal of Planning Education and Research*, 25 (2): 200–17.

Morales, A. (2007) "Law, Deviance, and Defining Vendors and Vending", in J. Cross and A. Morales (eds) *Street Entrepreneurs: People, Place and Politics in Local and Global Perspectives*, New York: Routledge.

Murray, M. (2011) *The City of Extremes: Spatial Politics in Johannesburg*, Durham, NC: Duke University Press.

Murray, M. (2008) *Taming the Disorderly City: the Spatial Landscape of Johannesburg after Apartheid*, Ithaca and London: Cornell University Press.

Myers, G. and Murray, M. (2006) "Introduction: Situating Contemporary Cities in Africa", in M. Murray and G. Myers (eds) *Cities in Contemporary Africa*, New York and London: Palgrave Macmillan.

Németh, J. (2009) "Defining a Public: the Management of Privately Owned Public Space", *Urban Studies*, 46 (1): 2463–90.

Németh, J. and Hollander, J. (2010) 'Security Zones and New York City's Shrinking Public Space', *International Journal of Urban and Regional Research* 34 (1): 20–34.

Németh, J. and Schmidt, S. (2007) "Toward a Methodology for Measuring the Security of Publicly Accessible Spaces", *Journal of the American Planning Association*, 73 (3): 283–97.

Neuwirth, R. (2006) *Shadow Cities: a Billion Squatters, a New Urban World*, New York: Routledge.

Nijman J. (2008) "Against the Odds: Slum Rehabilitation in Neoliberal Mumbai", *Cities*, 25: 73–85.

Ong, A. (2006) *Neoliberalism as Exception: Mutations in Citizenship and Sovereignty*, Durham, NC: Duke University Press.

Ong, A. (1999) *Flexible Citizenship: the Cultural Logics of Transnationality*, Durham, NC: Duke University Press.

Ong, A. and Roy, A. (2011) *Worlding Cities: Asian Experiments and the Art of Being Global*, Oxford: Wiley-Blackwell.

Parnell, S. and Pieterse, E. (2010) "The 'Right to the City': Institutional Imperatives of a Developmental State", *International Journal of Urban and Regional Research*, 34 (1): 146–62.

Perlman, J. (2010) *Favela: Four Decades of Living on the Edge in Rio de Janeiro*, New York: Oxford University Press.

Pieterse, E. (2008) *City Futures: Confronting the Crisis of Urban Development*, London: Zed Books.

Pírez, P. (2002) "Buenos Aires: Fragmentation and Privatization of the Metropolitan City", *Environment & Urbanization*, 14 (1): 145–58.

Pithouse, R. (2008) "Review of Mike Davis, *Planet of Slums*", *Journal of Asian and African Studies*, 43: 567–74.

Prouty, R. (2008) "Buying Generic: the Generic City in Dubai", *The London Consortium Static*, 8: 1–8.

Rao, V. (2006) "Slum as Theory: the South/Asian City and Globalization", *International Journal of Urban and Regional Research*, 30 (1): 225–32.

Read, S. (2005) "The Form of the Future", in S. Read, J. Rosemann and J. van Eldijk (eds) *Future City*, New York: Spon Press.

Robinson, J. (2011a) "Cities in a World of Cities: the Comparative Gesture", *International Journal of Urban and Regional Research*, 35 (1): 1–23.

Robinson, J. (2011b) "The Spaces of Circulating Knowledge: City Strategies and Global Urban Governmentality", in E. McCann and K. Ward (eds) *Mobile Urbanism*, Minnesota, MN: University of Minnesota Press.

Robinson, J. (2010) "Living in Dystopia: Past, Present and Future", in G. Prakash (ed.) *Noir Urbanisms*, Princeton: Princeton University Press.

Robinson, J. (2008) "Developing Ordinary Cities: City Visioning Processes in Durban and Johannesburg", *Environment and Planning A*, 40: 74–87.

Robinson, J. (2006) *Ordinary Cities: Between Modernity and Development*, London and New York: Routledge.

Robinson, J. (2005) "Urban Geography: World Cities, or a World of Cities", *Progress in Human Geography*, 29 (6): 757–65.

Robinson, J. (2002) "Global and World Cities: a View from off the Map", *International Journal of Urban and Regional Research*, 26 (3): 531–54.

Robinson, J. and Parnell, S. (2011) "Travelling Theory: Embracing Post-Neoliberalism through Southern Cities", in G. Bridge and S. Watson (eds) *New Companion to Urban Studies*, Oxford: Blackwell.

Rodgers, D. (2004) "'Disembedding' the City: Crime, Insecurity and Spatial Organization in Managua, Nicaragua", *Environment & Urbanization*, 16 (2): 113–24.

Rothman, H. and Davis, M. (eds) (2002) *The Grit Beneath the Glitter: Tales from the Real Las Vegas*, Berkeley and Los Angeles: University of California Press.

Roy, A. (2011a) "Postcolonial Urbanism: Speed, Hysteria, Mass Dreams", in A. Roy and A. Ong (eds) *Worlding Cities: Asian Experiments and the Art of Being Global*, New York: Wiley.

Roy, A. (2011b) "Slumdog Cities: Rethinking Subaltern Urbanism", *International Journal of Urban and Regional Research*, 35 (4): 223–38.

Roy, A. (2009a) "The 21st Century Metropolis: New Geographies of Theory", *Regional Studies*, 43: 819–30.

Roy, A. (2009b) "Civic Governmentality: the Politics of Inclusion in Beirut and Mumbai', *Antipode*, 41 (1): 159–79.

Roy, A. (2005) "Urban Informality: Toward an Epistemology of Planning", *Journal of the American Planning Association*, 71 (2): 147–58.

Sassen, S. (2001) *The Global City: New York, London, Tokyo*, [2nd Edition], Princeton, NJ: Princeton University Press.

Scott, J. (1985) *Weapons of the Weak: Everyday Forms of Peasant Resistance*, New Haven, CT: Yale University Press.

Shatkin, G. (2008) "The City and the Bottom Line: Urban Megaprojects and the Privatization of Planning in Southeast Asia", *Environment and Planning A*, 40: 383–401.
Silva, R. T. (2000) "The Connectivity of Infrastructure Networks and the Urban Space of São Paulo in the 1990s", *International Journal of Urban and Regional Research*, 24 (1): 139–64.
Silver, C. (2008) *Planning the Megacity: Jakarta in the Twentieth Century*, New York: Routledge.
Simon, D. (2010) "Situating Slums", *City*, 15 (6): 674–85.
Simone, A.-M. (2009) *City Life from Jakarta to Dakar: Movements at the Crossroads*, New York: Routledge.
Simone, A.-M. (2006) "Pirate Towns: Reworking Social and Symbolic Infrastructures in Johannesburg and Douala", *Urban Studies*, 43 (2): 357–70.
Simone, A.-M. (2005) "People as Infrastructure: Intersecting Fragments in Johannesburg", *Public Culture*, 16 (3): 407–29.
Simone, A.-M. (2004) *For the City Yet to Come: Urban Life in Four African Cities*, Durham, NC: Duke University Press.
Simone, A.-M. (2003) "Resource of Intersection: Remaking Social Collaboration in Urban Africa", *Canadian Journal of African Studies*, 37 (2–3): 513–38.
Simone, A.-M. (2001a) "On the Worlding of African Cities", *African Studies Review*, 44 (2): 15–41.
Simone, A.-M. (2001b) "Straddling the Divides: Remaking Associational Life in the Informal African City", *International Journal of Urban and Regional Research*, 25 (1): 102–17.
Simone, A-M. and Rao, V. (2012) "Securing the Majority: Living through Uncertainty in Jakarta", *International Journal of Urban and Regional Research*, 36 (2): 315–35.
Smith, N. (2002) "New Globalism, New Urbanism: Gentrification as Global Urban Strategy", *Antipode*, 34 (3): 427–50.
Smith, N. (1996) *The New Urban Frontier: Gentrification and the Revanchist City*, London: Routledge.
Srinivasan, R. and Fish, A. (2011) "Revolutionary Tactics, Media Ecologies, and Repressive States", *Public Culture*, 23 (3): 505–10.
Sundaram, R. (2010a) *Pirate Modernity: Delhi's Media Urbanism*, London and Delhi: Routledge.
Sundaram, R. (2010b) "Externalities, Urbanism and Pirate Modernities: India", Rising Powers working paper, Goldsmiths College, University of London.
Taylor, P. (2004) *World City Network*, London: Routledge.
Therborn, G. (2011) "End of a Paradigm: the Current Crisis and the Idea of Stateless Cities", *Environment and Planning A*, 43 (2): 272–85.
Trudeau, D. (2012) "Constructing Citizenship in the Shadow State", *Geoforum*, 43 (3): 442–52.
Von Schnitzler, A. (2008) "Citizenship Prepaid: Water, Calculability, and Techno-Politics in South Africa", *Journal of Southern African Studies*, 34 (4): 899–917.
Watson, V. (2009) "Seeing from the South: Refocusing Urban Planning on the Globe's Central Urban Issues", *Urban Studies*, 46: 2259–75.
Wacquant, L. (2008) "The Militarization of Urban Marginality: Lessons from the Brazilian Metropolis", *International Political Sociology*, 2 (1): 56–74.
Weinstein, L. (2008) "Mumbai's Development Mafias: Globalization, Organized Crime and Land Development", *International Journal of Urban and Regional Research*, 32 (1): 22–39.
Wright, G. (2002) "Building Global Modernisms", *Grey Room*, 7: 124–34.

Wright, H. (2008) *Instant Cities*, London: Black Dog Publishing.

Wu, F. (ed.) (2010) *China's Emerging Cities: the Making of a New Urbanism*, London and New York: Routledge.

Yeoh, B. (2005) "The Global Cultural City: Spatial Imagineering and Politics in the (Multi)cultural Marketplace of Southeast Asia", *Urban Studies*, 42 (5, 6): 945–58.

Yeoh, B. (2001) "Postcolonial Cities", *Progress in Human Geography*, 25 (3): 456–68.

Zérah, M.-H. (2008) "Splintering Urbanism in Mumbai: Contrasting Trends in a Multi-layered Society", *Geoforum*, 39: 1922–32.

Zérah, M.-H. (2009) "Participatory Governance in Urban Management and the Shifting Geometry of Power in Mumbai", *Development and Change*, 40 (5): 853–77.

Zukin, S. (1995) *The Cultures of Cities*, Malden, MA: Blackwell.

Index

Page numbers in *italics* denote tables, those in **bold** denote figures.